Differential Equation Based Solutions for Emerging Real-Time Problems

Modeling with differential equations is an effective tool to provide methodical and quantitative solutions to real-world phenomena including investigating measurable features, consolidation and processing of data, and designing and developing complex engineering systems. This book describes differential equations' correlation with qualitative and quantitative analysis, and mathematical modeling in the engineering and applied sciences. Given equations are explained from multidimensional characterizations with MATLAB® codes.

Features:

- Addresses differential equation-based approaches to solving varied engineering problems.
- Discusses derivation and solution of major equations of engineering and applied science.
- Reviews qualitative and quantitative (numerical) analysis and mathematical modeling.
- Includes mathematical models of the discussed problems.
- Discusses MATLAB® code and online materials related to differential equations.

This book is aimed at researchers and graduate students in electrical and electronics engineering, control systems, electron devices society, applied physics, and engineering design.

Computational Intelligence in Engineering Problem Solving

Series Editor:
Nilanjan Dey

Computational Intelligence (CI) can be framed as a heterogeneous domain that harmonized and coordinated several technologies, such as probabilistic reasoning, artificial life, multi-agent systems, neuro-computing, fuzzy systems, and evolutionary algorithms. Integrating several disciplines, such as Machine Learning (ML), Artificial Intelligence (AI), Decision Support Systems (DSS), and Database Management Systems (DBMS) increases the CI power and has an impact on several engineering applications. This book series provides a well-standing forum to discuss the characteristics of CI systems in engineering. It emphasizes the development of CI techniques and their role as well as the state-of-the-art solutions in different real-world engineering applications. The book series is proposed for researchers, academics, scientists, engineers, and professionals who are involved in the new techniques of CI. CI techniques including artificial fuzzy logic and neural networks are presented for biomedical image processing, power systems, and reactor applications.

IoT Security Paradigms and Applications
Research and Practices
Sudhir Kumar Sharma, Bharat Bhushan, and Narayan C. Debnath

Applied Intelligent Decision-Making in Machine Learning
Himansu Das, Jitendra Kumar Rout, Suresh Chandra Moharana, and Nilanjan Dey

Machine Learning and IoT for Intelligent Systems and Smart Applications
Madhumathy P, M Vinoth Kumar, and R. Umamaheswari

Industrial Power Systems
Evolutionary Aspects
Amitava Sil and Saikat Maity

Fuzzy Optimization Techniques in the Areas of Science and Management
Edited by Santosh Kumar Das and Massimiliano Giacalone

Semantic Web Technologies
Edited by Archana Patel, Narayan C. Debnath, and Bharat Bhushan

Differential Equation Based Solutions for Emerging Real-Time Problems
Edited by Papiya Debnath, Biswajit Sarkar, and Manash Chanda

For more information about this series, please visit:
http://www.routledge.com/Computational-Intelligence-in-Engineering-Problem-Solving/book-series/CIEPS

Differential Equation Based Solutions for Emerging Real-Time Problems

Edited by
Papiya Debnath, Biswajit Sarkar, and
Manash Chanda

CRC Press
Taylor & Francis Group
Boca Raton London New York

CRC Press is an imprint of the
Taylor & Francis Group, an **informa** business

First edition published 2024
by CRC Press
6000 Broken Sound Parkway NW, Suite 300, Boca Raton, FL 33487-2742

and by CRC Press
4 Park Square, Milton Park, Abingdon, Oxon, OX14 4RN

CRC Press is an imprint of Taylor & Francis Group, LLC

ISBN: 978-1-032-13138-2 (hbk)
ISBN: 978-1-032-13139-9 (pbk)
ISBN: 978-1-003-22784-7 (ebk)

DOI: 10.1201/9781003227847

Typeset in Times
by codeMantra

Contents

Preface

Differential equations and their wide approach in different engineering and science disciplines are inspired by engineering applications. To solve practical engineering problems, several types of theories and techniques of differential equations can be used. In science and engineering, one of the widely used effective tools is mathematical modeling. These models are systematic and provide quantitative descriptions of different real-world problems which may be Physical, Chemical, or Biological. These models help us to understand and analyze the data and select quality features to process the data. Based on features design is created and we develop complex engineering systems. To resolve specific practical problems, several types and complexity of numerous models have been considered.

There exist several mathematical models in different engineering disciplines that have become standard computational procedures, for instance, in construction models, analysis of the strength of materials, traffic noise propagation, heat flows, population dynamics and control, resource consumption, bifurcation analysis, and other engineering activities. In this book, solutions to many real-time problems have been detailed using differential equations. Chapter 1 introduces the different aspects of the differential equation. Chapter 11 also discusses the application of differential equations in inventory control. The application of a differential equation in a flexible production of the deteriorating item under the trade-credit policy is discussed in Chapter 2. The dynamic problem of the generation of elastic waves by an explosion within a spherical cavity with decelerated velocity is given in Chapter 3. Chapter 4 depicts the differential equation-based models to calculate accurate inventory and different parameters related to inventory. Chapter 5 discussed the existence of positive solutions to a nonlinear third-order boundary value problem by using an extension of the Krasnosel'skii fixed-point theorem and the Avery-Peterson fixed-point theorem. Constitutive equations including stress equation of motion, stress-strain relation, and appropriate boundary conditions are required to formulate fault systems in seismically active regions. To formulate mathematical models of earthquake processes in seismically active regions, ordinary and/or partial differential equations (Maxwell type) are very much useful. These have been detailed in Chapters 6 and 12, respectively. A combined study on time-dependent deterioration and carbon emission for fixed lifetime substitutable/complementary products in sustainable supply chain management is given in Chapter 7. A mathematical model of TiO_2 nanofluid flow over a slanted flat sheet soaked in a nonuniform absorbent medium in the occurrence of solar radiation is given in Chapter 8. Chapter 9 discussed the advection-diffusion equations and their applications in sciences and engineering. Differential Equation-Based Analytical Modeling of the Characteristics Parameters of the Junctionless MOSFET-based Label-Free Biosensors is detailed in Chapter 10. Chapter 13 explains the applications of fixed-point theory in differential equations. Finally, Chapter 14 shows the differential equation-based compact 2-D Modeling of asymmetric gate oxide hetero-junction Tunnel FET.

Key features of the book involve,

1. The book will be a must-read for a large group of readers like Professionals, Students, and Researchers in the fields of Science and Engineering, Health Informatics, and related fields.
2. Differential equation-based approaches to solving engineering problems.
3. It has clear explanations and examples and is very comprehensive.
4. State of art technological review and strong literature survey on the problems.
5. Not related to specific problems of a certain field, plenty of problems from different engineering fields have been discussed.
6. Discussions on the future scope to encourage and motivate the researchers.

The editors of this book have enormous expertise in the field of mathematical modeling. Dr. Papiya Debnath is presently working as an Assistant Professor in the Department of Basic Science and Humanities (Mathematics) at Techno International New Town (under Techno India Group), Kolkata, India. She has almost 13 years of experience in teaching, research, and administration. She has published more than 25 papers in national, international journals and conference proceedings. Her research interest lies in Mathematical Modeling and Simulation, Mathematical Computing, Earthquake Prediction, ML, and Data Science. Dr. Biswajit Sarkar is currently an Associate Professor in the Department of Industrial Engineering at Yonsei University, South Korea. He has dedicated his teaching and research abilities to various universities including Hanyang University, South Korea (2014–2019), Vidyasagar University, India (2010–2014), and Darjeeling Government College, India (2009–2010). He has worked as a Visiting Researcher in Tecnológico de Monterrey, Mexico in 2018. He has received 11 research projects. Under his supervision, 16 students have been awarded their PhDs and three students have been awarded their master's. He has published 180 journal articles in reputed journals of *Applied Mathematics and Industrial Engineering*, and he has published three books. Dr. Manash Chanda is working as Head of the Department of ECE at Meghnad Saha Institute of Technology. At present, he is the Chapter Advisor of the ED Meghnad Saha Institute of Technology Student Branch Chapter. Also, he is the Chairman of ED Kolkata Chapter. His current research interest spans the study of analytical modeling of sub-100-nm MOS-FETs and nanodevices considering quantum mechanical effects, low power VLSI designs, SPICE modeling of nano-scale devices, memory designs, etc. He has published more than 75 papers in refereed international journals of reputed publishers like IEEE, Elsevier, IET, Springer, Wiley, to name a few, and IEEE conferences.

Editors would like to thank the authors for their contribution to improving the quality of the book. We are indebted to the researchers for their valuable suggestions. We very much hope that this book will be efficacious to solve many research problems using the differential approaches mentioned here. Finally, the editors are especially grateful to their families for their tolerance, patience, and support throughout this very time-consuming project.

Dr. Papiya Debnath, Editor, Techno International New Town, Kolkata, India
Dr. Biswajit Sarkar, Editor, Yonsei University, South Korea.
Dr. Manash Chanda, Editor, Meghnad Saha Institute of Technology, Kolkata, India

April, 2022
MATLAB is a registered trademark of The MathWorks,
Inc. For product information, please contact:
The MathWorks, Inc.
3 Apple Hill Drive
Natick, MA 01760-2098 USA
Tel: 508-647-7000
Fax: 508-647-7001
E-mail: info@mathworks.com
Web: http://www.mathworks.com

Editors

Dr. Papiya Debnath completed Master of Science in Applied Mathematics from the Indian Institute of Engineering Science and Technology, Shibpur in 2006. She received an M.Phil degree in Applied Mathematics from the University of Calcutta, India in 2009 and PhD in Applied Mathematics from the University of Calcutta in 2019. Presently Dr. Debnath is working as an Assistant Professor in the Department of Basic Science and Humanities (Mathematics), at Techno International New Town (under Techno India Group), Kolkata, India. She has almost 15 years of experience in teaching, research, and administration. She has published more than 25 papers in national and international journals and conference proceedings. Dr. Debnath is also serving as a Guest Editor of the journal *WSEAS Transactions on Mathematics* (SCOPUS Indexed). Dr. Debnath has chaired different international conferences in India. Her research interest lies in Mathematical Modeling and Simulation, Mathematical Computing, Earthquake Prediction, ML, and Data Science. She is active in reviewing *Open Mathematics* (SCIE, SCOPUS Indexed), *Geomate* (SCOPUS Indexed), *Circuit World* (SCOPUS Indexed), and many other national/international journals. Recently she is actively engaged with a few IEEE conferences in IEEE R10 regions.

Dr. Biswajit Sarkar is currently an Associate Professor in the Department of Industrial Engineering at Yonsei University, South Korea. He completed his Bachelor's and Master's in Applied Mathematics in 2002 and 2004, respectively from Jadavpur University, India. He received his Master of Philosophy in the application of Boolean Polynomials from Annamalai University, India in 2008, his Doctor of Philosophy from Jadavpur University, India in 2010 in Operations Research, and his Post-Doctorate from the Pusan National University, South Korea (2012–2013). He has dedicated his teaching and research abilities to various universities including Hanyang University, South Korea (2014–2019), Vidyasagar University, India (2010–2014), and Darjeeling Government College, India (2009–2010). He worked as a Visiting Researcher in Tecnológico de Monterrey, Mexico in 2018. He has received 11 research projects. Between them, two projects are ongoing, and the rest have been completed. Under his supervision, 16 students have been awarded their PhDs and three students have been awarded their master's. Five students are continuing their PhD under his supervision. Under his guidance, four PhD theses have received the best theses award from Hanyang University, South Korea during 2018–2019. Since 2010, he has published 180 journal articles in reputed journals of *Applied Mathematics and Industrial Engineering*, and he has published three books. According to the SciVal 2020, Prof. Sarkar was the best active author in the topic cluster of the Supply Chain, Supply Chain Management, Industry during 2017–2020. According to SCOPUS, Prof. Sarkar is also the best active author in the sub-topic cluster in the field of Vendor Managed Inventory, Integrated Inventory Model, and Consignment Stock. According to Google Scholar, his total citation is 4788 and his h-index is 41.

He is the editorial board member of some reputed international journals of *Applied Mathematics and Industrial Engineering*. He is the Topic Editor of the SCIE-indexed journal *Energies*. He has served as the Guest Editor of five special issues of some SCIE-indexed journals. He is a member of several learned societies such as the Korean Society of Industrial Engineering, the Calcutta Mathematical Society, and many more. In 2014, his paper was selected as the best research paper in an international conference in South Korea. He has presented several research papers at international conferences as an Invited Speaker and chaired several sessions at several international conferences. He received a bronze medal for his capstone achievement from Hanyang University in 2016. He is the recipient of the Bharat Vikash Award as a young scientist from India in 2016. He received an International award from the Korean Institute of Industrial Engineers in 2017 at KAIST, Daejeon, South Korea. He is the recipient of the Hanyang University Academic Award as one of the most productive researchers in 2017 and 2018 consecutively. He has received the Top 100 International Distinguished Researchers 2020, Top 50 International Distinguished Young Researchers 2020, and Top 100 International Distinguished Educators 2020 from India in 2020.

Dr. Manash Chanda graduated in Electronics and Communication Engineering from the Kalyani Govt. Engineering College in 2005. He obtained his M.Tech degree in VLSI and Microelectronics from Jadavpur University. He completed his PhD in Engineering from ETCE Dept., Jadavpur University in 2018. At present; he is working as an Assistant Professor in the Department of ECE, Meghnad Saha Institute of Technology since February 2006. He is a Member of IEEE and is currently a Member of the IEEE Electron Device Society and Solid State Circuit Society. Dr. Chanda is a co-founder of the IEEE Student Branch and ED MSIT Student Branch Chapter. At present, he is the Chapter Advisor of the ED Meghnad Saha Institute of Technology Student Branch Chapter. Also, he is the Vice-Chairman of the ED Kolkata Chapter. He served as the Secretary of IEEE ED MSIT SBC from January 2018 to December 2019. His current research interest spans around the study of analytical modeling of sub-100-nm MOSFETs and nanodevices considering quantum mechanical effects, low power VLSI designs, SPICE modeling of nano-scale devices, memory designs, etc. He has published more than 75 papers in refereed international journals of reputed publishers and IEEE conferences. He is the reviewer of many reputed international journals and conferences. He is the recipient of a University Gold medal in M.Tech from Jadavpur University in 2008. One of his projects was selected in the Top 10 VLSI project design category (including B. Tech and M.Tech) all over India, organized by Cadence Design Contest, Bangalore, India in 2010.

Contributors

Nilkamal Bar received his Bachelor's and Master's in Applied Mathematics in 2013 and 2015, respectively, from Vidyasagar University, West Bengal India. He qualified for Junior Research Fellowship in the Joint CSIR-UGC Test in 2018. Now he is a research scholar at the Banasthali Vidyapith, India. His areas of research interest are Corporate Social Responsibility, Supply Chain Management, and Inventory.

Dr. Santanu Biswas is an IIT Madras alumni, with 7+ years of experience in teaching. Dr. Santanu Biswas joined Adamas University as an Assistant Professor in the Department of Mathematics (Science) in 2016. After obtaining a Master's degree in Mathematics from IIT Madras, Mr. Biswas pursued his Ph.D. from ISI- Kolkata. Presently, Dr. Biswas works as a post-doctoral research fellow at Jadavpur University. His research interests span several topics in mathematics, e.g., Mathematical Epidemiology, Nonlinear Dynamical Systems, Fractional Order Differential Equations, and Time Delay Differential Equation. Mr. Biswas has published 18 research articles in international journals.

Tanmoy Chakraborty is an Assistant Professor in the Department of Mathematics, College of Engineering and Technology, SRM Institute of Science and Technology, Kattankulathur, Chennai–603203, India.

Ruchi Chauhan is at the Department of Mathematics, Lovely Professional University, Phagwara, Punjab, 144 411, India.

Dr. Bikash Koli Dey is currently a post-doctorate fellow in the Department of Industrial & Data Engineering at Hongik University, Seoul, Republic of Korea. His research interests are production planning, inventory control, and smart manufacturing system. He received a B.S. and M.S. degree in mathematics and Applied mathematics from Vidyasagar University, India, and acquired his PhD in Mathematics from Banasthali Vidyapith, India. He published several articles in international journals. He served as a reviewer for several national and international journals. He is also a member of several learned societies.

Sudipta Ghosh is an Assistant Professor, Department of ECE, Meghnad Saha Institute of Technology, Kolkata, India.

R. T. Goswami is the Director at the Techno International New Town, Kolkata-700156, India.

Dr. Tapas Kumar Jana is currently an Assistant Professor in Mathematics at Ghatal Rabindra Satabarsiki Mahavidyalaya, Ghatal, West Bengal 721212, India. He obtained his B.Sc. with Honors in Mathematics and M.Sc. (1st rank in first class) degree in Applied Mathematics from Vidyasagar University, India, and PhD in Applied Mathematics from Jadavpur University, India. He is the recipient of the NBHM scholarship in M.Sc., as a Research Fellow of the Indian Statistical Institute. He has a teaching experience of more than 13 years. His research interests are in quantum mechanics, production planning, and inventory control. He published several articles in international journals.

Varun Joshi is an Assistant Professor in the Department of Mathematics, Lovely Professional University, Phagwara, Punjab, India (144411).

Mamta Kapoor works as an Assistant Professor in the Department of Mathematics, Lovely Professional University, Phagwara, Punjab, India (144411).

Dr. Asish Karmakar, Ph.D. was born in North 24 Parganas, W. B., India on 12th July of 1977. He completed his Master of Science and Doctor of Philosophy from the University of Calcutta in 2002 and 2017, respectively. He has service experience of 17 years as an Assistant Teacher in a high school. He started his carrier at Balti High School (H.S.), PO-Balti, North 24 Parganas, W.B., India, as an Assistant Teacher on the 27th December of 2003 and then he was transferred to Udairampur Pallisree Sikshayatan (H.S.), PO-Kanyanagar, PS-Bishnupur, Pin-743398, South 24 Parganas, W.B., India, on 5th November 2009. Also, the author has at least 10 years of research experience. His field of research work is basically confined to earthquake mechanisms. On this subject, the author has already published more than five research papers and he is interested to publish more research papers further in international journals.

Deepak Kumar is at the Department of Mathematics, Lovely Professional University, Phagwara, 144411, India.

Arunava Majumder is an Assistant Professor in the Department of Mathematics, Lovely Professional University, Phagwara 144411, Punjab, India.

Dr. Bula Mondal, Ph. D., was born in South 24 Parganas, West Bengal, India on 27th January 1979. She has achieved a Master of Science and a Doctor of Philosophy from the University of Calcutta in 2002 and 2017, respectively. She has been serving as an Assistant Teacher at Joka Bratachari Vidyasram Girls High School, Joka, Kolkata-104, West Bengal, India, since 2003. Also, the author has at least 10 years of research experience, which also includes Solid Mechanics. The author has published more than four research papers and is interested in publishing more in international journals.

Ashish Kumar Mondal received his bachelor's and master's degrees in applied mathematics from Jadavpur University, India, in 2002 and 2004, respectively. He has dedicated his teaching experiences in various schools and colleges from 2005 to 2018 in India. Now he is a research scholar at the Banasthali Vidyapith, India. His main research interests include supply chain management, inventory, and sustainability. He is also a member of several learned societies.

Richa Nandra is at the Department of Mathematics, Lovely Professional University, Phagwara, Punjab, 144 411, India.

Dr. Isha Sangal is an Associate Professor in the Department of Mathematics "& Statistics at Banasthali University, Rajasthan, India. She completed her Ph.D. (Inventory Management) in 2013. Earlier, she completed her M.Sc. in Mathematical Sciences with a specialization in Operations Research in 2008 from Banasthali University, Rajasthan. She has 14 years of teaching and research experience. Her research interests include inventory control and management, supply chain management, and fuzzy set theory. She has to her credit a number of publications in the national and international journals of repute.

Dr. Sharmila Saren is currently working as an Assistant Professor in the Department of Mathematics, Government General Degree College at Gopiballavpur-II, Jhargram, West Bengal, India. She completed her Bachelor's and Master's in Mathematics and Applied Mathematics in 2009 and 2011, respectively from Vidyasagar University, India. She has received a silver medal for claiming the second position in her Master's. She was passionately devoted to teaching at Raja Narendra Lal Khan Women's College, Midnapore, West Bengal, India (2011–2013). She received her Doctor of Philosophy from Vidyasagar University, India in 2020 titled Some Problems on Production Models. Her area of research interests are Inventory Management, Production Planning, Sustainable Development, Supply Chain Management, and Waste Management. She has published eight journal articles in reputed journals of *Applied Mathematics*. She has served as Guest Associate Editor and Review Editor of Sustainable Supply Chain Management. She is a lifetime member of the Operational Research Society of India (ORSI). She has presented various research papers at international conferences.

Arghyadeep Sarkar is at Department of ECE, Macmaster University, Hamilton, Canada.

Mitali Sarkar is at the Information Technology Research Centre, Chung-Ang University, Seoul, South Korea.

Avtar Singh is at Department of ECE, ASTU, Ethiopia.

1 Introduction

Papiya Debnath
Techno International New Town

Biswajit Sarkar
Yonsei University

Manash Chanda
Meghnad Saha Institute of Technology

Mathematics is one of the fundamental disciplines which has an immense influence on the advancement of modern society. Pure mathematics and applied mathematics are indispensable to developing the skill of mathematical modeling and problem-solving. However, applied mathematics is more essential, nowadays, to drive the advancement of science and technology. A plethora of mathematical formulas have been used to fulfill the demand for technology. Out of these, differential equations are the pivotal achievements without which the growth of the technology would not be possible. Over the last hundred years, Differential equations play a major role in mathematics. Besides academics, these differential equation-based approaches help scientists and engineers to solve the real problems arising from science and engineering. The book is intensely a pure high order of mathematics and very extensive in both depth and breadth. Connecting these equations in a single to multiple theories altogether and into applications in critical real-life problems from different engineering fields like electronic devices, Inventory control, biomedical engineering, aerodynamics, fluid mechanics, heat and thermodynamics, etc are the main theme of this book. An adequate literature survey will be detailed for understanding the problems clearly. A mathematical model to interpret the problem will be required also. As it is impossible to cover all the available techniques due to page limitations of the book, mainly Ordinary Differential Equations, Partial Differential Equations, Linear Differential Equations, Nonlinear Differential equations, Homogeneous Differential Equations, Non-homogenous Differential Equations, etc. will be used to solve the problems. MATLAB codes along with some online videos related to the differential methods will be provided as a reference to attract the readers. Not only researchers and scientists but also a large no of students and faculty members will also be benefited from the book because of the versatile problems and their solutions. Along with the solutions, some future scopes and the research ideas to proceed will also be detailed to motivate the researchers and to continue further studies on the issues if necessary. It also will help them to move into a higher order of mathematical reasoning on the basis of given equations from multidimensional characterizations.

DOI: 10.1201/9781003227847-1

To calculate the position of inventory for any product in the industry, the usefulness of the differential equation cannot be negligible. The differential equation shown in Chapter 2 helps interrogate the best value of the governing variable, which directly produces the optimal profit or cost for the inventory system. The current study focuses on a production-inventory model calculated for the products, which deteriorate with time. Since a product's demand can not always be constant, it is well-known that the selling price takes a critical role in determining the product's demand. Thus, considering the real applicability of this study, demand varies with the selling price. Owing to demand variability, the production rate is also considered controllable. Due to the deteriorating product production system, it is too essential to produce the exact amount of product; otherwise, the system may face a huge loss. To satisfy the consumer and earn more revenue, a credit period is provided through which some interest is generated. Finally, numerical studies are performed to show the study's necessity and awareness of the critical parameters. Numerical findings in Chapter 2 established that the current study is 77.04% beneficial over the traditional one due to variable production. The present study detailed in Chapter 2 proves the differential equation's applicability to solving the production-inventory model under a flexible production rate. In Chapter 4, an inventory model for deteriorating items is developed under the trade-credit policy. To ensure the actual applicability of this model, demand varies with the selling price of the item. Numerical results prove that the optimum result is obtained when the delay period exceeds the total cycle length. Numerical findings established that the current study is 29.30% beneficial over the traditional one due to demand variability. Sensitivity analysis for the valuable parameters is executed to establish the effect of the parameters on the system profit. In Chapter 11, the formation of differential equations with various demand structures is examined and how they assist to solve an optimization problem is studied. The study assumed an inventory model with Weibull deterioration under ramp-type demand. The model is studied with shortages at inventory starting and at the end of the inventory. The numerical solution is obtained for optimized decisions and minimum cost calculation.

Differential equations are very much useful to integrate the scientific theory and the mathematical model related to geological phenomena like geophysics, sedimentary geology, petrology and geochemistry, structural geology, etc. Geomathematics is used to study and model geological phenomena extensively. In general, when differential calculus is applied to geological problems, the parameters must be rigorous and should be set cautiously to consider nontrivial derivations. One of the pivotal issues is choosing the meaningful variable. The governing equations are PDE with suitable boundary conditions. These PDEs are solved by using suitable mathematical and numerical methods, involving integral transforms, Green's functions, correspondence principle, etc. MATLAB programs are used for numerical computations and graph plotting. This type of modeling of aseismic deformations has a major area of research in Geodynamics, Geophysics, Engineering and especially Seismology. In a viscoelastic half-space, tectonic force is generated due to mantle conversion and associated phenomenon. This tectonic force is the pivotal reason for the movements of the lithospheric plates which can lead to devastating earthquakes. In between two major seismic incidents, i.e., the aseismic period, gradually stresses will build up. When these accumulated stresses cross the threshold limit, some creeping movements are set in across the inclined strike-slip faults (SSF). For proper characterization of these interacting faults, mathematical modeling of the strain, stress, and displacements (SSDs) are indispensable. Using the differential equation, the model can be built up with better accuracy. However, with the help of numerical computations, one can estimate the rate of stress accumulation under certain boundary conditions, which will be efficacious for the characterization of the interacting faults. Chapter 3 models the explosion in a spherical cavity expanding with decelerated velocity. Chapter 6 details the modeling considering an inclined interred strike-slip fault which is located in the visco-elastic layer with inclination. A mathematical model of the SSDs is developed in Chapter 12 considering the behavior of interacting faults under increasing tectonic forces. This type of modeling of aseismic deformations has a major area of research in Geodynamics, Geophysics, Engineering, and especially Seismology. A rigorous analysis of the results thus obtained may lead us to a successful prediction of the next major event.

In Chapter 5, we have discussed the existence of positive solutions to a nonlinear third-order boundary value problem by using an extension of the Krasnosel'skii fixed point theorem and Avery - Peterson fixed point theorem. Also, numerical solutions for the third-order boundary value problem are analyzed by Bernoulli polynomials. The numerical scheme & the existing results may be helpful for Eco-epidemiology & food chain models.

In today's competitive business world, it is important to find out suitable strategies for supply chain managers when working with substitutable and complementary products to make the supply chain more profitable. In Chapter 7, the objective is to find out selling strategies for fixed lifetime deteriorated substitutable and complementary products holding sustainability of the supply chain. This chapter is structured with one common retailer and two manufacturers where the manufacturers produce either a substitutable product or complementary product under a flexible production system. Then the supply of those products to the common retailer follows the single setup multiple-equal delivery (SSMD) policy. The study maximizes the profit by getting the optimal values of the production rate, the cycle length of the retailer, and the number of shipments. The results explore that the system profit is maximum for the complementary product with low cross-price elasticity and for substitutable products the profit is maximum with high cross-price elasticity. Also, the result shows that with the increase in the retailer's cycle time, products' deterioration rate increases but the adaptation of SSMD strategy reduces the retailer's cycle time and carbon emission by 8% and increases the system total profit by 10%.

Nanofluidics and microfluidics are the most important part of computational fluid dynamics (CFD). Because of simplified geometrical presentation, incomplete mathematical models and numerical errors impose limitations on the acceptability of the computer results related to the CFDs. So, the formulation of the mathematical model is very much effective to accelerate the simulation of the CFDs. To formulate the model, differential equations and the relevant boundary conditions have been used extensively in Chapter 8. Here, a mathematical model of TiO_2 nanofluid flow has been addressed over a slanted flat sheet soaked in a non-uniform absorbent medium in the occurrence of solar radiation. Also, the consequence of suction/injection with velocity and thermal slips is considered at the solid-liquid boundary to understand the flow behavior in a porous medium. Similarity transformations are imposed on the leading partial differential equation. Numerical fallouts are gained through the Runge Kutta-4th order method coupled with shooting practice. Chapter 8, highlights the potential applications of the partial differential equations to solve the nanofluidic and microfluidic issues.

The model of advection-diffusion equation (ADE) plays a vital role in different fields, as it is treated as the model of different areas like dispersion, diffusion processes, transport, or intrusion in different media. Applications of such types of equations are helpful in tackling a large number of models in various areas of science, like biochemistry, aerospace sciences, ecology, and many more. In Chapter 9, an attempt has been presented to put light upon the origin of the ADE, the classification of ADEs, its applications in different fields of sciences as well as a deep review of the literature regarding such types of equations.

In recent decades, the design and analysis of biosensors have gained significant attention due to the COVID-19 pandemic worldwide. When we analyze the biosensors, the sensitivity, selectivity, limit of detection, linearity, etc need to be analyzed thoroughly. It can be done theoretically and experimentally. Experimental analysis can give more accurate data; however, the process is not cost-effective. Before the experimental approach, mathematical modeling of the system is highly required to estimate the figure of merit roughly. These models are basically based on the differential equation, which is solved considering some boundary conditions based on the device parameters and applications. Once the differential equation-based model has been formed, the model needs to be calibrated. If the experimental data is available then the calibration can be done more accurately. However, in the case of real-time complicated issues, the experimental data is not available in general. Then the mathematical models can be calibrated with the help of simulated data to enhance the accuracy of the models. Differential equation-based analytical modeling of the characteristics parameters of the junctionless MOSFET-based label-free biosensors has been discussed here in-depth in Chapter 10.

Fixed point theory is an old and rich branch of analysis. A point under a defined transformation is called a fixed point when it remains invariant, irrespective of the nature of the transformation. The theorem which guarantees that a function under certain conditions has at least one fixed point is called the fixed point theorem. It has immense applications in mathematics especially; the theory has gained a remarkable scope of research in nonlinear analysis. One of the famous theorems called Schauder's fixed point theorem is useful to check the existence of the solution of the differential equations under some boundary conditions. In Chapter 13, we gave two mathematical models for the flow of Casson fluid flow and Powell-Eyring fluid flow in the annulus of rotating concentric cylinders in the presence of a magnetic field, and the existence of the solutions of their non-dimensional governing equations are presented using Schauder's fixed point theorem.

Compact modeling of the asymmetric gate oxide heterojunction Tunnel FET has been discussed in-depth in Chapter 14. A variable separation method has been adopted in this work for modeling the asymmetric dual metal heterojunction TFET (A-DMDG-HTFET) and a comparative performance study is carried out with a single metal double gate hetero junction TFET (SMDG-HTFET) structure. Modeled data have been compared with SILVACO TCAD simulation results to validate the proposed analytical model.

The main highlight is on the derivation and solution of major equations of engineering and applied science for understanding and thinking mathematically while highlighting the supremacy and significance of mathematics in science and engineering. The purpose of this book is to describe comprehensive research in the recent expansion of differential equations correlating with qualitative and quantitative (numerical) analysis and mathematical modeling in the engineering and applied sciences.

2 Application of Differential Equation in a Flexible Production of Deteriorating Item under Trade Credit Policy

Bikash Koli Dey
Hongik University

Nilkamal Bar
Banasthali Vidyapith

Sharmila Saren
Government General Degree College

Tapas Kumar Jana
Ghatal Rabindra Satabarsiki Mahavidyalaya

Biswajit Sarkar
Yonsei University
Saveetha University

CONTENTS

DOI: 10.1201/9781003227847-2

2.1   INTRODUCTION

The application of differential equations in the inventory model is unavoidable. Without inventory level calculation, it is impossible to decide for the industry. Generally, most of the products that are use on day-to-day basis are deteriorating items, and each product has a particular expiry date. If the expiry date is over, the product is useless and may be used for a secondary product (Saxena et al., 2020). However, if the product starts to deteriorate and can not be sold within the expiry date, the company may face a huge loss. Thus, to optimize the loss of the company, it is necessary to take proper decision on the production rate (Sarkar et al., 2022a). In constant production rate, one major problem is if the amount of the product is much more than the demand, then excess products start to deteriorate, after that the industry faces a huge loss; on the other hand, if the production rate is much less then there is demand, and again the company faces a shortage situation and loss. To overcome the shortage situation and optimize the cost/profit, controllable production rate is too effective for any SCM (Sarkar and Chung, 2020).

It is always not possible for the consumer/retailers to pay the whole amount of the product. Thus, to satisfy the customers, a certain delay period is provided to the consumer; in that period, they can pay the amount of the product, and no extra amount will be charged for this delay period. But, if this delay period is over, a certain percentage of the amount will be charged as a late fine. The holding cost of the retailer and credit period are interlinked (Khan et al., 2021). By providing a credit period, the industry can save the holding cost of the product. Moreover, to attract more consumers, instead of a single-level credit period, a two-level credit period is considered in this study.

Trade credit satisfies the customer and plays a critical part in determining the demand for the product. Owing to different real situations, demand can not always be consistent. In this competitive business environment, determining the exact demand is very beneficial for any retailing strategy. Practically, the selling price of the products is inversely proportional to the demand for products. Thus, this current study tries to develop the optimal retailing strategy under the controllable production rate, where demand varies along with the credit period and selling price of the product.

To obtain an optimal retailing strategy, the following questions arise:

1. How much selling price is beneficial to obtain optimum profit?
2. How much cycle time is appropriate for an inventory model consisting of deteriorating items?
3. How much credit period is optimal to satisfy the customers and earn more revenue?
4. What is the rate at which the product can provide the best result?

By occupying the above questions, the current model focuses on the following research gaps

1. Demand variability for a production inventory model is widespread. In contrast, some models were developed along with price elasticity demand rate (Dey et al., 2019b), some inventory systems were credit period-dependent demand rate (Sarkar et al., 2020b). However, selling price and two-level credit period-dependent demand rates are infrequent.
2. In general, the rate of production for basic inventory model of deteriorating products is treated as constant (Sett et al., 2016; Li et al., 2019), whereas controllable production rate is more reliable for several SCM (Dey et al., 2021a). However, a flexible production rate for the retailing strategy of the deteriorating product is quite often in this path of research.
3. A single credit period was applied in most of the existing studies for the deteriorating item (Mashud et al., 2021; Pervin et al., 2019). However, the two-level credit period is rare in this research direction.

Based on the above discussion, it is found that a big research gap in the literature was there for the deteriorating item. The present study focuses on those research gaps and tries to provide the optimal retailing strategy for the deteriorating items under flexible production rate with the help of differential equations.

The rest of the manuscript goes as follows: gaps in the literature are discussed in the next Segment 2 as the literature review. The description of the problems, assumptions, which are considered to formulate this research, and notations used throughout the model is illustrated in Segment 3. Section 2.4 contains detailed modeling with the help of differential equations. Section 2.5 contains the methodology used to obtain the best result of the controlling variables. To establish the concavity and applicability of this model, some numerical cases along with comparison are provided in Section 2.6. The efficiency of the parameters is elaborated in Section 2.7. What are the findings? And what the industry managers can decide in this study? It is provided in Section 2.8. Finally, the study's research benefit, future research direction, and limitation are describe in the Conclusion part of Section 2.9.

2.2 PREVIOUS STUDIES AND LITERATURE GAP

In this section, the gaps in the literature and the previous studies in this direction are presented in detail

2.2.1 DEMAND VARIABILITY

The journey of the inventory model started several years ago. Researchers always try to incorporate different new strategies along with the inventory system so that the industry can optimize the system's cost or profit. Most of the products which are necessary for our daily life are deteriorating in nature. Simultaneously, deciding on the demand is crucial for smoothly maintaining the inventory situation. Traditional inventory systems for deteriorating items were developed under constant demand rate (Wu et al., 2014), whereas, in reality, the demand for the deteriorating product can not be constant (Pando et al., 2021). Different practical situations and issues were responsible for this demand variability (Dey et al., 2021a). Stock of any inventory system plays a vital role in determining the demand (Mandal et al., 2021). The selling price of the product is always much more responsible for demand variability of any product and selling price of the product is inversely proportional to the demand of the product (Sarkar et al., 2021a). Some customers care about the quality of the product in spite of the price of the product (Dey et al., 2021b).

However, in reality, most of the customers try to buy the product with lesser price (De and Sana, 2015; Sarkar and Saren, 2015). Bai et al. (2015) calculated the effect of the selling price for a deteriorating item under an SCM environment. An inventory system along with quality discount and price variability demand rate was constructed by Alfares and Ghaithan (2016). They also considered a time-varying holding cost. In the same year, Rabbani et al. (2016) discussed the policy for replenishment and pricing of deteriorating items. A pricing strategy for deteriorating items was formulated by Hsieh and Dey, (2017), under the consideration of the product's freshness. Feng et al. (2017) constructed an inventory policy in which they considered that demand varies with the product's stock level and expiry date. The effect of the freshness of the product and selling price in a lot size model was proved by Li and Teng (2018). They also focused on the stock level of the product. In the year 2019, an integrated system for retailing was proposed by Dey et al. (2019b) under the assumption of price elasticity demand. A single credit period variability demand was considered by Li et al. (2019b). The best pricing policy for perishable items was calculated by Hassan et al. (2020). They considered agriculture products for their study. Recently, Mandal et al. (2021) constructed their inventory strategy for deteriorating products along with credit-level dependency demand rate.

Many inventories or SCM models were constructed within selling price- dependent demand, or credit period dependent demand, or freshness- dependent demand. However, optimal retailing and pricing strategies for inventory systems, along with demand dependency on selling price, deterioration rate, and a two-level trade credit period are unique. Thus, an innovative approach through a differential equation is illustrated in this study to overcome these research gaps.

2.2.2 CONTROLLABLE PRODUCTION RATE

The variable production system is too effective due to demand variability and the deterioration of product production systems. Another necessity of the variable production system is to overcome the shortage situation. The main problem of the deteriorating product production system is that if the produced product is too high compared to demand, the industry faces a huge loss, as products are starting to deteriorate after a certain time period. On the other hand, if the amount of the produced product is less than the demand, then the company again faces shortages, leading to a huge loss for the company. Thus, the variable production rate is very much essential for deteriorating product production system.

The concept of the variable production rate for a production model was first considered by Khouja and Mehrez (1994). Khouja (1995) extended his own model by considering scheduling and lot-sizing approaches. Giri et al. (2005) designed a production inventory model along with the consideration of a controllable production rate. The concept of FPR and maintenance policy for a production model were considered by Ayed et al. (2012). An SCM for multi-supplier and single-manufacturer under controllable production rate was analyzed by Choi et al. (2013). An integrated system for single-vendor and single-manufacturer along with stochastic demand and flexible production was considered by Al Durgam et al. (2017). A single-vendor and multi-buyer SCM was considered by Sarkar et al. (2018b), in which they discussed the effect of variable production rate (VPR). Sarkar et al. (2018b) model was extended by Dey et al. (2019a). They considered two different strategies to reduce lead time along VPR. Recently, a smart SCM was constructed by Dey et al. (2021a), in which demand is dependent on the advertisement and the production rate was flexible. They also calculate the exact lead time with the help of lead time variance. A sustainable green product production model along with the concept of remanufacturing was incorporated by Sarkar et al. (2022b). In a similar direction, the smart production model along with VPR for biofuel manufacturing was considered by Sarkar et al. (2022a).

Different researchers in the literature developed several production models by considering VPR. However, a production inventory model and optimal retailing strategy for deteriorating items are still not considered by any existing literature.

2.2.3 DETERIORATION UNDER TRADE CREDIT SCENARIO

The daily life usable products are most of the deteriorating in nature, which has a maximum life period and everyone needs those items to spend their daily life. Sometimes the cost of those products is a little bit high, or when a consumer buys several deteriorated products in a single batch, then at a time, they have to expense a huge amount of money. To pay a large amount at a single time is quite a headache for the customers. On the other side, if the deteriorating products are not solid within the product's life span, then the company may face a huge disadvantage. Credit period or trade credit is the best option to overcome both situations. In trade credit, customers do not need to pay the whole amount at a time, which is a relief for the customers.

The company can save their holding cost and sell their deteriorate product smoothly (Sarkar and Sarkar, 2013).

Sarkar (2012) formed an inventory strategy to emphasize the effect of delayed payment under a variable decay rate. Sarkar et al. (2012) established an inventory model where both backorder and demand depended on time under shortage situations for deteriorating products. Sarkar (2013) considered a two-echelon SCM for deteriorating products and the rate of deterioration being probabilistic. In the same direction, another SCM was developed by Ghiami et al. (2013), where they cared about the capacity and partial backorder. Das et al. (2013) constructed a model for the deteriorating product with price-dependent credit period. The policy of credit financing and optimal retailing in an SCM for deteriorating products was proposed by Chen and Teng (2014). Some policies for inventory management involved deteriorating products and trade credit by Bhunia and Shaikh (2015). They utilized the PSO metaheuristic approach to solve their model. The optimal strategy for retailing under the consideration of deteriorating products was presented by Sarkar et al. (2016) and Sett et al. (2016), where the rate of deterioration was variable. The concept of advanced payment for a deteriorating product and the policy for lot size was evaluated by Teng et al. (2016). A two-warehouse model including deteriorating products and trade credit financing was developed by Tiwari et al. (2016). They also discussed the optimal retailing strategy for their model. Ordering and transportation policy for a deteriorating item under the consideration of warehouse was formulated by Sarkar and Saren (2017). Banerjee and Agrawal (2017) presented the optimal policy to order and discount. Mahata and De (2017) formulated an SC for deteriorating products under credit risk. Sarkar et al. (2018) proposed a SC to point out the effects of single-setup-multiple-delivery (SSMD) and carbon emission under multidelay-in-payments. Bhunia et al. (2018) illustrated an inventory system that included deteriorating items under different market strategies and variable demand. They used two separate metaheuristic approaches (genetic algorithm and particle swarm optimization) to solve their model and proved that the GA algorithm provided better results than the PSO algorithm. Li et al. (2019) proposed an inventory system in which replenishment policy and pricing decisions were introduced, also considered a deteriorating product and preservation policy for controlling the rate of deterioration. Tiwari et al. (2019) formulated a sustainable inventory system under multi-trade credit where the optimal policy for ordering was applied. A (Q, r) inventory system under random demand for deteriorating products was developed by Braglia et al. (2019) by considering positive lead time. With a similar assumption for deteriorating products and complete backlogging, Tiwari et al. (2020) developed an inventory system with an optimal ordering policy. Saren et al. (2020) considered a discount policy for deteriorating items, where they provided the effectiveness of a delay-in payment policy. Mashud et al. (2020) developed an inventory strategy that included deteriorating products considering advanced payment and credit period policy. In the environment of default risk and a two-stage trade credit policy (Mahata and Mahata, 2021), proposed an order quantity model with deteriorating items. A production inventory system was proposed by Das et al. (2021a) where partial trade credit was considered along with reliability for deteriorating products.

Table 2.1
Contribution in Existing Literature

Author(s)	Retailing	Item	Demand Dependency	Credit -Period	Production Rate
Ayed et al. (2012)	NA	ND	Constant	NA	Variable
Das et al. (2013)	Yes	Det.	Constant	Yes	Constant
Sarkar (2013)	SCM	Det.	Constant	NA	Constant
Wu et al. (2014)	INV.	Det.	Constant	TL	Constant
De and Sana (2015)	EOQ	ND	SP	NA	Constant
Rabbani et al. (2016)	Yes	Det.	SP	NA	Constant
Jaggi et al. (2017)	EOQ	Det.	SP	Yes	Constant
Sarkar et al. (2018b)	SCM	ND	Constant	NA	Variable
Li et al. (2019)	Yes	Det.	SP	NA	Constant
Tiwari et al. (2020)	Yes	Det.	Constant	NA	Constant
Sarkar et al. (2020b)	Yes	Det.	Credit period	TL	Constant
Mahata and Mahata (2021)	INV.	Det.	Constant	TL	Constant
This model	Yes	Det.	SP & TL	TL	Variable

NA, Not applicable; INV., Inventory; SCM, Supply chain management; Det., Deteriorating item; TL, Two-level credit period; SP, Selling price; ND, non-deteriorating product; EOQ, Economic order quantity.

Several inventories, manufacturing, or SCM were proposed for a deteriorating item under the consideration of trade credit policy or delay-in-payment strategies. Still, the major issue for deteriorated product production is CPR, where demand depends on selling price, rate of deterioration, and two-level credit financing which are still not present in earlier research. Thus, a pioneering experiment is done in this study to fill those research gaps. In Table 4.1, it is trying to show those gaps in the literature and the novelty of the present study compared to the existing literature.

2.3 DESCRIPTION OF PROBLEM, SYMBOLS, AND ASSUMPTIONS

In depth analysis of the problems and the necessity of the model are discussed in this segment.

2.3.1 PROBLEM DEFINITION

The replenishment decision for the retailer is one of the most important decisions in the industry. To apply this study in reality, the demand for deteriorating items varies with the price for selling and credit period. This study was developed for the deteriorating items, where the rate of deterioration is exponential and function of on-hand inventory (Sarkar et al., 2015). Due to demand variability and controlling the shortage situation rate of the production is considered as flexible, where the cost for unit production varies with ingredient cost, improvement cost, and tool/die cost. To earn more revenue supplier provides some delay period to the retailer, in which no extra amount will be charged, and after that, some percentage of the amount will be charged as a late fine. Three different cases are described based on the delay period. Finally, the entire profit of the retailing is determined based on the best production rate, delay period, entire time cycle, and price of the item for selling.

2.3.2 NOTATION

The symbols which are used throughout the model discussed in this segment

Decision	Variables
s_u	Price for sell the product (\$/unit)
M_r	Offered delay period (days)
Υ	Entire cycle of time (days)
P_r	Rate of the production (unit/cycle)
Parameters	
$D(s_u, M_r)$	Demand which depends on s_u and M_r (unit/cycle)
$\tau(s_u)$	Demand when $M_r = 0$
$\rho(s_u)$	Denotes maximum demand
γ	Saturation for the demand, such that $0 \leq \gamma < 1$
c_o	Cost for order per ordering (\$/order)
h_c	Cost to hold (\$/unit/unit time)
ω_1	Development cost (\$/unit)
ω_2	Tool/die cost (\$/unit)
c_p	Cost related to material per item (\$/item)
M_s	Permissible delay period provided to retailer through the supplier
$S(t)$	Level of the inventory at time t, where $0 \leq t \leq \Upsilon$
t_s	Production stoping time in a cycle
R_1	Earned interest (\$ /year)
α	Rate of deterioration, where $0 \leq \alpha < 1$
R_2	Charged interest through the supplier (\$/year)
Ω	Lifetime of the deteriorating products
$T_P(s_u, \Upsilon, M_r, P_r)$	Annual total profit (\$/cycle)

D is alter with $D(s_u, M_r)$ throughout this manuscript.

2.3.3 ASSUMPTIONS

With the help of the following hypothesis, this study is conducted:

1. The retailing model is constructed for deteriorating items, where items are packed in a container and sent to the market for selling.
2. Due to deteriorating item, the demand function i.e., $D(s_u, M_r)$ is directly proportional with M_r and inversely proportional with price s_u.
3. Owing to the demand variability, deterioration, and overcoming shortage situations, the rate of production is treated as flexible, which helps to enhance the revenue of the industry. The material cost, tool or die cost, and product improvement cost help to determine the unit cost of production. Thus, the cost

function for the production is obtained as $c_p + \omega_1 P_r + \frac{\omega_2}{P_r}$ (Sarkar and Seo, 2021).

4. A time-varying rate of deterioration of the product is adopted. Therefore, time-varying deterioration rate is considered as $\alpha = \frac{1}{1+\Omega-t}$; Ω denotes the maximum lifetime of the deteriorating item whereas $\alpha \to 1$ if the product reaches its maximum lifetime i.e., when, $t \to \Omega$. (See Sarkar (2012)).

5. Upto period of time M_s, revenue will be earned by the retailer at a yearly rate R_1. After completing the period of credit $t = M_s$, the retailer has to pay a certain amount of interest with an annual rate R_2 for reaming items.

6. Each client enjoy some credit period M_r to complete the remaining payment. The skyline of the time is boundless.

7. Cost for unit holding is calculated based on the sum of robust stock. The shortage situation is neglected and the time difference between placing and receiving items is negligible.

2.4 MATHEMATICAL MODEL

On the basis of such assumptions, that for a positive selling price $s_u > 0$, the demand of deteriorating products vary proportionally with the credit period (See Figure (2.1)). Now, by using the concept of differential equation the demand pattern is obtained as

$$\frac{\partial D(s_u, M_r)}{\partial M_r} = \gamma[\rho(s_u) - D(s_u, M_r)]. \tag{2.1}$$

Then we have

$$\frac{\partial D(s_u, M_r)}{\partial M_r} + \gamma D(s_u, M_r) = \gamma\rho(s_u). \tag{2.2}$$

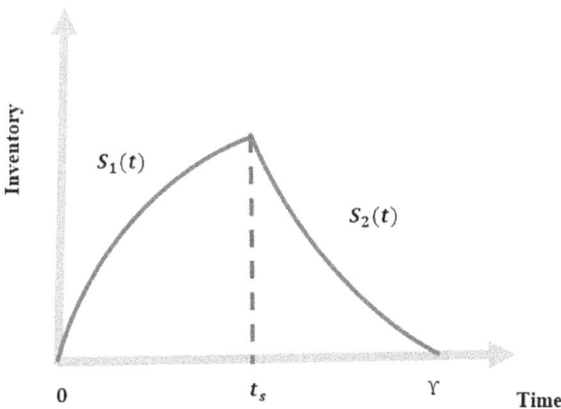

Figure 2.1 Inventory position of the deteriorating items.

With the help of the initial condition $M_r = 0$, $D(s_u, M_r) = \tau(s_u)$ demand can obtain from equation (2.2) as:

$$D(s_u, M_r) = \rho(s_u) - [\rho(s_u) - \tau(s_u)]e^{-\gamma M_r}. \tag{2.3}$$

where $\rho(s_u)$ denotes maximum demand, and saturation for the demand denoted by γ, such that $0 \leq \gamma < 1$.

Again, one can obtain the demand as follows:

$$D(s_u, M_r + 1) - D(s_u, M_r) = \gamma[\rho(s_u) - D(s_u, M_r)]. \tag{2.4}$$

Combining equations (2.3) and (2.4),

$$D(s_u, M_r) = \rho(s_u) - [\rho(s_u) - \tau(s_u)]e^{-\gamma M_r}. \tag{2.5}$$

$$\text{and } D(s_u, M_r) = \rho(s_u)[1 - (1-\gamma)^{M_r}] + \tau(s_u)(1-\gamma)^{M_r}. \tag{2.6}$$

Graphical representation of the level of inventory $S(t)$ provided in Figure (2.1) which obtained by following differential equation:

Within time interval $0 \leq t \leq t_s$,

$$\frac{dS(t)}{dt} + \alpha S(t) = P_r - D(s_u, M_r) \text{ if } 0 \leq t \leq t_s, \text{ where } \alpha = \frac{1}{1 + \Omega - t}. \tag{2.7}$$

With the help of deterioration rate α, one can rewrite equation (2.7) as:

$$\frac{dS(t)}{dt} + \frac{1}{1 + \Omega - t} S(t) = P_r - D(s_u, M_r). \tag{2.8}$$

By the help of trivial conditions that mean $t = 0$, and $S(0) = 0$, equation (2.8) can be rewritten after integration as:

$$S(t) = (P_r - D(s_u, M_r))(1 + \Omega - t) \log\left(\frac{1 + \Omega}{1 + \Omega - t}\right). \tag{2.9}$$

During interval of $t_s \leq t \leq \Upsilon$, one can formulated the differential equation as:

$$\frac{dS(t)}{dt} + \alpha S(t) = -D(s_u, M_r) \text{ if } t_s \leq t \leq \Upsilon, \text{ where } \alpha = \frac{1}{1 + \Omega - t}. \tag{2.10}$$

Again by the help of deterioration rate α, equation (2.10), can be modified as:

$$\frac{dS(t)}{dt} + \frac{1}{1 + \Omega - t} S(t) = -D(s_u, M_r). \tag{2.11}$$

After integrating, equation (2.11) rewritten as:

$$S(t) = (1 + \Omega - t)D(s_u, M_r) \log\left(\frac{1 + \Omega - t}{1 + \Omega - \Upsilon}\right) \tag{2.12}$$

Thus, with the help of the concept of the differential equation, the inventory level can be determined as

$$S(t) = \begin{cases} S_1(t) & \text{if } 0 \leq t \leq t_s \\ S_2(t) & \text{if } t_s \leq t \leq \Upsilon \end{cases} \tag{2.13}$$

where,

$$S_1(t) = (P_r - D(s_u, M_r))(1 + \Omega - t)\log\left(\frac{1+\Omega}{1+\Omega-t}\right) \tag{2.14}$$

$$S_2(t) = (1 + \Omega - t)D(s_u, M_r)\log\left(\frac{1+\Omega-t}{1+\Omega-\Upsilon}\right). \tag{2.15}$$

With help of $S_1(t) = S_2(t)$, at $t = t_s$,

$$t_s = (1 + \Omega)\left[1 - e^{-\frac{D(s_u, M_r)}{P_r} \log \frac{(1+\Omega)}{(1+\Omega-\Upsilon)}}\right] \tag{2.16}$$

2.4.1 HOLDING COST (HC)

To keep the deterioration product some associated cost is required, which is known as the holding cost. For the present study, the annual holding cost without interest charged is provided as:

$$\begin{aligned}
HC &= \frac{h_c}{\Upsilon} \int_0^\Upsilon S(t)dt \\
&= \frac{h_c}{2\Upsilon}(P_r - D(s_u, M_r))\log(1 + \Omega)[(1 + \Omega)^2 - (1 + \Omega - t_s)^2] \\
&\quad + \frac{h_c}{\Upsilon}(P_r - D(s_u, M_r))\left[\frac{(1+\Omega-t_s)^2}{2}\log(1+\Omega-t_s)\right. \\
&\quad \left. - \frac{(1+\Omega-t_s)^2}{4} - \frac{(1+\Omega)^2}{2}\log(1+\Omega) + \frac{(1+\Omega)^2}{4}\right] \\
&\quad - \frac{h_c}{\Upsilon}D(s_u, M_r)\left[\frac{(1+\Omega-\Upsilon)^2}{2}\log(1+\Omega-\Upsilon)\right. \\
&\quad \left. - \frac{(1+\Omega-\Upsilon)^2}{4} - \frac{(1+\Omega-t_s)^2}{2}\log(1+\Omega-t_s) + \frac{(1+\Omega-t_s)^2}{4}\right] \\
&\quad + \frac{h_c}{2\Upsilon}D(s_u, M_r)\log(1 + \Omega - \Upsilon)[(1 + \Omega - \Upsilon)^2 - (1 + \Omega - t_s)^2] \tag{2.17}
\end{aligned}$$

2.4.2 ORDERING COST (OC)

To order something, some cost is essential, which may be variable or fixed and known as ordering cost. In this present study, ordering costs are obtained as follows:

$$OC = \frac{c_o}{\Upsilon}$$

2.4.3 DETERIORATING COST (DC)

This cost is also known as the cost of degradation. To consider the degradation of a commodity which effects diminishing the usefulness from the original situation, the deteriorating cost is applied. The deteriorating cost is determined as follows:

$$\text{DC} = \frac{(c_p + \omega_1 P_r + \frac{\omega_2}{P_r}) P_r t_s}{\Upsilon} + (c_p + \omega_1 P_r + \frac{\omega_2}{P_r}) D(s_u, M_r)$$

2.4.4 UNIT PRODUCTION COST (UPC)

To produce the deteriorating item it is necessary to invest some cost, which is known as production cost, and the unit production cost is provided as

$$\text{UPC} = c_p + \omega_1 P_r + \frac{\omega_2}{P_r}$$

2.4.5 PAYABLE INTEREST (PI)

Different cases for the payable interest are calculated on the basis of the value of Υ, M_r, and M_s

Case I: $M_r \leq M_s \leq \Upsilon + M_r$

$$\text{PI} = (c_p + (\omega_1 P_r + \frac{\omega_2}{P_r})) R_2 \frac{D(s_u, M_r)}{\alpha \Upsilon} \int_{M_s}^{\Upsilon + M_r} \left(e^{\alpha(\Upsilon + M_r - t)} - 1 \right) dt$$

$$= (c_p + (\omega_1 P_r + \frac{\omega_2}{P_r})) R_2 \frac{D(s_u, M_r)}{\Upsilon} (1 + \Omega - t)^2 \left[e^{\frac{(\Upsilon + M_r - M_s)}{(1 + \Omega - t)}} - \frac{(\Upsilon + M_r - M_s)}{(1 + \Omega - t)} - 1 \right]$$

(2.18)

Case II: $M_r \leq \Upsilon + M_r \leq M_s$
No interest is charged for this scenario.

Case III: $M_s \leq M_r \leq \Upsilon + M_r$

$$\text{PI} = \frac{(c_p + (\omega_1 P_r + \frac{\omega_2}{P_r})) R_2}{\Upsilon} \left[\int_{M_s}^{M_r} P_r t_s dt + \int_{M_r}^{\Upsilon + M_r} \frac{D(s_u, M_r)}{\alpha (e^{\alpha(\Upsilon + M_r - t)} - 1) dt} \right]$$

$$= (c_p + (\omega_1 P_r + \frac{\omega_2}{P_r})) R_2 \frac{D(s_u, M_r)}{\Upsilon} (1 + \Omega - t)^2 \left[(\frac{1}{1 + \Omega - t})^2 P_r t_s (M_r - M_s) \right.$$

$$+ D(s_u, M_r) \left(e^{\frac{\Upsilon}{(1 + \Omega - t)}} - \frac{\Upsilon}{1 + \Omega - t} - 1 \right) \right]$$

(2.19)

2.4.6 EARNED INTEREST (EI)

Similarly, different cases for earned interest are calculated:

Case I: $M_r \leq M_s \leq \Upsilon + M_r$ (See Figure 2.2)

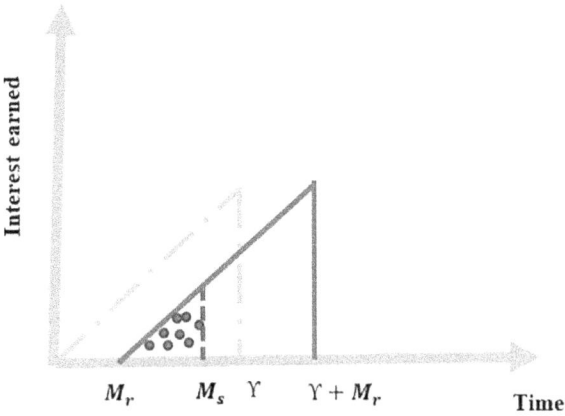

Figure 2.2   Earned interest in the during $M_r \le M_s \le \Upsilon + M_r$.

For this scenario, earned interest throughout the year is provided as:

$$\text{EI} = \frac{s_u R_1 D(s_u, M_r)(M_s - M_r)^2}{2\Upsilon} \tag{2.20}$$

Case II: $M_r \le \Upsilon + M_r \le M_s$ (See Figure 2.3)

The annual interest earned is

$$\text{EI} = \frac{s_u R_1}{\Upsilon} \left[\frac{D(s_u, M_r)\Upsilon^2}{2} + D(s_u, M_r)\Upsilon(M_s - \Upsilon - M_r) \right]$$

$$= s_u R_1 D(s_u, M_r)[M_s - M_r - \frac{\Upsilon}{2}] \tag{2.21}$$

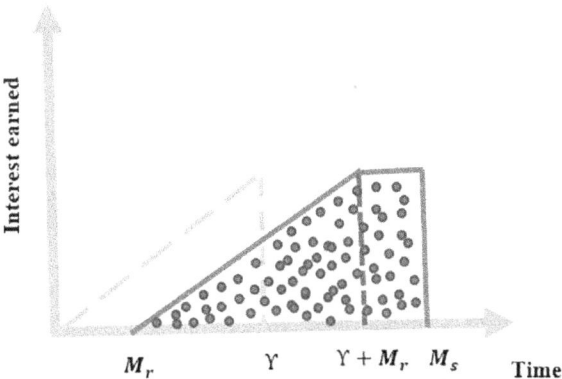

Figure 2.3   Earned interest in the during $M_r \le \Upsilon + M_r \le M_s$.

Case III: $M_s \leq M_r \leq \Upsilon + M_r$

Since, offered delay period is grater then provided permissible period, no extra interest will be earned in this scenario.

2.4.7   SALES REVENUE (SR)

Revenue earned through product sales is

$$\left(s_u - c_p - (\omega_1 P_r + \frac{\omega_2}{P_r}) \right) D(s_u, M_r).$$

Now, the entire profit for this retailing system provided as

$$T_P(s_u, \Upsilon, M_r, P_r) \quad = SR - HC - OC - DC - PI + EI$$

Then, the annual profit for different cases are

$$T_P(s_u, \Upsilon, M_r, P_r) = \left\{ \begin{array}{ll} T_{P_1}(s_u, \Upsilon, M_r, P_r), & \text{if } M_r \leq M_s \leq \Upsilon + M_r \\ T_{P_2}(s_u, \Upsilon, M_r, P_r), & \text{if } M_r \leq \Upsilon + M_r \leq M_s \\ T_{P_3}(s_u, \Upsilon, M_r, P_r), & \text{if } M_s \leq M_r \leq \Upsilon + M_r \end{array} \right\} \qquad (2.22)$$

2.5   SOLUTION METHODOLOGY

To find out the analytic solutions of the managing variables, one can take the help of differential equations. First, take the partial derivative of the equation (2.22) with regards to managing factors, and independently equating to zero. After that, a lemma and algorithm are constructed as follows

2.5.1   NECESSARY CONDITION

To obtain singularities partial derivative of the equation (2.22) with regards to managing factors are taken as follows

$$\frac{\partial T_{P_i}}{\partial s_u} = 0, \ \frac{\partial T_{P_i}}{\partial M_r} = 0, \ \frac{\partial T_{P_i}}{\partial \Upsilon} = 0, \ \frac{\partial T_{P_i}}{\partial P_r} = 0, \ \text{where, i=1,2,3.} \qquad (2.23)$$

A two-variable function g is concave in nature in a closed interval under the stated condition:

$$g(\gamma \tau_1 + (1 - \gamma)\tau_2) \quad \geq \quad \gamma g(\tau_1) + (1 - \gamma)g(\tau_2)$$

where, $\tau_1, \tau_2 \in (\varepsilon_1, \varepsilon_2)$ and each γ, $0 \leq \gamma \leq 1$, the taking after is fulfilled: Intermediate value theorem states that if a continuous function ζ is defined in a closed interval $[\xi_1, \xi_2]$ as its domain and $\zeta(\varepsilon_1), \zeta(\varepsilon_2)$ are opposite in sign, then there exists a point σ within $[\varepsilon_1, \varepsilon_2]$, such that $\zeta(\sigma) = 0$

Lemma 2.1. *If a function $\zeta(t)$ is continuous on $(\varepsilon_1, \varepsilon_2)$ and differentiable with a non increasing function (i.e, $\frac{d\zeta}{dt}$ is non-increasing) then ζ is concave.*

Proof. Details calculation are provided in Appendix A.   □

Case 1: $M_r \leq M_s \leq \Upsilon + M_r$
For given values of s_u and M_r, the first derivative of T_{P_1} with respect to Υ was given earlier in this report. The optimal value can be solved from equation $\frac{dT_{P_1}}{d\Upsilon} = 0$, it is easy to show that

$$
\begin{aligned}
\frac{\partial^2 T_{P_1}}{\partial \Upsilon^2} = &-\Big[h_c(P_r - D(s_u, M_r)) \Big[\frac{1}{2}(1+\Omega-\Upsilon)^2 \log(1+\Omega-\Upsilon) - \frac{1}{4}(1+\Omega-\Upsilon)^2 \\
&+ \frac{1}{2}(1+\Omega-t_s)^2 \log(1+\Omega-t_s) + \frac{1}{4}(1+\Omega-t_s)^2 \Big] \\
&+ h_c D(s_u, M_r) \log(1+\Omega-\Upsilon)(1+\Omega-\Upsilon) \\
&+ \frac{1}{2} h_c D(s_u, M_r)[\log(1+\Omega-\Upsilon)][(1+\Omega-\Upsilon)^2 \\
&- (1+\Omega-t_s)^2][-2-2t+2\Upsilon] \\
&+ \frac{1}{2} h_c D(s_u, M_r) \log(1+\Omega-\Upsilon)[(1+\Omega-\Upsilon)^2 - (1+\Omega-t_s)^2] \\
&+ \Big(c_p + \omega_1 P_r + \frac{\omega_2}{P_r} \Big) P_r t_s + \frac{1}{2} s_u R_1 D(s_u, M_r)(M_s - M_r)^2 \Big] < 0 \qquad (2.24)
\end{aligned}
$$

As, $\frac{\partial^2 T_{P_1}}{\partial \Upsilon^2} < 0$, the function T_{P_1} is non-increasing on $(0, \infty)$ and the Lemma T_{P_1} is a concave function on $(0, \infty)$. At zero, the following is true.

$$
\begin{aligned}
T_{P_1}(0) = &\Big(c_o - \frac{s_u R_1 D(s_u, M_r)(M_s - M_r)^2}{2} \Big) \\
&- \Big(c_p + \omega_1 P_r + \frac{\omega_2}{P_r} \Big) R_1 D(s_u, M_r)(1+\Omega-t)^2 \Big[e^{\frac{(\Upsilon+M_r-M_s)}{(1+\Omega-t)}} \\
&- \frac{(\Upsilon+M_r-M_s)}{(1+\Omega-t)} - 1 \Big] \qquad (2.25)
\end{aligned}
$$

Since, $e^{\frac{(\Upsilon+M_r-M_s)}{(1+\Omega-t)}} > \frac{(\Upsilon+M_r-M_s)}{(1+\Omega-t)} - 1$, $T_{P_1}(0) > 0$, if $\Big(c_o - \frac{s_u R_1 D(s_u, M_r)(M_s-M_r)^2}{2} \Big) > 0$.
$T_{P_1} = -\infty < 0$ whereas Υ leads to ∞, and $T_{P_1}(0) > 0$
   Thus, by the intermediate value theorem, the result is unique as well as optimum.
   By similar arguments and using the following algorithms, one can obtain other decision variables s_u and M_r along with optimal profit.

Algorithm 2.1. *To obtain the best solution of s_u P_r and M_r, the following algorithm is structured*

Step 1 Set $m = 1$. With the help of equation $\left(s_u - c_p - \omega_1 P_r - \frac{\omega_2}{P_r}\right)\frac{\partial D}{\partial s_u} + D = 0$, and value of the parameters, obtained s_{u_j}.

Step 2 By obtained value of s_{u_j} and help of equations (2.24) and (2.25), calculate Υ_{im} ($i = 1,2,3$) now set $s_{im} = s_{u_j}$.

Step 3 By equation (2.23), i.e., $\frac{\partial T_{P_i}}{\partial s_u} = 0$ and using the value of Υ solve s_u. Calculate $s_{u_i}^*$ in such a way that $\frac{\partial^2 T_{P_i}}{\partial s_u^2} \geq 0$ for $(\Upsilon = \Upsilon_{im}, s_u + s_{u_i}^*)$ and considered $s_{i,m+1} = s_{u_i}^*$.

Step 4 Now by using equation (2.23) i.e., $\frac{\partial T_{P_i}}{\partial \Upsilon} = 0$ one can find Υ with help of $s_u = s_{i,m+1}$ and obtained solution considered as $\Upsilon_{i,m+1}$.

Step 5 Using the value of Υ^*, s_u^*, and M_r^* and by the help of the equation (2.23) i.e., $\frac{\partial T_{P_i}}{\partial P_r} = 0$ for P_r, one can obtained the solution as $P_{r_{i,m+1}}$.

Step 6 For a fixed $\varepsilon > 0$ $|\Upsilon_{im} - \Upsilon_{im+1}| < \varepsilon$, $|P_{r_{im}} - P_{r_{im+1}}| < \varepsilon$ and $|s_{im} - s_{im+1}| < \varepsilon$, then $s_u^* = s_{im+1}, P_r^* = P_{r_{im+1}}$, and $\Upsilon^* = \Upsilon_{im+1}$. If the condition does not satisfy then put $m = m + 1$ and shifted to step 3.

2.6 NUMERICAL ANALYSIS

For explanation based on numerical value discussed in this segment, the parametric values similar to Sarkar et al. (2020b) model and use the computer software Mathematica 11.0, one can obtain the optimum profit under the best value of the managing variables. Different parametric values and different demand structure are considered as follows:

Example 2.1 The demand is $\rho(s_u) - [\rho(s_u) - \tau(s_u)]e^{-\gamma M_r}$ and other required values are: $\gamma = 0.5$; $\omega_1 = 1/300$; ordering cost $c_o = \$100/\text{order}$, interest earned $R_1 = 25\%$ per year, cost to hold $h_c = \$4.5$ per unit time, $\omega_2 = 300$; cost for metrical $c_p = \$15$ per unit, permissible delay period $M_s = 30$ days, lifetime $\Omega = 100$ days, the interest charged $R_2 = 15\%$ per year, $\rho(s_u) = 80 - 1.21s_u$, $\tau(s_u) = 30 - 1.21s_u$, By utilizing above parameters, optimum profit with respect to all variables are determined for three cases are:

(a) For case 1: The annual profit function T_{P_1} is $\$8018.18$ for production rate (P_r^*) is 130.58 unit/cycle, unit price for selling (s_u^*) is $\$312.40$ /unit, cycle time (Υ^*) is 42.26 days, and provided delay period (M_r^*) is 28.75 days.

(b) For case 2: The annual profit function T_{P_2} is $\$7922.96$ for production rate (P_r^*) is 170.42 unit/cycle, unit price for selling (s_u^*) is $\$243.48$ /unit, cycle time (Υ^*) is 18.16 days, and provided delay period (M_r^*) is 11.09 days.

(c) For case 3: The annual profit function T_{P_3} is $\$7936.77$ for production rate (P_r^*) is 161.53 unit/cycle, unit price for selling (s_u^*) is $\$221.78$ /unit, cycle time (Υ^*) is 34.21 days, and provided delay period (M_r^*) is 30.46 days.

Table 2.2

Optimum Result along with Optimum Values of Decision Variables

	Production Rate (unit/cycle)	Selling Price s_u ($/unit)	Delay-Period M_r (days)	Cycle Time Υ (days)	Profit ($)
T_{P_1}	**130.58**	**312.40**	**28.75**	**42.26**	**8018.18**
T_{P_2}	170.42	243.48	11.09	18.16	7922.96
T_{P_3}	161.53	221.78	30.46	34.21	7936.77

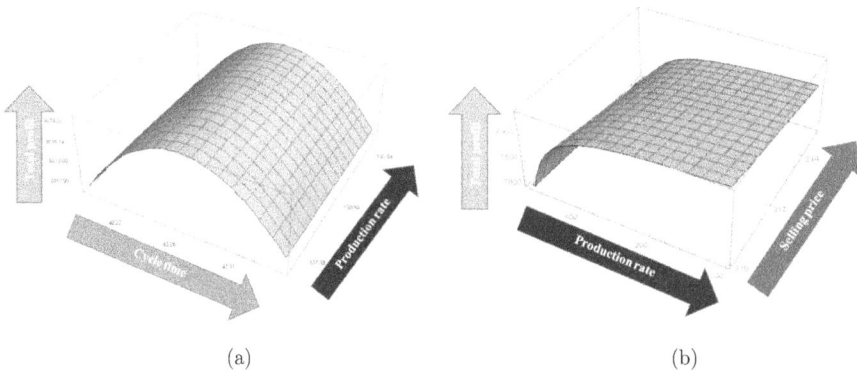

(a)   (b)

Figure 2.4 Graphical representation of concavity with respect to decision variables. (a) Graphical illustration of concavity with respect to cycle time and production rate. (b) Graphical illustration of concavity with respect to selling price and production rate.

After observing all three cases, the best result is obtained when permissible delay period is larger then the credit period, which was $8018.18 (see Table 2.2) and the related unit price for selling is (s_u^*) is $312.40 /unit. The best result is achieved when cycle time (Υ^*) is 42.26 days, and provided delay period (M_r^*) is 28.75 days, whereas the production rate is 130 units (Figure 2.4). The bold values in Table 2.2 represents the best value of this study.

2.6.1 NUMERICAL COMPARISON WITH EXISTING LITERATURE

Since the numerical values are calculated with the help of the values of Sarkar et al. (2020b) model, thus in this section, a comparison is made along with the Sarkar et al. (2020b) model and Mandal et al. (2021). In Sarkar et al. (2020b), they considered deteriorating items under a two-level trade credit policy. But due to demand variability, measurement in production is very crucial, which is proved through this study. Due to consideration of controllable production rate present study provide fifteen times more profit compare to Sarkar et al. (2020b) model. On the other hand, Mandal et al. (2021) model considered stock and advertisement dependent demand pattern for a deteriorating item under constant production rate and their model provided the optimal profit as $1840.41 per year, where the current model is beneficial

Table 2.3

Comparison with Previous Study

	Sarkar et al. (2020b)	Mandal et al. (2021)	This Study
Total Profit ($)	520.53	1840.41	8018.18

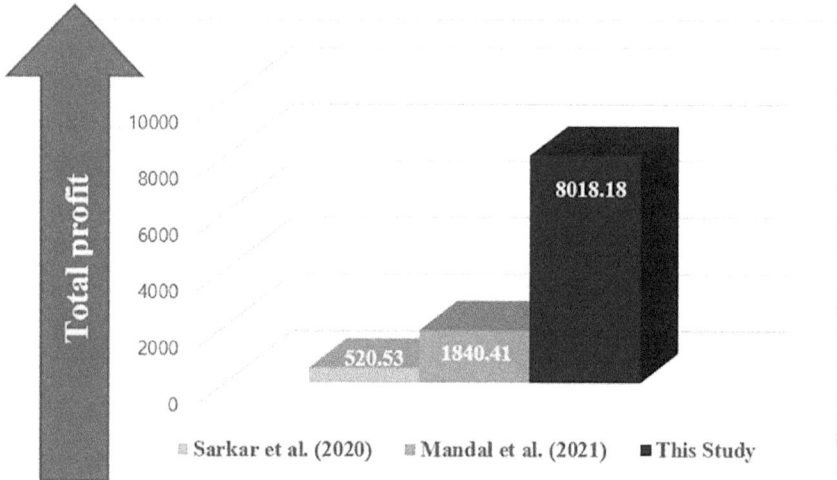

Figure 2.5   Graphical representation of comparison with different existing studies.

nearly about four times compared to Mandal et al. (2021). Thus implementation of VPR in an inventory system for deteriorating products under demand variability is very beneficial. In the following Table 4.3 the comparison is made and a graphical representation is provided in Figure 2.5.

2.7   SENSITIVITY ANALYSIS

It is found from the numerical analysis section that system profit is optimized for Case I. Therefore in this segment, the impact of crucial parameters on the total system profit for Case I are presented, numeric value of the sensitivity is illustrated in Table 2.4 and a graphical representation is structured in Figure 2.6. The following discussion can be made from the Sensitivity Analysis Table 2.4:

1. Holding cost is always very much crucial for any business sector. Since the current study focuses on deteriorating items, the unit holding cost for the deteriorating items is found to be much essential for this retailing strategy. A fifty percent reduction in unit holding cost decreases the system profit up to 46% and a reduction in unit holding cost up to 25% reduces the total

Table 2.4
Numerical Representation of Sensitivity Analysis

Parameters	Percentage Changes (%)	T_{P_1}	Parameters	Percentage Changes (%)	T_{P_1}
	−50	+0.015		−50	−72.45
	−25	+0.007		−25	−37.08
c_o	+25	−0.007	Ω	+25	+39.07
	+50	−0.015		+50	+79.74
	−50	+46.92		−50	+0.979
	−25	+23.46		−25	+0.489
h_c	+25	−23.46	R_2	+25	−0.489
	+50	−46.92		+50	−0.979
	−50	+1.615		−50	−0.023
	−25	+0.807		−50	−0.012
c_p	+25	−0.807	R_1	+25	+0.012
	+50	−1.615		+50	+0.023

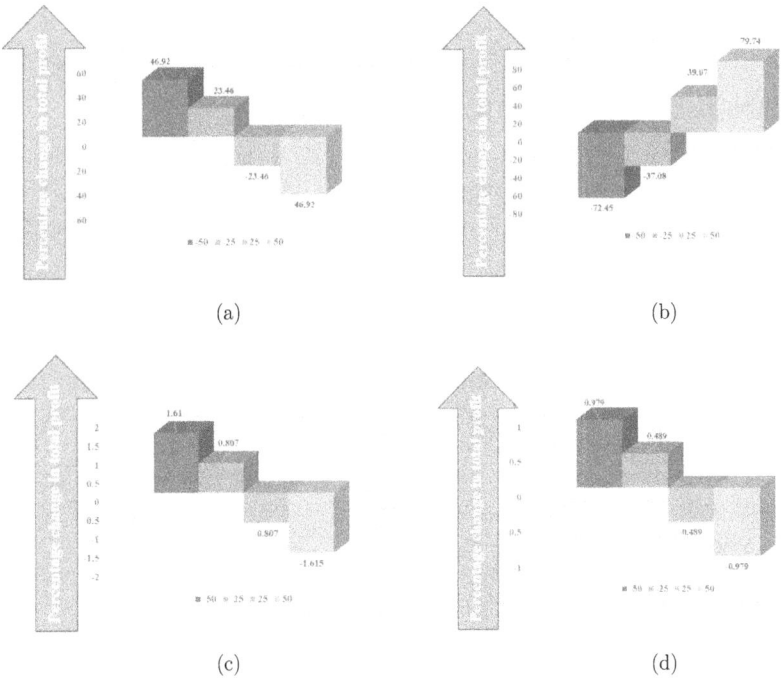

Figure 2.6 Graphical representation of sensitivity of the key parameters. (a) Sensitivity of the unit holding on total profit, (b) sensitivity of the lifetime of the product on total profit, (c) sensitivity of the unit metrical cost on total profit, and (d) sensitivity of charged interest on total profit.

system profit up to 23%. Similarly, increases unit holding cost 50% the system profit reduces upto 46%, whereas increase unit holding cost 25%, the profit reduces 23%.

2. The maximum lifetime of the deteriorating product is another major issue for any retailing strategy. Reduction in the lifetime of the product is harmful to optimize the system profit. With fifty percent reduction in the lifetime of the deteriorating items, the system profit decreases by 72% and with the reduction of 25% in maximum lifetime of the product, the system profit is reduced to nearly about 37%. Simultaneously,a 25% increase in the maximum lifetime of the product increase the system profit by nearly about 39% and an increase in the maximum lifetime by 50% enhances the system profit by 79%.

3. The unit material cost for the unit production cost is quite sensitive. Reduction in unit material cost is beneficial for the industry. Fifty percent reduction in material cost, system profit increases 1.615% and reduction 25%, system profit increases 0.807%. In addition to, an increase in material cost reduces the system profit.

4. This study was conducted for a two-level credit period, and that's why the interest charged for the delay period is a little bit sensitive in optimizing the total system profit.

5. Ordering cost and earned interest are a little bit sensitive for this study.

2.8 MANAGERIAL IMPLICATION

The real applicability of this research and the benefit of this study in the industry are illustrated in detail in this segment. The decision maker or the managers of the industry can take several major decisions through this study, which are elaborated as follows:

1. The calculation of demand is very essential for the industry. In this study it is considered that demand is variable, thus managers can take decisions on the demand and calculate how much demand is beneficial for the industry.

2. Demand for the deteriorating product varies with selling price and credit period. Thus, the decision maker takes a decision regarding the selling price of the product which precisely comforts to increment the profit of the industry. Simultaneously, by this study, one can take the decision on credit period, which is another major finding of the research.

3. How much credit period can provide maximum interest for a deteriorating product can also be found through this study.

4. Owing to demand variability, it is necessary to calculate the exact amount of production. Since, this study is conducted for a deteriorating item, if the rate of production is too high, then the company may face a huge loss due to deterioration. Thus, to control the loss and shortages, production variability is very essential. Through the current study, managers of the industry can take the right decision on production rate. Moreover, flexible production rate makes the system more reliable and smart compare to traditional one.

2.9 CONCLUSIONS

The optimal strategy for retailing is always beneficial for any industry. Nowadays, most of the daily life product is a perishable product. Thus, an optimal retailing strategy for perishable products is studied in this study, where the demand for the products vary with selling price and two-level delay period. To increase the profit of the industry, trade credit is one of the best policies through which the industry can earn interest. Calculation of the exact credit period is always necessary to optimize the profit of the system. Due to consideration of demand variability, it is necessary to calculate the exact amount for production rate, otherwise the industry may face a big loss due to deterioration. Thus, a flexible production rate helps to minimize system costs and increase the revenue of the industry. A solution procedure along with an algorithm is provided to find an excellent solution for this model. The optimal result is obtained for the first case for the optimal values of production rate, delay period, cycle time, and selling price of the deteriorating products.

Fixed ordering cost, no shortage, and negligible lead time are the limitations of the current study. Instead of constant ordering cost one can use flexible ordering cost along with some investment to reduce ordering cost (Dey et al., 2019a). Recycling or the reproduction of the deteriorating product is another limitation of this study (Iqbal and Sarkar, 2019). By considering different environmental effects and preservation technology one can extend the current study (Sepehri et al., 2021a). One can consider the effect of advertisement on demand to extend this type of model for deteriorating items Dey et al. (2021a). In spite of two-level trade credit, one can take multi-level credit period for deteriorating item (Ahmed et al., 2021). One can assume more than two players i.e., three-echelon SCM to extend this type of study Ullah et al. (2021). Some smart technology can be applied to identify the lifetime of the deteriorating product (Dey et al., 2021b; Ullah and Sarkar, 2020).

Appendix A: Proof of Lemma

To prove lemma, one can consider function ζ within close interval $[0,1]$ in the form:

$$\zeta(t) = g(t\tau_2 + (1-t)\tau_1) - tg(\tau_2) - (1-t)g(\tau_1)$$

in which, $\varepsilon_1 < \tau_1 < \tau_2 < \varepsilon_2$

Now, it is required to prove that ζ is non-negative within $[0,1]$. Based on assumption ζ is continuous and $\zeta(0) = \zeta(1) = 0$ then,

$$\frac{d\zeta(t)}{dt} = (\tau_2 - \tau_1)\frac{dg}{dt} - g(\tau_2) + g(\tau_1)$$

however, $t + f > t$, and one can obtain

$$\frac{d\zeta(t+f)}{dt} - \frac{d\zeta(t)}{dt} = (\tau_2 - \tau_1)\left[\frac{dg(t+f)}{dt} - \frac{dg(t)}{dt}\right]$$

Due to non-increasing nature of the function $\frac{dg}{dt}$,

$$\frac{dg(t+f)}{dt} - \frac{dg(t)}{dt} < 0$$

Then $\frac{d\zeta(t)}{dt}$ is non-increasing in nature with in $[0,1]$. Again, if one consider that ζ obtained its minimum at the point $\sigma \in [0,1]$. When, $\sigma = 1$, $\zeta(t) \geq \zeta(1) = 0$ within $[0,1]$, which indicate ζ takes a local minimum at σ, and $\frac{d\zeta(\sigma)}{dt} \geq 0$. $\frac{d\zeta}{dt}$ is again non-increasing in nature, thus, $\frac{d\zeta}{dt} \geq 0$ on $[0,\sigma]$. Again one can prove that ζ is non-decreasing on $[0,\sigma]$, then the minimum of ζ in $[0,1]$ is non-negative and $\zeta \geq 0$ is on $[0,1]$.

REFERENCES

Ahmed, W., Moazzam, M., Sarkar, B., and Rehman, S. U. (2021). Synergic effect of reworking for imperfect quality items with the integration of multi-period delay-in-payment and partial backordering in global supply chains. *Engineering*, 7(2):260–271.

Al Durgam, M., Adegbola, K., and Glock, C. H. (2017). A single-vendor single-manufacturer integrated inventory model with stochastic demand and variable production rate. *International Journal of Production Economics*, 191:335–350.

Alfares, H. K. and Ghaithan, A. M. (2016). Inventory and pricing model with price-dependent demand, time-varying holding cost, and quantity discounts. *Computers & Industrial Engineering*, 94:170–177.

Ayed, S., Sofiene, D., and Nidhal, R. (2012). Joint optimization of maintenance and production policies considering random demand and variable production rate. *International Journal of Production Research*, 50(23):6870–6885.

Bai, Q.-G., Xu, X.-H., Chen, M.-Y., and Luo, Q. (2015). A two-echelon supply chain coordination for deteriorating item with a multi-variable continuous demand function. *International Journal of Systems Science: Operations & Logistics*, 2(1): 49–62.

Banerjee, S. and Agrawal, S. (2017). Inventory model for deteriorating items with freshness and price-dependent demand: Optimal discounting and ordering policies. *Applied Mathematical Modeling*, 52:53–64.

Bhunia, A. K. and Shaikh, A. A. (2015). An application of PSO in a two-warehouse inventory model for deteriorating items under permissible delay in payment with different inventory policies.*Applied Mathematics and Computation*, 256:831–850.

Bhunia, A. K., Shaikh, A. A., Dhaka, V., Pareek, S., and Cárdenas-Barrón, L. E. (2018). An application of genetic algorithm and PSO in an inventory model for a single deteriorating item with variable demand dependent on marketing strategy and displayed stock level. *Scientia Iranica*, 25(3):1641–1655.

Braglia, M., Castellano, D., Marrazzini, L., and Song, D. (2019). A continuous review, (q,r) inventory model for a deteriorating item with random demand and positive lead time. *Computers & Operations Research*, 109:102–121.

Chen, S.-C. and Teng, J.-T. (2014). Retailer's optimal ordering policy for deteriorating items with maximum lifetime under supplier's trade credit financing. *Applied Mathematical Modeling*, 38(15–16):4049–4061.

Choi, T.-M., Yeung, W.-K., and Cheng, T. (2013). Scheduling and coordination of multi- suppliers single-warehouse-operator single-manufacturer supply chains with VPR and storage costs. *International Journal of Production Research*, 51(9):2593–2601.

Das, B. C., Das, B., and Mondal, S. K. (2013). Integrated supply chain model for a deteriorating item with procurement cost-dependent credit period. *Computers & Industrial Engineering*, 64(3):788–796.

Das, S., Khan, M. A.-A., Mahmoud, E. E., Abdel-Aty, A.-H., Abualnaja, K. M., and Shaikh, A. A. (2021). A production inventory model with partial trade credit policy and reliability. *Alexandria Engineering Journal*, 60(1):1325–1338.

De, S. K. and Sana, S. S. (2015). Backlogging EOQ model for promotional effort and selling price sensitive demand-an intuitionistic fuzzy approach. *Annals of Operations Research*, 233(1):57–76.

Dey, B. K., Bhuniya, S., and Sarkar, B. (2021a). Involvement of controllable lead time and variable demand for a smart manufacturing system under supply chain management. *Expert Systems with Applications*, 184:115464.

Dey, B. K., Pareek, S., Tayyab, M., and Sarkar, B. (2021b). Autonomation policy to control work-in-process inventory in a smart production system. *International Journal of Production Research*, 59(4):1258–1280.

Dey, B. K., Sarkar, B., and Pareek, S. (2019a). A two-echelon supply chain management with setup time and cost reduction, quality improvement, and variable production rate. *Mathematics*, 7(4):328.

Dey, B. K., Sarkar, B., Sarkar, M., and Pareek, S. (2019b). An integrated inventory model involving discrete setup cost reduction, variable safety factor, selling price-dependent demand, and investment. *RAIRO-Operations Research*, 53(1):39–57.

Feng, L., Chan, Y.-L., and Cárdenas-Barrón, L. E. (2017). Pricing and lot-sizing policies for perishable goods when the demand depends on selling price, displayed stocks, and expiration date. *International Journal of Production Economics*, 185:11–20.

Ghiami, Y., Williams, T., and Wu, Y. (2013). A two-echelon inventory model for a deteriorating item with stock-dependent demand, partial backlogging, and capacity constraints. *European Journal of Operational Research*, 231(3):587–597.

Giri, B. C., Yun, W., and Dohi, T. (2005). Optimal design of unreliable production inventory systems with VPR. *European Journal of Operational Research*, 162(2):372–386.

Hasan, M. R., Mashud, A. H. M., Daryanto, Y., and Wee, H. M. (2020). A non-instantaneous inventory model of agricultural products considering deteriorating impacts and pricing policies. *Kybernetes*, 50(8):2264–2288.

Hsieh, T.-P. and Dye, C.-Y. (2017). Optimal dynamic pricing for deteriorating items with reference price effects when inventories stimulate demand. *European Journal of Operational Research*, 262(1):136–150.

Iqbal, M. W. and Sarkar, B. (2019). Recycling of lifetime-dependent deteriorated products through different supply chains. *RAIRO-Operations Research*, 53(1):129–156.

Jaggi, C. K., Tiwari, S., and Goel, S. K. (2017). Credit financing in economic ordering policies for non-instantaneous deteriorating items with price-dependent demand and two storage facilities. *Annals of Operations Research*, 248(1–2): 253–280.

Khan, M. A.-A., Shaikh, A. A., and Cárdenas-Barrón, L. E. (2021). An inventory model under linked-to-order hybrid partial advance payment, partial credit policy, all-units discount, and partial backlogging with capacity constraint. *Omega*, 103:102418.

Khouja, M. (1995). The economic production lot size model under volume exibility. *Computers & Operations Research*, 22(5):515–523.

Khouja, M. and Mehrez, A. (1994). Economic production lot size model with variable production rate and imperfect quality. *Journal of the Operational Research Society*, 45(12):1405–1417.

Li, G., He, X., Zhou, J., and Wu, H. (2019a). Pricing, replenishment and preservation technology investment decisions for non-instantaneous deteriorating items. *Omega*, 84:114–126.

Li, R. and Teng, J.-T. (2018). Pricing and lot-sizing decisions for perishable goods when demand depends on selling price, reference price, product freshness, and displayed stocks. *European Journal of Operational Research*, 270(3): 1099–1108.

Li, R., Teng, J.-T., and Zheng, Y. (2019b). Optimal credit term, order quantity, and selling price for perishable products when demand depends on selling price, expiration date, and credit period. *Annals of Operations Research*, 280(1): 377–405.

Mahata, G. C. and De, S. K. (2017). Supply chain inventory model for deteriorating items with maximum lifetime and partial trade credit to credit risk customers. *International Journal of Management Science and Engineering Management*, 12(1):21–32.

Mahata, P. and Mahata, G. C. (2021). Two-echelon trade credit with default risk in an EOQ model for deteriorating items under dynamic demand. *Journal of Industrial & Management Optimization*, 17(6):3659.

Mandal, B., Dey, B. K., Khanra, S., and Sarkar, B. (2021). Advance sustainable inventory management through advertisement and trade credit policy. *RAIRO-Operations Research*, 55(1):261–284.

Mashud, A. H. M., Hui-Ming, W., and Huang, C.-V. (2021). Preservation technology investment, trade credit, and partial backordering model for a non-instantaneous deteriorating inventory. *RAIRO: Recherche Opérationnelle*, 55:51.

Mashud, A. H. M., Wee, H.-M., Sarkar, B., and Li, Y.-H. C. (2020). A sustainable inventory system with the advanced payment policy and trade credit strategy for a two-warehouse inventory system. *Kybernetes*, 50(5), 1321–1348.

Pando, V., San-José, L. A., Sicilia, J., and Alcaide-Lopez-de Pablo, D. (2021). Maximization of the return on inventory management expense in a system with price-and stock-dependent demand rates. *Computers & Operations Research*, 127:105134.

Pervin, M., Roy, S. K., and Weber, G. W. (2019). Multi-item deteriorating two-echelon inventory model with price-and stock-dependent demand: A trade credit policy. *Journal of Industrial & Management Optimization*, 15(3):1345.

Rabbani, M., Zia, N. P., and Rafei, H. (2016). Joint optimal dynamic pricing and replenishment policies for items with simultaneous quality and physical quantity deterioration. *Applied Mathematics and Computation*, 287:149–160.

Saren, S., Sarkar, B., and Bachar, R. K. (2020). Application of various price-discount policies for deteriorated products and delay in payments in an advanced inventory model. *Inventions*, 5(3):50.

Sarkar, B. (2012). An EOQ model with delay in payments and time-varying deterioration rates. *Mathematical and Computer Modeling*, 55(3-4):367–377.

Sarkar, B. (2013). A production inventory model with probabilistic deterioration in two-echelon supply chain management. *Applied Mathematical Modeling*, 37(5):3138–3151.

Sarkar, B., Ahmed, W., and Kim, N. (2018a). Joint effects of variable carbon emission cost and multi-delay in payments under the single-setup-multiple-delivery policy in a global sustainable supply chain. *Journal of Cleaner Production*, 185:421–445.

Sarkar, B., Dey, B. K., Sarkar, M., and AlArjani, A. (2021). A sustainable online-to-offine (O2O) retailing strategy for supply chain management under controllable lead time and variable demand. *Sustainability*, 13(4):1756.

Sarkar, B., Dey, B. K., Sarkar, M., Hur, S., Mandal, B., and Dhaka, V. (2020). Optimal replenishment decision for retailers with variable demand for deteriorating products under a trade credit policy. *RAIRO-Operations Research*, 54(6):1685–1701.

Sarkar, B., Majumder, A., Sarkar, M., Kim, N., and Ullah, M. (2018b). Effects of variable production rate on quality of products in a single-vendor multi-buyer supply chain management. *The International Journal of Advanced Manufacturing Technology*, 99(1):567–581.

Sarkar, B., Mridha, B., and Pareek, S. (2022a). A sustainable smart multi-type biofuel manufacturing with optimum energy utilization under flexible production. *Journal of Cleaner Production*, 332:129869.

Sarkar, B. and Saren, S. (2015). Partial trade credit policy of retailers with exponentially deteriorating items. *International Journal of Applied and Computational Mathematics*, 1(3):343–368.

Sarkar, B. and Saren, S. (2017). Ordering and transfer policy and variable deterioration for a warehouse model. *Hacettepe Journal of Mathematics and Statistics*, 46(5):985–1014.

Sarkar, B., Saren, S., and Cárdenas-Barrón, L. E. (2015). An inventory model with trade credit policy and variable deterioration for fixed lifetime products. *Annals of Operations Research*, 229(1):677–702.

Sarkar, B. and Sarkar, S. (2013). Variable deterioration and demand—an inventory model. *Economic Modeling*, 31:548–556.

Sarkar, B., Sarkar, S., and Yun, W. Y. (2016). Retailer's optimal strategy for fixed lifetime products. *International Journal of Machine Learning and Cybernetics*, 7(1):121–133.

Sarkar, B., Ullah, M., and Sarkar, M. (2022b). Environmental and economic sustainability through innovative green products by remanufacturing. *Journal of Cleaner Production*, 332:129813.

Sarkar, M. and Chung, B. D. (2020). The flexible work-in-process production system in supply chain management under quality improvement. *International Journal of Production Research*, 58(13):3821–3838.

Sarkar, M. and Seo, Y. W. (2021). Renewable energy supply chain management with exibility and automation in a production system. *Journal of Cleaner Production*, 324:129149.

Sarkar, T., Ghosh, S. K., and Chaudhuri, K. (2012). An optimal inventory replenishment policy for a deteriorating item with time-quadratic demand and time-dependent partial backlogging with shortages in all cycles. *Applied Mathematics and Computation*, 218(18):9147–9155.

Saxena, N., Sarkar, B., and Singh, S. (2020). Selection of remanufacturing/production cycles with an alternative market: A perspective on waste management. *Journal of Cleaner Production*, 245:118935.

Sepehri, A., Mishra, U., and Sarkar, B. (2021). A sustainable production inventory model with imperfect quality under preservation technology and quality improvement investment. *Journal of Cleaner Production*, 310:127332.

Sett, B. K., Sarkar, S., Sarkar, B., and Young Yun, W. (2016). Optimal replenishment policy with variable deterioration for fixed lifetime products. *Scientia Iranica*, 23(5):2318–2329.

Teng, J.-T., Cárdenas-Barrón, L. E., Chang, H.-J., Wu, J., and Hu, Y. (2016). Inventory lot size policies for deteriorating items with expiration dates and advance payments. *Applied Mathematical Modeling*, 40(19–20):8605–8616.

Tiwari, S., Ahmed, W., and Sarkar, B. (2019). Sustainable ordering policies for non-instantaneous deteriorating items under carbon emission and multi-trade credit policies. *Journal of Cleaner Production*, 240:118183.

Tiwari, S., Cárdenas-Barrón, L. E., Khanna, A., and Jaggi, C. K. (2016). Impact of trade credit and inflation on retailer's ordering policies for non-instantaneous deteriorating items in a two-warehouse environment. *International Journal of Production Economics*, 176:154–169.

Tiwari, S., Cárdenas-Barrón, L. E., Shaikh, A. A., and Goh, M. (2020). Retailer's optimal ordering policy for deteriorating items under order-size dependent trade credit and complete backlogging. *Computers & Industrial Engineering*, 139:105559.

Ullah, M., Asghar, I., Zahid, M., Omair, M., AlArjani, A., and Sarkar, B. (2021). The ramification of remanufacturing in a sustainable three-echelon closed-loop supply chain management for returnable products. *Journal of Cleaner Production*, 290:125609.

Ullah, M. and Sarkar, B. (2020). Recovery-channel selection in a hybrid manufacturing remanufacturing production model with RFID and product quality. *International Journal of Production Economics*, 219:360–374.

Wu, J., Ouyang, L.-Y., Cárdenas-Barrón, L. E., and Goyal, S. K. (2014). Optimal credit period and lot size for deteriorating items with expiration dates under two-level trade credit financing. *European Journal of Operational Research*, 237(3):898–908.

3 Explosion in a Spherical Cavity Expanding with Decelerated Velocity

Asish Karmakar
Udairampur Pallisree Sikshayatan (H.S.)

Bula Mondal
J.B.V. Girls' High School

CONTENTS

3.1 INTRODUCTION

The investigation of elastic waves generated by spherical sources situated in an infinite elastic medium have been continuing for a long time. Such spherical sources can represent some of the main features of deep, underground explosion, including nuclear explosion. Many theoretical models have been developed to study the effects of time-dependent stress acting on the boundary of a spherical cavity situated in an infinite elastic medium and the nature of disturbances thus produced in the medium. Various forms of such theoretical models have been considered by Jeffreys (1931), Honda (1959), Vodicva (1963), Gurvich and Yanovskii (1967), and Gurvich (1968). Different techniques for large underground explosions are considered by Ghosh (1969), Archambeau and Sammis (1970), Mukhopadhyay (1971), Aboudi (1971), Aboudi (1972), Andrews (1973), Archambeau (1972), and Minster and Suteau (1977) and others. Earlier researchers considered the models

DOI: 10.1201/9781003227847-3

with the outer boundary of the cavity either fixed or moving with a uniform velocity. But in our model, we consider that the outer boundary is expanding with a high but decelerated velocity and finally comes to a halt after a finite length of time.

In an explosion, the build-up of high temperature and pressure falls off with the expansion of the cavity. The deep focus earthquake may be generated by the sudden phase transition in a volume material, rather than fault slip was discussed by Randal (1964) and Randal (1966). The effect of sudden dynamical failure or sudden phase transition in a volume of material in the interior of the Earth would be expected. The expansion starts in a small region and then expands into a large volume of material, and hence such dynamical failure or phase transition would stop after some time. This type of symmetrical expanding volume source has been considered in this chapter.

In recent years, to understand the prediction model for amplitude-frequency characteristics, natural hazards due to explosion, the effects of underground explosions on soil and structure, pressure distribution from explosives buried have been studied. Such work was done by Rigby et al. (2016), Chenglong et al. (2018), Daniel and Bibiana (2019), and Min et al. (2022). Also spherical cavity expansion (SCE) theory has been extensively utilized to model dynamic deformation processes related to indentation and penetration problems in many fields. Such work was done by Mario and Alejandro (2020).

We consider the large explosion within a spherical cavity situated in an infinite, homogeneous, isotropic, and perfectly elastic medium. We also assume that the entire medium including the boundary of the cavity is at rest just before the explosion. We assume that the equations of elasticity hold everywhere in the medium except within the cavity. The region in which the equations of elasticity hold will be called elastic region, while the region in which the equations of elasticity do not hold, will be called nonelastic region. A sufficiently large motion is generated by the explosion. So the effect of explosion produced melting, vaporization, crushing of the material, and very high stresses. Hence the condition of elasticity ceases to hold in a part of the medium which extends, with expansions of the outward boundary of the cavity. The outer boundary of the nonelastic region, which is also the inner boundary of the elastic regions would start expanding immediately after the explosion. After a sufficient time, the pressure generated within the cavity due to the explosion gets reduced and becomes sufficiently small to permit the material to remain in the elastic state. During the explosion, the elastic region would be under large outward stresses across the elastic-nonelastic boundary.

It may be noted that under suitable circumstances the displacements in the elastic region for all explosive or non-explosive expanding volume sources depend on the initial and final states of the elastic-non-elastic boundary and also on the velocity and mode of expansion of the boundary and the stress acting at the boundary of the elastic region. In our model we obtained an exact solution in closed form for the displacement in the elastic medium for a spherical source, expanding with a decelerated velocity.

3.2 FORMULATION AND CONSTITUTIVE EQUATIONS

We introduce spherical polar co-ordinates (r, θ, ϕ) with the pole at the center of the cavity of radius 'a'. We assume that the outer boundary of the nonelastic regions, which is given by $r = a$ before the explosion. Phase transition or other failure processes, starts expanding at the time $t = 0$, which is taken to be the instant at which the explosion or failure starts. The outer boundary of the nonelastic region is taken to remain spherical throughout the process of expansion, the rate of expansion of the spherical boundary being taken to be decreasing and say it $v(t)$. We assume that after the elastic-nonelastic boundary attains a sufficiently large radius $b > a$, where the cause of failure of the elastic properties becomes sufficiently small to permit the material to remain in the elastic state, further expansion of the elastic-nonelastic boundary ceases. We take the elastic-nonelastic boundary to be subjected to long normal stress that may be taken to be the limiting pressure for the material, beyond which elastic conditions cease to hold. When the elastic-nonelastic boundary finally comes to rest after attaining a maximum radius $b > a$, we assume that the applied stress at the boundary falls off subsequently with time.

We assume complete spherical symmetry of motion in the elastic region and take the displacement in the elastic region to be independent of θ and ϕ, and a function of r and t alone. Under these conditions, the equation of motion for the radial displacement u reduces to the form,

$$u = \frac{\partial \phi}{\partial r} \tag{3.1}$$

where displacement potential ϕ satisfies

$$\left.\begin{array}{c} \nabla^2 \phi = \dfrac{1}{\alpha^2} \dfrac{\partial^2 \phi}{\partial t^2} \\[2mm] \text{where } \alpha^2 = \dfrac{\lambda + 2\mu}{\rho} \end{array}\right\} \tag{3.2}$$

where α is the velocity of propagation of the dilatational wave in the elastic medium. If β is the velocity of propagation of transverse wave, we have $\beta^2 = \dfrac{\mu}{\rho}$. So that $\dfrac{\beta^2}{\alpha^2} = \dfrac{\mu}{\lambda + 2\mu}$.

The stress component τ_{rr}, which is taken to have prescribed values at the elastic-nonelastic boundary is taken to be related to ϕ by the equation

$$\tau_{rr} = \rho\alpha^2 \frac{\partial \phi^2}{\partial r^2} + \frac{2\lambda}{r}\frac{\partial \phi}{\partial r} \tag{3.3}$$

where λ is the Lame' parameter and ρ is the density.

The boundary condition at the elastic-nonelastic boundary

$$\tau_{rr} = 0 \text{ for } t < 0 \text{ on } r = a \tag{3.4}$$

$$\tau_{rr} = P_0 + \frac{P_1 a^n}{(a+s)^n} \text{ for } 0 \le t \le T \text{ on } r = a+s, \text{ where } s = ut - \frac{1}{2}ft^2 \qquad (3.5)$$

and

$$\tau_{rr} = P_2 e^{-k(t-T)} \text{ for } t > T \text{ on } r = a+s(T) \qquad (3.6)$$

where $P_2 = \left\{ P_0 + \dfrac{P_1 a^n}{(a+s(T))^n} \right\}$ and f is uniform deceleration and u is the initial velocity. In equations (3.5) and (3.6), $T \, ¿0$ is the duration of expansion of the elastic-nonelastic boundary and P_0, P_1, P_2 are finite constants which would be positive for an explosive source, and k is a positive constant. We note that the continuity of the normal stress at the elastic-nonelastic boundary when the boundary comes to rest can be secured by taking $P_0 + \dfrac{P_1 a^n}{(a+s)^n} = \left\{ P_0 + \dfrac{P_1 a^n}{(a+s)^n} \right\} e^{-k(t-T)}$ at $t = T$. However, the solution which we obtain would remain valid even if this condition is not satisfied. We also note that increasing or decreasing stress at the elastic-nonelastic boundary can be obtained by taking $n < 0$ and $n > 0$ respectively, while $n = 0$ corresponds to a constant stress at the elastic-nonelastic boundary. In equation (3.6), $k = 0$ would correspond to a constant stress at the elastic-nonelastic boundary after the boundary comes to rest. The boundary conditions (3.4)–(3.6) reduces to

$$\frac{\partial^2 \phi}{\partial r^2} + \frac{1}{r} \left(2 - \frac{4\beta^2}{\alpha^2} \right) \frac{\partial \phi}{\partial r} = 0, \text{ for } t < 0 \text{ on } r = a \qquad (3.7)$$

$$\frac{\partial^2 \phi}{\partial r^2} + \frac{1}{r} \left(2 - \frac{4\beta^2}{\alpha^2} \right) \frac{\partial \phi}{\partial r} = \frac{P_0}{\rho \alpha^2} + \frac{P_1 a^n}{\rho \alpha^2 (a+s)^n}, \text{ for } 0 \le t \le T \text{ on } r = a+s \qquad (3.8)$$

and

$$\frac{\partial^2 \phi}{\partial r^2} + \frac{1}{r} \left(2 - \frac{4\beta^2}{\alpha^2} \right) \frac{\partial \phi}{\partial r} = \frac{P_2}{\rho \alpha^2}, \text{ for } t > T \text{ on } r = a+s(T) = b \qquad (3.9)$$

Here ρ is the density of the elastic region and β is the velocity of propagation of shear waves in the region.

We note that the equation of motion (3.2), which holds in the elastic region, has a solution representing a wave radiating outwards in the form,

$$\left. \begin{array}{l} \phi = \dfrac{F\left(t - \frac{r}{\alpha}\right)}{r} \\ r \ge a, \, t < 0; \, r \ge a+s, 0 \le t \le T; \, r \ge b, t > T \end{array} \right\} \qquad (3.10)$$

We try to determine the function $F\left(t - \dfrac{r}{\alpha}\right)$ in such a way that ϕ satisfies the boundary conditions (3.7)–(3.9), as also the initial condition $u = 0$ and $\dfrac{\partial u}{\partial t} = 0$ for

$t \leq 0$, $r \geq a$. We first try to determine the form of $F\left(t - \dfrac{r}{\alpha}\right)$ for which the boundary conditions (3.7)–(3.9) are satisfied. We consider the boundary condition (3.8). We find that equation (3.8) will be satisfied by ϕ, given by equation (3.10), if

$$
\left.
\begin{aligned}
&\frac{(a+s)^2}{\alpha^2}F''\left(t - \frac{a+s}{\alpha}\right) + \frac{4\beta^2(a+s)}{\alpha^3}F'\left(t - \frac{a+s}{\alpha}\right) + \frac{4\beta^2}{\alpha^2}F\left(t - \frac{a+s}{\alpha}\right) = \\
&\frac{P_0(a+s)^3}{\alpha^2\rho}H(t) + \frac{P_1 a^n (a+s)^{(3-n)}}{\alpha^2\rho}H(t) \\
&\text{for } (0 \leq t \leq T)
\end{aligned}
\right\}
$$

(3.11)

In equation (3.11) primes denote differentiation with respect to the argument.
Now

$$
\left.
\begin{aligned}
a+s &= a + v_H t - \frac{1}{2}\frac{v_H t^2}{T} \\
&= a + v_H t - \frac{v_H t^2}{T} + \frac{1}{2}\frac{v_H t^2}{T} \\
&= \left(a + \frac{1}{2}\frac{v_H t^2}{T}\right) + v_H t\left(1 - \frac{t}{T}\right) \\
&= a_1 + vt
\end{aligned}
\right\}
$$

where v_H = initial velocity of the expansion of the cavity $(t = 0)$, $a_1 = a + \dfrac{1}{2}\dfrac{v_H t^2}{T}$, $v = v_H\left(1 - \frac{t}{T}\right)$

Write,

$$
r_1 = t - \frac{a+s}{\alpha} = t - \frac{a_1 + vt}{\alpha} = t\left(1 - \frac{v}{\alpha}\right) - \frac{a_1}{\alpha} = tv_1 - \frac{a_1}{\alpha}
$$

where $v_1 = 1 - \dfrac{v}{\alpha}$

Therefore,

$$
\left.
\begin{aligned}
a + s &= a_1 + vt \\
&= \frac{a_1 v_1 + vv_1 t}{v_1} \\
&= \frac{a_1 + vr_1}{v_1}
\end{aligned}
\right\}
$$

Hence equation (3.11) reduce to

$$
\left.
\begin{aligned}
&\frac{(a_1 + vr_1)^2}{\alpha^2 v_1^2}\frac{d^2 F(r_1)}{dr_1^2} + \frac{4\beta^2}{\alpha^3}\frac{(a_1 + vr_1)}{v_1}\frac{dF(r_1)}{dr_1} + \frac{4\beta^2}{\alpha^2}F(r_1) = \\
&\frac{P_0(a_1 + vr_1)^3}{\alpha^2\rho v_1^3} + \frac{P_1 a^n (a_1 + vr_1)^{(3-n)}}{\alpha^2\rho v_1^{(3-n)}}
\end{aligned}
\right\}
$$

(3.12)

for

$$\left[-\frac{a_1}{\alpha} \le r_1 \le T\left(1-\frac{v}{\alpha}\right)-\frac{a_1}{\alpha}\right] \tag{3.13}$$

Now equation (3.12) can be solved by using $a_1 + vr_1 = av_1 e^y$, where, y is a new variable. It is found, on detail investigation, that the nature of the solution depends on $\frac{v}{\alpha}$. We finally obtain for equation (3.13) and the solution of equation (3.12) is given below

$$\left.\begin{aligned}
F(r_1) &= A\left(\frac{a_1+vr_1}{av_1}\right)^{m_1+m_2} + B\left(\frac{a_1+vr_1}{av_1}\right)^{m_1-m_2} \\
&\quad + P_{01}\left(\frac{a_1+vr_1}{v_1}\right)^3 + P_{11}a^n\left(\frac{a_1+vr_1}{v_1}\right)^{(3-n)} \\
&\quad \text{for } \frac{\alpha}{S_1} < v < \alpha
\end{aligned}\right\} \tag{3.14}$$

$$\left.\begin{aligned}
F(r_1) &= \left[A' + B'\log\left(\frac{a_1+vr_1}{av_1}\right)\right]\left(\frac{a_1+vr_1}{av_1}\right)^{m_1} \\
&\quad + P_{01}\left(\frac{a_1+vr_1}{v_1}\right)^3 + P_{11}a^n\left(\frac{a_1+vr_1}{v_1}\right)^{(3-n)} \\
&\quad \text{for } v = \frac{\alpha}{S_1}
\end{aligned}\right\} \tag{3.15}$$

$$\left.\begin{aligned}
F(r_1) &= \left(\frac{a_1+vr_1}{av_1}\right)^{m_1}\left[A_1\cos\left\{m_2'\log\left(\frac{a_1+vr_1}{av_1}\right)\right\}\right. \\
&\quad \left. + B_1\sin\left\{m_2'\log\left(\frac{a_1+vr_1}{av_1}\right)\right\}\right] \\
&\quad + P_{01}\left(\frac{a_1+vr_1}{v_1}\right)^3 + P_{11}\left(\frac{a_1+vr_1}{v_1}\right)^{(3-n)} \\
&\quad \text{for } 0 \le v < \frac{\alpha}{S_1}
\end{aligned}\right\} \tag{3.16}$$

In equations (3.14)–(3.16), A, B, A', B', A_1, B_1 are constants. Also

$$S_1 = 1 + \frac{\beta_1 - \sqrt{\beta_1}}{4(\beta_1^2 - \beta_1)} \tag{3.17}$$

where $\beta_1 = \dfrac{\beta^2}{\alpha^2} < 1$ and $1 < S_1 < 1.125$.

Again,

$$
\left.\begin{aligned}
m_1 &= \frac{1}{2}\left(1 - 4\alpha_1\beta_1\right) \\[2mm]
m_2 &= \frac{1}{2}\left[16\alpha_1^2(\beta_1^2 - \beta_1) - 8\alpha_1\beta_1 + 1\right]^{\frac{1}{2}} \\[2mm]
m_2' &= \frac{1}{2}\left[16\alpha_1^2(\beta_1 - \beta_1^2) + 8\alpha_1\beta_1 - 1\right]^{\frac{1}{2}}
\end{aligned}\right\}
\tag{3.18}
$$

where

$$
\alpha_1 = \frac{\alpha}{v} - 1
\tag{3.19}
$$

and

$$
P_{01} = \frac{P_0}{v_2^2\rho(4\alpha_1^2\beta_1 + 12\alpha_1\beta_1 + 6)}
\tag{3.20}
$$

$$
P_{11} = \frac{P_1}{v_2^2\rho\left[(3-n)(4\alpha_1\beta_1 - n + 2) + 4\alpha_1^2\beta_1\right]}
\tag{3.21}
$$

where

$$
v_2 = \frac{v}{v_1} = \frac{\alpha}{\alpha_1}
\tag{3.22}
$$

We therefore find from equation (3.10) that for,

$$
-\frac{a_1}{\alpha} \le r_1 \le T\left(1 - \frac{v}{\alpha}\right) - \frac{a_1}{\alpha}
$$

$$
i.e.,\quad -\frac{a_1}{\alpha} \le t - \frac{r}{\alpha} \le T\left(1 - \frac{v}{\alpha}\right) - \frac{a_1}{\alpha}
$$

$$
i.e.,\quad \frac{r - a_1}{\alpha} \le t \le T + \frac{r - (a_1 + vT)}{\alpha}
$$

We have

$$
\left.\begin{aligned}
\phi &= \frac{F\left(t - \frac{r}{\alpha}\right)}{r} \\[2mm]
&= \frac{A}{r}\left(\frac{t_1}{av_1}\right)^{m_1 + m_2} + \frac{B}{r}\left(\frac{t_1}{av_1}\right)^{m_1 - m_2} + \frac{P_{01}}{r}\left(\frac{t_1}{v_1}\right)^3 \\[2mm]
&\quad + \frac{P_{11}a^n}{r}\left(\frac{t_1}{v_1}\right)^{(3-n)} \\[2mm]
&\qquad \text{if } \frac{\alpha}{S_1} < v < \alpha
\end{aligned}\right\}
\tag{3.23}
$$

$$
\left.
\phi = \frac{1}{r}\left(\frac{t_1}{av_1}\right)^{m_1}\left[A' + B'\log\left(\frac{t_1}{av_1}\right)\right] + \frac{P_{01}}{r}\left(\frac{t_1}{v_1}\right)^3 + \frac{P_{11}a^n}{r}\left(\frac{t_1}{v_1}\right)^{(3-n)}
\right\}
$$

$$
\text{if } v = \frac{\alpha}{S_1}
$$

(3.24)

and

$$
\left.
\begin{aligned}
\phi = {}& \frac{1}{r}\left(\frac{t_1}{av_1}\right)^{m_1}\left[A_1\cos\left\{m_2'\log\left(\frac{t_1}{av_1}\right)\right\} + B_1\sin\left\{m_2'\log\left(\frac{t_1}{av_1}\right)\right\}\right] \\
& + \frac{P_{01}}{r}\left(\frac{t_1}{v_1}\right)^3 + \frac{P_{11}a^n}{r}\left(\frac{t_1}{v_1}\right)^{(3-n)} \\
& \text{if } 0 \le v < \frac{\alpha}{S_1}
\end{aligned}
\right\}
$$

(3.25)

where $r \ge (a+s)$ for $0 \le t \le T$ and $r \ge b$ for $t > T$ and

$$
t_1 = a_1 + v\left(t - \frac{r}{\alpha}\right)
$$

(3.26)

On taking ϕ in the form given in equations (3.23)–(3.25), the boundary condition (3.8) is satisfied.

Proceeding in a similar way, we find that the boundary condition (3.7) is satisfied if the function $F\left(t - \frac{r}{\alpha}\right)$ in equation (3.10) satisfies the relation

$$
\left.
\begin{aligned}
& \frac{a^2}{\alpha^2}F''\left(t - \frac{a}{\alpha}\right) + \frac{4\beta^2 a}{\alpha^3}F'\left(t - \frac{a}{\alpha}\right) + \frac{4\beta^2}{\alpha^2}F\left(t - \frac{a}{\alpha}\right) = 0 \\
& \text{for } (t < 0)
\end{aligned}
\right\}
$$

(3.27)

On writing $r_2 = t - \dfrac{a}{\alpha}$, we get,

$$
\left.
\begin{aligned}
& \frac{a^2}{\alpha^2}F''(r_2) + \frac{4\beta^2 a}{\alpha^3}F'(r_2) + \frac{4\beta^2}{\alpha^2}F(r_2) = 0 \\
& \text{for } (t < 0)
\end{aligned}
\right\}
$$

Hence

$$
F(r_2) = e^{-\frac{2\beta^2}{\alpha a}r_2}\left[A_{11}\cos\left(\frac{2\beta}{a}\sqrt{1-\beta_1}\,r_2\right) + A_{12}\sin\left(\frac{2\beta}{a}\sqrt{1-\beta_1}\,r_2\right)\right]
$$

$$
\text{for } r_2 < -\frac{a}{\alpha}
$$

(3.28)

So that, from equation (3.10) we find that the boundary condition (3.7) is satisfied on taking

$$
\phi = e^{-\frac{2\beta^2}{\alpha a}\left(t - \frac{r}{\alpha}\right)} \left[A_{11}\cos\left(\frac{2\beta}{a}\sqrt{1-\beta_1}\left(t - \frac{r}{\alpha}\right)\right) \right.
$$
$$
\left. + A_{12}\sin\left(\frac{2\beta}{a}\sqrt{1-\beta_1}\left(t - \frac{r}{\alpha}\right)\right) \right]
$$

(3.29)

for $t < \dfrac{r-a}{\alpha}$, where $r \geq a$ for $t < 0$, $r \geq a+s$ for $0 \leq t \leq T$ and for $r \geq b$ for $t > T$. We note, however, that the entire medium $r \geq a$ is at rest for all $t \leq 0$. So, that, for all $t \leq 0$, we have, $\dfrac{\partial \phi}{\partial r} = 0$ and $\dfrac{\partial^2 \phi}{\partial r \partial t} = 0$ for all $r \geq a$. Hence we have $A_{11} = A_{12} = 0$, so that we obtain

$$
\phi = 0 \text{ for } t < \frac{r-a}{\alpha}
$$

(3.30)

satisfying equation (3.7), where $r \geq a$ for $t < 0$, $r \geq a+s$ for $0 \leq t \leq T$ and $r \geq b$ for $t > T$.

We now try to find out the constants A, B, A', B', A_1, B_1 in equations (3.23)–(3.25). These are determined, for the respective ranges of value v, from the continuity of u and $\left(\tau_{rr} + \rho\alpha\dfrac{\partial u}{\partial t}\right)$ across $r = a + \alpha t$. We obtain the values of these constants, using the relations (3.1) and (3.3) for u and τ_{rr}, where ϕ is given by equations (3.23)–(3.25) for the respective ranges of value of v. These values are given in later, while giving the final expression for the displacement for the entire elastic region.

We consider the next boundary condition (3.9). From equation (3.10), we find that this boundary condition is satisfied if

$$
\frac{b^2}{\alpha^2}F''\left(t - \frac{b}{\alpha}\right) + \frac{4\beta^2 b}{\alpha^3}F'\left(t - \frac{b}{\alpha}\right) + \frac{4\beta^2}{\alpha^2}F\left(t - \frac{b}{\alpha}\right) = \frac{P_2}{\rho\alpha^2}e^{-k(t-T)}
$$
$$
\text{for } (t > T)
$$

(3.31)

Writing $r_3 = t - \dfrac{b}{\alpha}$, we find that equation (3.31) satisfied if

$$
\frac{b^2}{\alpha^2}\frac{d^2 F(r_3)}{dr_3^2} + \frac{4\beta^2 b}{\alpha^3}\frac{dF(r_3)}{dr_3} + \frac{4\beta^2}{\alpha^2}F(r_3) = \frac{P_2}{\rho\alpha^2}e^{-k\left(r_3 + \frac{b}{\alpha} - T\right)}
$$
$$
\text{for }\left(r_3 > T - \frac{b}{\alpha}\right)
$$

(3.32)

Solving equation (3.32) and using equation (3.10), we find that the boundary condition (3.9) is satisfied on taking

$$
\left.
\begin{aligned}
\phi = \frac{1}{r}F\left(t - \frac{r}{\alpha}\right) = \quad & \frac{e^{-\frac{2\beta^2}{\alpha b}\left(t - T - \frac{r-b}{\alpha}\right)}}{r} \\
& \times \left[A_3 \cos\left\{\beta_2\left(t - T - \frac{r-b}{\alpha}\right)\right\} + B_3 \sin\left\{\beta_2\left(t - T - \frac{r-b}{\alpha}\right)\right\}\right] \\
& + \frac{P_{21}}{r} e^{-k\left(t - T - \frac{r-b}{\alpha}\right)} \\
& \qquad\qquad \text{for } \left(r_3 > T - \frac{b}{\alpha}\right)
\end{aligned}
\right\}
\tag{3.33}
$$

for, $t > T + \dfrac{r-b}{\alpha}$, where $r \geq b$ and

$$
\left.
\beta_2 = \frac{2\beta}{b}\sqrt{1 - \beta_1}
\right\}
\tag{3.34}
$$

$$
\left.
\begin{aligned}
P_{21} &= \frac{P_3}{\rho\left[b^2k^2 + 4\beta^2\left(1 + \dfrac{kb}{\alpha}\right)\right]} \\
&\text{where } P_2 = P_0 + \frac{P_1 a^n}{(a + s(T))^n}
\end{aligned}
\right\}
\tag{3.35}
$$

and A_3, B_3 are constants. The constants A_3 and B_3 are determined from the continuity of u and $\left(\tau_{rr} + \rho\alpha\dfrac{\partial u}{\partial t}\right)$ across $r = a + \alpha(t - T)$, in terms of A_1, B_1 or A', B' or A, B for the respective ranges of v. Since A_1, B_1, etc. are determined as explained earlier, from a continuity condition, A_3, B_3 are determined. Finally, this gives the expression for potential ϕ, satisfying the initial condition, boundary condition, and equation of motion, for the entire elastic region for all values of time. Finally, we obtain the displacement at all points of the elastic region for all values of time, for all values of $v < \alpha$. After some simplification, it is found that, for values of v,

$$
u = 0 \text{ for } t < \frac{r-a}{\alpha}
\tag{3.36}
$$

where, $r \geq a$ for $t < 0$, $r \geq a_1 + vt$ for $0 \leq t \leq T$, and $r \geq b$ for $t > T$.

For $t \geq \dfrac{r-a}{\alpha}$, it is found that the expression for u depends on the values of v. It is found that for $\dfrac{r-a}{\alpha} \leq t \leq T + \dfrac{r-b}{\alpha}$.

$$u = \frac{\partial \phi}{\partial r} \text{ in equation (3.23)}$$

$$= -\frac{At_1^{m_1+m_2}}{\alpha r^2 (av_1)^{m_1+m_2}} \left[\alpha + \frac{vr(m_1+m_2)}{t_1} \right]$$

$$- \frac{Bt_1^{m_1-m_2}}{\alpha r^2 (av_1)^{m_1-m_2}} \left[\alpha + \frac{vr(m_1-m_2)}{t_1} \right] \tag{3.37}$$

$$- \frac{P_{01}t_1^3}{\alpha r^2 v_1^3} \left[\alpha + \frac{3vr}{t_1} \right] - \frac{P_{11}a^n t_1^{3-n}}{\alpha r^2 v_1^{3-n}} \left[\alpha + \frac{(3-n)vr}{t_1} \right]$$

$$\text{if } \frac{\alpha}{S_1} < v < \alpha$$

and

$$u = \frac{\partial \phi}{\partial r} \text{ in equation (3.24)}$$

$$= -\frac{A't_1^{m_1}}{\alpha r^2 (av_1)^{m_1}} \left[\alpha + \frac{vrm_1}{t_1} \right]$$

$$- \frac{B't_1^{m_1}}{\alpha r^2 (av_1)^{m_1}} \left[\left\{ \alpha + \frac{vrm_1}{t_1} \right\} \log \left(\frac{t_1}{av_1} \right) + \frac{rv}{t_1} \right] \tag{3.38}$$

$$- \frac{P_{01}t_1^3}{\alpha r^2 v_1^3} \left[\alpha + \frac{3vr}{t_1} \right] - \frac{P_{11}a^n t_1^{3-n}}{\alpha r^2 v_1^{3-n}} \left[\alpha + \frac{(3-n)vr}{t_1} \right]$$

$$\text{if } v = \frac{\alpha}{S_1}$$

and

$$u = \frac{\partial \phi}{\partial r} \text{ in equation (3.25)}$$

$$= \frac{t_1^{m_1-1}}{r(av_1)^{m_1}} \left[\left\{ \frac{vm_2'A_1}{\alpha} - \left(\frac{t_1}{r} + \frac{m_1 v}{\alpha} \right) B_1 \right\} \sin\left\{ m_2' \log\left(\frac{t_1}{av_1} \right) \right\} \right.$$

$$\left. - \left\{ \left(\frac{t_1}{r} + \frac{m_1 v}{\alpha} \right) A_1 + \frac{vm_2'B_1}{\alpha} \right\} \cos\left\{ m_2' \log\left(\frac{t_1}{av_1} \right) \right\} \right]$$

$$- \frac{P_{01}t_1^2}{rv_1^3} \left[\frac{3v}{\alpha} + \frac{t_1}{r} \right] + \frac{P_{11}a^n t_1^{2-n}}{rv_1^{3-n}} \left[\frac{(n-3)v}{\alpha} - \frac{t_1}{r} \right] \qquad (3.39)$$

$$\text{if } 0 \le v \le \frac{\alpha}{S_1}$$

where $r \ge a_1 + vt$ for $0 \le t \le T$ and $r \ge b$ for $t > T$.

For $t > T + \dfrac{r-b}{\alpha}$, we obtain,

$$u = \frac{\partial \phi}{\partial r} \text{ in equation (3.33)}$$

$$= \frac{e^{-\frac{2\beta^2 t_2'}{\alpha b}}}{r^2} \left[\left\{ \frac{r\beta_2}{\alpha} A_3 + \left(\frac{2r\beta_1}{b} - 1 \right) B_3 \right\} \sin\left(\beta_2 t_2' \right) \right.$$

$$\left. + \left\{ \left(\frac{2r\beta_1}{b} - 1 \right) A_3 - \frac{r\beta_2}{\alpha} B_3 \right\} \cos\left(\beta_2 t_2' \right) \right] + \frac{P_{21}e^{-kt_2'}}{r^2} \left(\frac{kr}{\alpha} - 1 \right) \qquad (3.40)$$

where $t_2' = t - T - \dfrac{r-b}{\alpha}$ and $r \ge b$ and P_{21} is given by equation (3.35).

The constants A_1, B_1, A', B', A, B, determined as described earlier and given by

$$
\left.
\begin{aligned}
A_1 &= \frac{F_1(v)}{F_2(v)}, B_1 = \frac{F_3(v)}{F_2(v)} \text{ for } \left(0 \le v < \frac{\alpha}{S_1} \right) \\[2mm]
A' &= \frac{F_4(v)}{F_5(v)}, B' = \frac{F_6(v)}{F_5(v)} \text{ for } \left(v = \frac{\alpha}{S_1} \right) \\[2mm]
\text{and } A &= \frac{F_7(v)}{F_8(v)}, B = \frac{F_9(v)}{F_8(v)} \text{ for } \left(\frac{\alpha}{S_1} < v < \alpha \right)
\end{aligned}
\right\}
\tag{3.41}
$$

where

$$
\left.
\begin{aligned}
F_1(v) &= \chi_1(v)\psi_2(v) - \psi_1(v)\chi_2(v), \\
F_2(v) &= \phi_1(v)\psi_2(v) - \psi_1(v)\phi_2(v), \\
F_3(v) &= \phi_1(v)\chi_2(v) - \chi_1(v)\phi_2(v), \\
&\quad \text{for } \left(0 \le v < \frac{\alpha}{S_1} \right) \\[3mm]
F_4(v) &= \chi_3(v)\psi_4(v) - \psi_3(v)\chi_4(v), \\
F_5(v) &= \phi_3(v)\psi_4(v) - \psi_3(v)\phi_4(v), \\
F_6(v) &= \phi_3(v)\chi_4(v) - \chi_3(v)\phi_4(v), \\
&\quad \text{for } \left(v = \frac{\alpha}{S_1} \right) \\[3mm]
F_7(v) &= \chi_5(v)\psi_6(v) - \psi_5(v)\chi_6(v), \\
F_8(v) &= \phi_5(v)\psi_6(v) - \psi_5(v)\phi_6(v), \\
F_9(v) &= \phi_5(v)\chi_6(v) - \chi_5(v)\phi_6(v), \\
&\quad \left(\frac{\alpha}{S_1} < v < \alpha \right)
\end{aligned}
\right\}
\tag{3.42}
$$

In equation (3.42),

$$\phi_1(v) = \left\{ \frac{vm_2'X}{\alpha} - \left(\frac{t_2}{a} + \frac{vm_1}{\alpha} \right) Y \right\}$$

$$\phi_2(v) = \left\{ \frac{vm_2'X}{\alpha} - \left(\frac{t_2}{a} + \frac{vm_1}{\alpha} \right) Y \right\} t_4 + \left\{ -\frac{v_H m_2'X}{T\alpha} + \frac{vm_2'^2 t_5 Y}{\alpha t_2} \right.$$

$$\left. - \left(\frac{t_3}{a} - \frac{v_H m_1}{T\alpha} \right) Y + \left(\frac{t_2}{a} + \frac{vm_1}{\alpha} \right) X \frac{m_2' t_5}{t_2} \right\} t_2$$

$$\psi_1(v) = \left\{ -\left(\frac{t_2}{a} + \frac{vm_1}{\alpha} \right) X - \frac{vm_2'}{\alpha} Y \right\}$$

$$\psi_2(v) = \left\{ -\left(\frac{t_2}{a} + \frac{m_1 v}{\alpha} \right) X - \frac{vm_2'}{\alpha} Y \right\} t_4 + \left\{ -\left(\frac{t_3}{a} - \frac{v_H m_1}{T\alpha} \right) X - \right.$$

$$\left. \left(\frac{t_2}{a} + \frac{m_1 v}{\alpha} \right) \frac{Y m_2' t_5}{t_2} + \frac{v_H m_2' Y}{T\alpha} + \frac{vm_2'^2 t_5 X}{T\alpha} \right\} t_2$$

$$\chi_1(v) = \frac{a(av_1)^{m_1}}{t_2^{m_1-1}} \left[\frac{t_2^2 P_{01}}{av_1^3} \left(\frac{3v}{\alpha} + \frac{t_2}{a} \right) - \frac{P_{11} t_2^{2-n} a^n}{av_1^{3-n}} \left\{ \frac{(n-3)v}{\alpha} - \frac{t_2}{a} \right\} \right]$$

$$\chi_2(v) = \frac{a(av_1)^{m_1}}{t_2^{m_1-2}}$$

$$\times \left[\left\{ \frac{2t_2 t_3 v_1^3 - 3t_2^2 v_1^2 \frac{v_H}{\alpha t}}{av_1^6} \left(\frac{3v}{\alpha} + \frac{t_2}{a} \right) + \frac{t_2^2}{av_1^3} \left(-\frac{3v_H}{\alpha T} + \frac{t_3}{a} \right) \right\} P_{01} \right.$$

$$- \left\{ \frac{a^n(2-n)t_2^{1-n} t_3 v_1^{3-n} - t_2^{2-n}(3-n)v_1^{2-n} \frac{v_H}{\alpha t}}{av_1^{2(3-n)}} \left\{ \frac{(n-3)v}{\alpha} - \frac{t_2}{a} \right\} \right.$$

$$\left. - \frac{a^n t_2^{2-n}}{av_1^{3-n}} \left\{ \frac{(n-3)v_H}{\alpha T} + \frac{t_3}{a} \right\} \right\} P_{11} \right]$$

$$\text{for } \left(0 \leq v < \frac{\alpha}{S_1} \right)$$

(3.43)

where

$$X = \sin\left\{ m_2' \log\left(\frac{t_2}{av_1}\right) \right\}$$

$$Y = \cos\left\{ m_2' \log\left(\frac{t_2}{av_1}\right) \right\}$$

$$t_2 = av_1 + \frac{1}{2}\frac{v_H t^2}{T}$$

and

$$\phi_3(v) = \left(1 + \frac{vm_1 a}{\alpha t_2}\right)$$

$$\phi_4(v) = -\frac{m_1 t_2^{m_1}}{(av_1)^{m_1}}\left\{ \left(\frac{t_3}{t_2} - \frac{v_H}{\alpha T v_1}\right)\left(1 + \frac{avm_1}{\alpha t_2}\right) + \frac{a}{\alpha}\frac{\frac{t_2 v_H}{T} + vt_3}{t_2^2} \right\}$$

$$\psi_3(v) = \left[\frac{av}{\alpha t_2} + \left\{1 + \frac{vm_1 a}{\alpha t_2}\right\}\log\left(\frac{t_2}{v_1}\right)\right]$$

$$\psi_4(v) = -\frac{t_2^{m_1}}{(av_1)^{m_1}}\left\{ \left(\frac{t_3}{t_2} - \frac{v_H}{\alpha T v_1}\right)\frac{av}{\alpha t_2} + \left(1 + \frac{m_1 va}{\alpha t_2}\right)\log\left(\frac{t_2}{av_1}\right) \right\}$$

$$+ \left\{ \frac{a}{\alpha}\frac{\frac{v_H t_2}{T} + vt_3}{t_2^2} + \frac{am_1}{\alpha}\frac{\frac{v_H t_2}{T} + vt_3}{t_2^2}\log\left(\frac{t_2}{av_1}\right) + \left(1 + \frac{m_1 va}{\alpha t_2}\right)\frac{t_3 v_1 - \frac{v_H t_2}{\alpha T}}{t_2 v_1} \right\}$$

$$\chi_3(v) = -\frac{a^2 v_1^{m_1}}{t_2^{m_1}}\left[\frac{t_2^3 P_{01}}{a^2 v_1^3}\left(1 + \frac{3va}{\alpha t_2}\right) + \frac{P_{11} t_2^{3-n} a^n}{a^2 v_1^{3-n}}\left\{ 1 + \frac{(3-n)va}{\alpha t_2} \right\} \right]$$

$$\chi_4(v) = \frac{3t_2^3 P_{01}}{v_1^3}\left[\left(\frac{t_3}{t_2} - \frac{v_H}{\alpha T v_1}\right)\left(1 + \frac{3va}{\alpha t_2}\right) - \frac{a}{\alpha}\frac{\frac{v_H t_2}{T} + vt_3}{t_2^2} \right]$$

$$+ \frac{(3-n)a^n t_2^{3-n} P_{11}}{v_1^{3-n}}\left[\left(\frac{t_3}{t_2} - \frac{v_H}{\alpha T v_1}\right)\left\{ 1 + \frac{(3-n)va}{\alpha t_2} \right\} - \frac{a}{\alpha}\frac{\frac{v_H t_2}{T} + vt_3}{t_2^2} \right]$$

$$\text{for } v = \frac{\alpha}{S_1}$$

(3.44)

where

$$t_3 = v_H \left\{ 1 - \frac{1}{T} \left(t - \frac{r}{\alpha} \right) \right\}$$

$$t_4 = (m_1 - 1)t_3 - \frac{m_1 t_2}{v_1} \frac{v_H}{\alpha T}$$

$$t_5 = t_3 - \frac{a v_H t_2}{a v_1 \alpha T}$$

$$\phi_5(v) = 1 + \frac{va(m_1 + m_2)}{\alpha t_2}$$

$$\phi_6(v) = -\frac{(m_1 + m_2)t_2^{m_1 + m_2}}{(av_1)^{m_1 + m_2}} \left[\left(\frac{t_3}{t_2} - \frac{v_H}{\alpha T v_1} \right) \left\{ 1 + \frac{va(m_1 + m_2)}{\alpha t_2} \right\} \right.$$

$$\left. + \frac{a}{\alpha} \frac{\frac{v_H t_2}{T} + vt_3}{t_2^2} \right]$$

$$\psi_5(v) = \frac{(av_1)^{2m_2}}{t_2^{2m_2}} \left\{ 1 + \frac{va(m_1 - m_2)}{\alpha t_2} \right\}$$

$$\psi_6(v) = -\frac{(m_1 - m_2)t_2^{m_1 - m_2}}{(av_1)^{m_1 - m_2}} \left[\left(\frac{t_3}{t_2} - \frac{v_H}{\alpha T v_1} \right) \left\{ 1 + \frac{(m_1 - m_2)va}{\alpha t_2} \right\} \right.$$

$$\left. + \frac{a}{\alpha} \frac{\frac{v_H t_2}{T} + vt_3}{t_2^2} \right]$$

$$\chi_5(v) = -\frac{(av_1)^{m_1 + m_2}}{t_2^{m_1 + m_2}}$$

$$\times \left[\frac{P_{01} t_2^3}{v_1^3} \left(1 + \frac{3va}{\alpha t_2} \right) + \frac{P_{11} a^n t_2^{(3-n)}}{v_1^{(3-n)}} \left\{ 1 + \frac{(3-n)va}{\alpha t_2} \right\} \right]$$

$$\chi_6(v) = \frac{3 t_2^3 P_{01}}{v_1^3} \left[\left(\frac{t_3}{t_2} - \frac{v_H}{\alpha T v_1} \right) \left(1 + \frac{3va}{\alpha t_2} \right) - \frac{a}{\alpha} \frac{\frac{v_H t_2}{T} + vt_3}{t_2^2} \right]$$

$$+ \frac{(3-n)a^n t_2^{3-n} P_{11}}{v_1^{3-n}}$$

$$\times \left[\left(\frac{t_3}{t_2} - \frac{v_H}{\alpha T v_1} \right) \left\{ 1 + \frac{(3-n)va}{\alpha t_2} \right\} - \frac{a}{\alpha} \frac{\frac{v_H t_2}{T} + vt_3}{t_2^2} \right]$$

$$\text{for } \frac{\alpha}{S_1} < v < \alpha$$

(3.45)

The constants A_3 and B_3 are found to have different values for different ranges of values of v. Proceeding in similar way stated earlier, using the continuity condition of u and $\dfrac{\partial u}{\partial t}$ at $t = T$, $r = b$, we get

$$
\left.
\begin{aligned}
A_3 &= \frac{F_9(v)}{F_{10}(v)}, B_3 = \frac{F_{11}(v)}{F_{10}(v)} \quad \text{for} \quad \left(0 \le v < \frac{\alpha}{S_1}\right) \\[2mm]
A_3 &= \frac{F_{12}(v)}{F_{10}(v)}, B_3 = \frac{F_{13}(v)}{F_{10}(v)} \quad \text{for} \quad \left(v = \frac{\alpha}{S_1}\right) \\[2mm]
\text{and} & \\[1mm]
A_3 &= \frac{F_{14}(v)}{F_{10}(v)}, B_3 = \frac{F_{15}(v)}{F_8(v)} \quad \text{for} \quad \left(\frac{\alpha}{S_1} < v < \alpha\right)
\end{aligned}
\right\}
\tag{3.46}
$$

where

$$
\left.
\begin{aligned}
F_9(v) &= \chi_7(v)\psi_8(v) - \psi_7(v)\chi_8(v), \\
F_{10}(v) &= \phi_7(v)\psi_8(v) - \psi_7(v)\phi_8(v), \\
F_{11}(v) &= \phi_7(v)\chi_8(v) - \chi_7(v)\phi_8(v), \\
F_{12}(v) &= \chi_9(v)\psi_8(v) - \psi_7(v)\chi_{10}(v), \\
F_{13}(v) &= \phi_7(v)\chi_{10}(v) - \chi_9(v)\phi_8(v), \\
F_{14}(v) &= \chi_{11}(v)\psi_8(v) - \psi_7(v)\chi_{12}(v), \\
F_{15}(v) &= \chi_{12}(v)\phi_7(v) - \phi_8(v)\chi_{11}(v),
\end{aligned}
\right\}
\tag{3.47}
$$

In equation (3.47),

$$
\left.
\begin{aligned}
\phi_7 &= \frac{2\beta_1 - 1}{b^2}, \quad \psi_7 = -\frac{\beta_2}{b\alpha} \\[3mm]
\phi_8 &= \frac{1}{\alpha b^2}\left[b\beta_2^2 - \frac{2\beta^2}{b}(2\beta_1 - 1)\right], \quad \psi_8 = \frac{\beta_2}{b^2}(4\beta_1 - 1)
\end{aligned}
\right\}
\tag{3.48}
$$

Also in equation (3.47),

$$
\left.
\begin{aligned}
\chi_7 =\ & \frac{P_{21}}{b^2}\left(1 - \frac{kb}{\alpha}\right) \\[2mm]
& + \frac{T_1^{m_1-1}}{b(av_1)^{m_1} F_2}\left[\left\{\frac{vm_2'F_1}{\alpha} - \left(\frac{T_1}{b} + \frac{m_1 v}{\alpha}\right)F_3\right\}\sin\left\{m_2'\log\left(\frac{T_1}{av_1}\right)\right\} \right. \\[2mm]
& \left. - \left\{\frac{vm_2'F_3}{\alpha} + \left(\frac{T_1}{b} + \frac{m_1 v}{\alpha}\right)F_1\right\}\cos\left\{m_2'\log\left(\frac{T_1}{av_1}\right)\right\}\right] \\[2mm]
& - \frac{T_1^2 P_{01}}{v_1^3 b}\left(\frac{3v}{T_1} + \frac{T_1}{b}\right) + \frac{a^n T_1^{2-n} P_{11}}{v_1^{3-n} b}\left\{\frac{(n-3)v}{\alpha} - \frac{T_1}{b}\right\}
\end{aligned}
\right\}
\tag{3.49}
$$

$$\chi_8 = -\frac{kP_{21}}{b^2}\left(1 - \frac{kb}{\alpha}\right)$$

$$+\frac{T_1^{m_1-2}}{b(av_1)^{m_1}F_2}\left[\left[-\left\{\frac{vm_2'F_3}{\alpha} + \left(\frac{T_1}{b} + \frac{m_1v}{\alpha}\right)F_1\right\}t_4\right.\right.$$

$$+T_1\left\{\frac{vm_2'F_1}{\alpha} - \left(\frac{T_1}{b} + \frac{m_1v}{\alpha}\right)F_3\right\}\frac{m_2't_5}{T_1}$$

$$-\left\{F_1\left(\frac{t_3}{b} - \frac{v_Hm_1}{T\alpha}\right) - \frac{v_Hm_2'F_3}{T\alpha}\right\}T_1\right]\cos\left\{m_2'\log\left(\frac{T_1}{av_1}\right)\right\}$$

$$+\left[\left\{\frac{vm_2'F_1}{\alpha} - \left(\frac{T_1}{b} + \frac{m_1v}{\alpha}\right)F_3\right\}t_4\right.$$

$$+T_1\left\{\frac{vm_2'F_3}{\alpha} + \left(\frac{T_1}{b} + \frac{m_1v}{\alpha}\right)F_1\right\}\frac{m_2't_5}{T_1}$$

$$+\left.\left\{-F_3\left(\frac{t_3}{b} - \frac{v_Hm_1}{T\alpha}\right) - \frac{v_Hm_2'F_1}{T\alpha}\right\}T_1\right]\sin\left\{m_2'\log\left(\frac{T_1}{av_1}\right)\right\}\right]$$

$$-\frac{T_1^2P_{01}}{v_1^3b}\left(\frac{2t_3}{T_1} - \frac{3v_H}{v_1T\alpha}\right)\left(\frac{3v}{\alpha} + \frac{T_1}{b}\right) - \frac{P_{01}T_1^2}{v_1^3b}\left(-\frac{3v_H}{T\alpha} + \frac{t_3}{b}\right)$$

$$+\frac{P_{11}a^nT_1^{2-n}}{v_1^{3-n}b}\left\{\frac{(2-n)t_3}{T_1} - \frac{(3-n)v_H}{v_1T\alpha}\right\}\left\{\frac{(n-3)v}{\alpha} - \frac{T_1}{b}\right\}$$

$$+\frac{a^nT_1^{2-n}P_{11}}{v_1^{3-n}b}\left\{-\frac{(n-3)v_H}{\alpha} - \frac{t_3}{b}\right\}$$

$$\left.\right\} \quad (3.50)$$

$$\chi_9 = \frac{P_{21}}{b^2}\left(1 - \frac{kb}{\alpha}\right)$$

$$-\frac{T_1^{m_1}}{\alpha b^2(av_1)^{m_1}F_5}$$

$$\times\left[\left(\alpha + \frac{vbm_1}{T_1}\right)F_4 + \left\{\frac{vb}{T_1} + \left(\alpha + \frac{vbm_1}{T_1}\right)\log\left(\frac{T_1}{av_1}\right)\right\}F_6\right]$$

$$-\frac{T_1^3P_{01}}{\alpha b^2v_1^3}\left(\alpha + \frac{3vb}{T_1}\right) - \frac{P_{11}a^nT_1^{3-n}}{\alpha b^2v_1^{3-n}}\left\{\alpha + \frac{(3-n)vb}{T_1}\right\}$$

$$\left.\right\} \quad (3.51)$$

$$\chi_{10} = -\frac{kP_{21}}{b^2}\left(1 - \frac{kb}{\alpha}\right) - \frac{T_1^{m_1-1}}{\alpha b^2 (av_1)^{m_1} F_5}$$

$$\times \left[\left\{m_1\left(t_3 - \frac{v_H T_1}{\alpha T v_1}\right)\left(\alpha + \frac{vbm_1}{T_1}\right) + \frac{m_1 b}{T_1}\left(-\frac{T_1 v_H}{T} - vt_3\right)\right\}F_4\right.$$

$$+ \left\{\frac{b}{T_1}\left(-\frac{v_H T_1}{T} - vt_3\right) + \frac{m_1 b}{T_1}\left(-\frac{v_H T_1}{T} - vt_3\right)\log\left(\frac{T_1}{av_1}\right)\right.$$

$$+ \left(\alpha + \frac{vbm_1}{T_1}\right)\left(t_3 - \frac{v_H T_1}{\alpha T v_1}\right)\right\}F_6$$

$$\left.+ m_1\left(t_3 - \frac{v_H T_1}{\alpha T v_1}\right)\left\{\frac{vb}{T_1} + \left(\alpha + \frac{vbm_1}{T_1}\right)\log\left(\frac{T_1}{av_1}\right)\right\}F_6\right]$$

$$- \frac{3T_1^2 P_{01}}{\alpha b^2 v_1^3}\left[\left(t_3 - \frac{v_H T_1}{\alpha T v_1}\right)\left(\alpha + \frac{3vb}{T_1}\right) + \frac{b}{T_1}\left(-\frac{v_H T_1}{T} - vt_3\right)\right]$$

$$- \frac{(3-n)P_{11}a^n T_1^{2-n}}{\alpha b^2 v_1^{3-n}}\left[\left(t_3 - \frac{v_H T_1}{\alpha T v_1}\right)\left\{\alpha + \frac{(3-n)vb}{T_1}\right\}\right.$$

$$\left.+ \frac{b}{T_1}\left(-\frac{v_H T_1}{T} - vt_3\right)\right]$$

(3.52)

$$\chi_{11} = \frac{P_{21}}{b^2}\left(1 - \frac{kb}{\alpha}\right) - \frac{1}{\alpha b^2 F_8}\left[\left\{\alpha + \frac{vb(m_1+m_2)}{T_1}\right\}\frac{T_1^{m_1+m_2}F_7}{(av_1)^{m_1+m_2}}\right.$$

$$\left.+ \left\{\alpha + \frac{vb(m_1-m_2)}{T_1}\right\}\frac{T_1^{m_1-m_2}F_9}{(av_1)^{m_1-m_2}}\right]$$

$$- \frac{T_1^3 P_{01}}{\alpha b^2 v_1^3}\left(\alpha + \frac{3vb}{T_1}\right) - \frac{P_{11}a^n T_1^{3-n}}{\alpha b^2 v_1^{3-n}}\left\{\alpha + \frac{(3-n)vb}{T_1}\right\}$$

(3.53)

$$\chi_{12} = -\frac{kP_{21}}{b^2}\left(1 - \frac{kb}{\alpha}\right) - \frac{1}{\alpha b^2 F_8}\left[\frac{(m_1+m_2)T_1^{m_1+m_2-1}F_7}{(av_1)^{m_1+m_2}}\right.$$

$$\times \left\{\left(t_3 - \frac{v_H T_1}{\alpha T v_1}\right)\left(\alpha + \frac{vb(m_1+m_2)}{T_1}\right) + \frac{b}{T_1}\left(-\frac{T_1 v_H}{T} - vt_3\right)\right\}$$

$$+ \frac{(m_1-m_2)T_1^{m_1-m_2-1}F_9}{(av_1)^{m_1-m_2}}$$

$$\times \left\{\left(t_3 - \frac{v_H T_1}{\alpha T v_1}\right)\left(\alpha + \frac{vb(m_1-m_2)}{T_1}\right) + \frac{b}{T_1}\left(-\frac{T_1 v_H}{T} - vt_3\right)\right\}\right]$$

$$- \frac{3T_1^2 P_{01}}{\alpha b^2 v_1^3}\left[\left(t_3 - \frac{v_H T_1}{\alpha T v_1}\right)\left(\alpha + \frac{3vb}{T_1}\right) + \frac{b}{T_1}\left(-\frac{v_H T_1}{T} - vt_3\right)\right]$$

$$- \frac{(3-n)P_{11}a^n T_1^{2-n}}{\alpha b^2 v_1^{3-n}}\left[\left(t_3 - \frac{v_H T_1}{\alpha T v_1}\right)\left\{\alpha + \frac{(3-n)vb}{T_1}\right\}\right.$$

$$\left.+ \frac{b}{T_1}\left(-\frac{v_H T_1}{T} - vt_3\right)\right]$$

$$\tag{3.54}$$

where $T_1 = t_1$ at $r = b, t = T$, and t_1 is given by equation (3.26).

3.3 NUMERICAL COMPUTATIONS, RESULTS, AND DISCUSSIONS

We note here that the initial velocity v_H of the wall of the cavity and final radius b of the cavity essentially depend upon the model parameters including the magnitude of P_0, P_1, P_2 and also on the values of λ, μ and ρ which characterizes the elastic medium. The initial radius of the cavity is taken to be 10 m in all possible cases. We assume here $P_0 = 0$ and take different values of P_1–higher values of P_1 induce higher values of v_H as well as higher values of v, keeping the values of λ and μ unchanged. We assume here $\lambda = \mu$ (Poisson condition) which is satisfied in a major part of the Earth. So that $\alpha = \sqrt{3}\beta$. The velocity of the longitudinal wave (α) and that of shear waves (β) are taken to 7.8 and 4.5 km/s. respectively. In all the cases we take $n = 1$.

3.3.1 CASE I

We take $P_1 = 0.1$ bar, exerted on the wall of the cavity by the explosion, producing an initial velocity of $v_H = 1$ km/s. We take $b = 300$ m with $a = 10$ m. The time T during which the cavity expands is found to be 0.58 seconds. So the magnitude of deceleration is 1.72 km/s². The displacements against time at a point outside of the cavity for which $r = 500$ m have been shown in the following Figure 3.1.

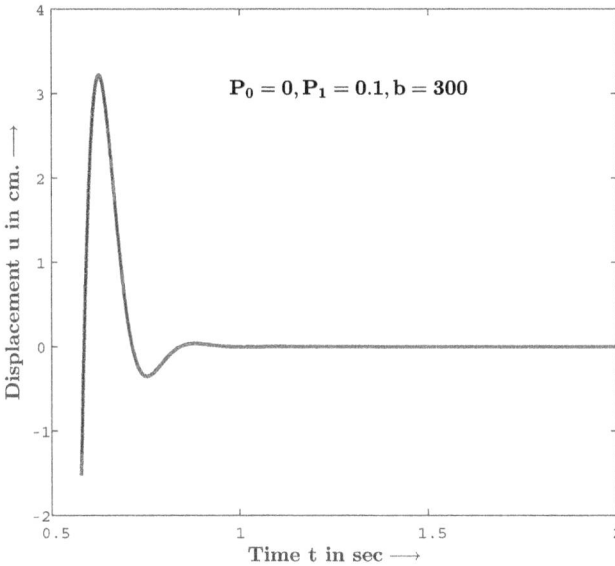

Figure 3.1   Time-displacement curve at $r = 500$ m (with $P_0 = 0.1$ bar and $v_H = 1$ km/s).

It shows that the point undergoes an oscillatory motion with a maximum amplitude of 3.21 cm. and the oscillation last for 1.1 seconds. after which the oscillatory nature ceased.

3.3.2   CASE II

We consider a slightly higher explosion within the cavity producing $P_1 = 0.2$ bar and an initial outward velocity $v_H = 2$ km/s. We construct the time-displacement curve at the same point. The highest magnitude of the amplitude is found to be 3.64 cm. but the oscillation takes about 1.07 seconds. to die out. Figure 3.2 is shown below

3.3.3   CASE III

Next, we consider a higher explosion with respect to previous one within the cavity producing $P_1 = 0.3$ bar and initial velocity $v_H = 3$ km/s. In this case, we take $b = 450$ m with $a = 10$ m. The expansion of the cavity stops with time $T = 0.29$ seconds. We compute the displacement at a point outside the cavity for which $r = 500$ m. In this case, it is found that the highest amplitude is 12.38 cm. and oscillation last for about 1.06 seconds and after that it dies out (Figure 3.3).

3.3.4   CASE IV

Lastly, we consider much higher expansion within the cavity with respect to the previous three cases, such that it produces $P_1 = 0.4$ bar and initial velocity $v_H = 4$ km/s. Here we take $b = 455$ m with $a = 10$ meter and the displacement points $r = 500$ m. The expansion of the cavity would stop after time $T = 0.22$ s. In this case,

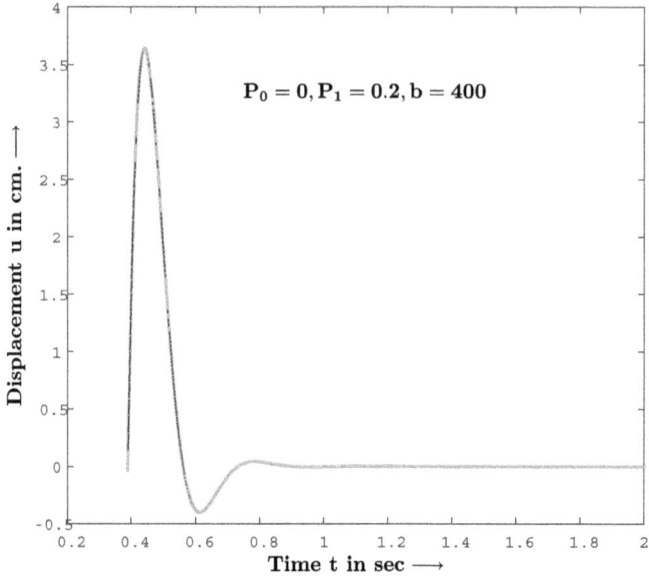

Figure 3.2   Time-displacement curve at $r = 500$ m (with $P_0 = 0.2$ bar and $v_H = 2$ km/s).

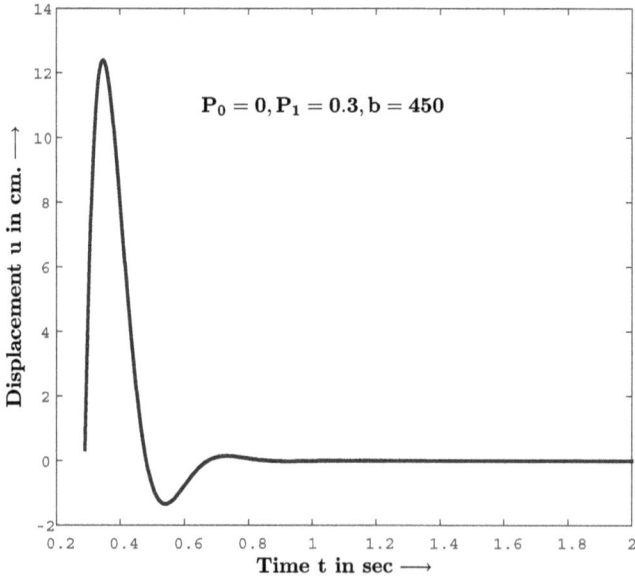

Figure 3.3   Time-displacement curve at $r = 500$ m (with $P_0 = 0.3$ bar and $v_H = 3$ km/s).

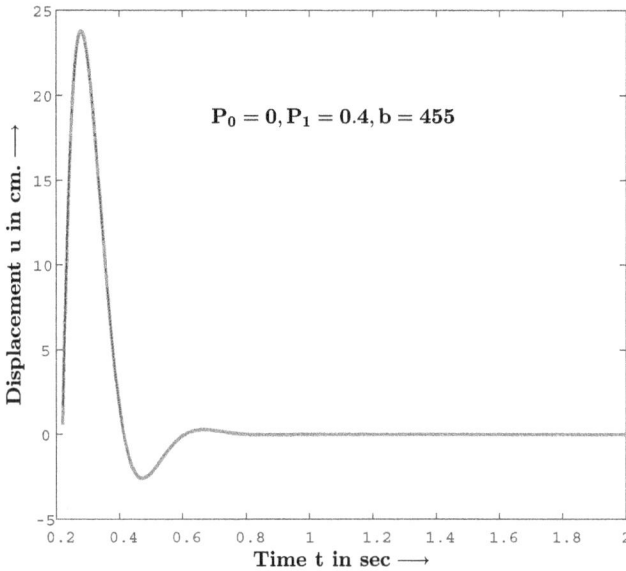

$P_0 = 0, P_1 = 0.4, b = 455$

Figure 3.4 Time-displacement curve at $r = 500$ m (with $P_0 = 0.4$ bar and $v_H = 4$ km/s.).

the highest magnitude of the amplitude is 23.75 cm. and the oscillation time takes about 1.00 seconds and thereafter it goes to zero. Figure 3.4 is shown below

3.4 REMARKS AND CONCLUSION

In most of the earlier papers dealing with the explosion within embedded cavities, the boundary of the cavity has been assumed to be either fixed or moves with a uniform velocity, under the impulse of the force due to the explosion. However, it is expected that boundary of the cavity should move in an outward direction with a velocity which decreases gradually as the pressure exerted on the boundary decreases with time. Further, with the same value of P_0 and P_1 with constants values of model parameters such as λ, μ, ρ, etc. the velocity of expansion of the boundary can not be different as suggested in the earlier publications. To generate higher velocities of expansion of the boundary of the cavity with the same value of the model parameter it is essential that P_0 and P_1 will be of higher magnitude. This concept has been followed in this model of expanding cavities with gradually decreasing velocity.

REFERENCES

Aboudi, J. (1971). The motion excited by an impulsive source in an elastic half-space with a surface obstacle. *Bulletin of the Seismological Society of America*, 61:747–763.

Aboudi, J. (1972). The response of an elastic half-space to the dynamic expansion of an embedded spherical cavity. *Bulletin of the Seismological Society of America*, 62(1):115–127.

Andrews, D. J. (1973). A numerical study of tectonic stress by an underground explosion. *Bulletin of the Seismological Society of America*, 63(4):1375–1391.

Archambeau, C. B. (1972). The theory of stress wave radiation from explosions in prestressed media. *Geophysical Journal of the Royal Astronomical Society*, 29:329.

Archambeau, C. B. and Sammis, C. (1970). Seismic radiation from an explosion in prestressed media and the measurement of tectonic stress in the Earth. *Reviews of Geophysics*, 8:473.

Chenglong, Y., Zhongqi, W., and Wengong, H. (2018). A prediction model for amplitude-frequency characteristics of blast induced seismic waves. *Geophysics*, 83(3):159–173.

Daniel, A. and Bibiana, L. (2019). Effects of underground explosions on soil and structures. *Science Direct*, 5:324–338.

Ghosh, M. L. (1969). On the propagation of spherical waves due to large underground explosion. *Pure and Applied Geophysics*, 72:22–34.

Gurvich, I. I. (1968): Theory of spherical emitter of transverse seismic waves; Izv. Acad. Sci. U.S.S.R., *Phys. Solid. Earth*. 6, 22–29.

Gurvich, I. I.and Yanovskii, V. (1967): Seismic impulses of an explosion in a homogeneous absorbing medium; Izv. Acad. Sci. U.S.S.R., *Phys. Solid. Earth.*, 5, 634–641.

Honda, H. (1959). The elastic waves generated from a spherical source. *Science Reports of the Tohoku University*, 5:11–78.

Jeffreys, H. (1931). On the cause of oscillatory movements in seismograms. *Monthly Notices of the Royal Astronomical Society (MNRAS)*, 2:329–334.

Mario, B. and Alejandro, M. (2020). Spherical cavity expansion approach for the study of rigid-penetrator's impact problems. *Applied Mechanics*, 1:20–26.

Min, R., Guanwen, C., Wancheng, Z., Wen, N., Kai, G., and Tianhong, Y. (2022). A prediction model for surface deformation caused by undergrpound mining based on spatio–temporal associations. *Geomatics, Natural Hazards and Risk*, 13:94–122.

Minster, J. B. and Suteau, A. M. (1977). Far-field wave from an arbitrarily expanding, transparent spherical cavity in a prestressed media. *Geophysical Journal of the Royal Astronomical Society*, 50:215–233.

Mukhopadhyay, A. (1971). Thesis entitled " source of finite dimensions in elastic media ".

Randal, M. J. (1964). On the mechanism of earthquakes. *Bulletin of the Seismological Society of America*, 54:1283–1289.

Randal, M. J. (1966). Seismin radaiation from a sudden phase transition. *Journal of Geophysical Research.*, 71:5297–5302.

Rigby, S., Fay, S., Clarke, S., Tyas, A., Reay, J., Warren, J., M. Gant, M., and Elgy, I. (2016). Measuring spatial pressure distribution from explosives buried in dry leighton buzzard sand. *International Journal of Impact Engineering*, 96:89–104.

Vodicva, V. (1963). Radial vibration of an infinite medium with spherical cavity. *ZAMP*, 14(6):745–748.

4 Importance of Differential Equations in a Retailing Strategy under Credit Period Consideration

Bikash Koli Dey
Hongik University

Ashish Kumar Mondal
Banasthali Vidyapith

Tapas Kumar Jana
Ghatal Rabindra Satabarsiki Mahavidyalaya

Biswajit Sarkar
Yonsei University
Saveetha University

CONTENTS

DOI: 10.1201/9781003227847-4

4.1   INTRODUCTION

Deciding on inventory for any industry is a very much crucial decision. The inventory of any business is one of the major indicators for running the business smoothly with optimizing profit. If the price of a particular product is high, then the industry can earn more revenue. But on the other hand, a huge amount of selling price has a critical impact on the demand for the product. In reality, high selling prices reduce demand, which leads to the company's loss. Therefore, taking a decision on the selling price of the product plays a vital role in determining the demand of the product and profit of the product (Sarkar et al., 2021a).

Nowadays, customers want their desired product as early as possible, but due to the high price of the product, it is not possible all time that they pay a whole amount at a time for a particular product. To resolve this problem and satisfy customers' demands, the industry provided a certain period to pay the amount, called delay. Customers can take the products home by paying a percentage or without paying any amount. The product's company provides a certain time period during which customers can pay the amount for their product. No extra amount will be deducted on that period, or some time company charges a very small amount for this delay time. However, if this delay period is over, customers have to pay a certain percentage of the extra amount as a fine (Mashud et al., 2020).

The concept of delay in payment is beneficial for customers and retailers simultaneously. The customer's benefit is that they can take their desired product to the home at an affordable price and not handle the pressure of massive payment at a time. On the other hand, the retailer is benefited in several ways; first of all, a retailer can save their holding cost; second, the retailer can feel relief from the anxiety of deterioration of the product; third, the retailer can earn some interest after the delay period. Thus, the concept of delay in payment is beneficial for both players.

This study deals with a single deteriorate item; thus, the concept of trade credit is beneficial for both players. The time when deterioration is started, and the total length of the cycle is too important. Moreover, customers always attract by the items shown in the retail shop. Thus, the stock of the retailer is a major component to determine the demand for the product. Traditional inventory deals with fixed demand; however, in reality, demand can not always be constant; it varies with a different system component. Thus, demand variability is another trending topic in inventory research.

The following research gaps and research questions are fulfilled through the current study, along with the help of differential equations.

1. Several inventory models were developed in the literature for deteriorating items, but few deal with trade-credit or delay in payment policy.

2. Traditional economic order quantity (EOQ) deals with fixed demand; however, stock level, time, and selling price are dependent on the demand for a deteriorating item under trade credit policy, which is very rare.

The background of the research or the existing literature is discussed in the next Sections 4.2 and 4.3 containing the idea, symbols, and assumptions of the research. The main modeling along with the help of differential equation is presented in Section 4.4. Solution methodology which was established with the help of a differential equation is provided in Section 4.5. The model was validated by some numerical examples in Section 4.6. The effect of the parameters is shown in Section 4.7, as a sensitivity analysis. Industrial implementations and suggestions for the managers of the industry are provided in Section 4.8. Finally, some concluding remarks along with how the feature researcher will be benefited from this study are discussed in the Section 4.9.

4.2 LITERATURE REVIEW

In-depth elaboration of the previous studies is discussed in this section as follows:

4.2.1 STOCK AND PRICE DEPENDENT DEMAND

The basic requirement of any business sector is the perfect decision on the inventory. The necessity of the EOQ model is to calculate the exact inventory and make a smooth decision on inventory and cycle time. In general EOQ models have two different categories, one continuous review model, in which it is known when to order but it is not known how much to order and another is the periodic review model, in which quantity is known but when to order i.e., decision made on the basis of cycle time. The journey of EOQ models started about hundreds of years ago, and after that several models were developed on the basis of different concepts and parameters Cárdenas-Barrón et al. (2014).

The demand for the product is crucial for the inventory system. It is quite obvious that customers will attract by the stock of the product, in other words, the stock of a particular product increases the demand Sana and Chaudhuri (2008). An EOQ model was considered by Choi et al. (2008), where demand was constant but the rate of replenishment was partial. The effect of imperfectness in the production process was another major issue in optimizing the profit of a stock- dependent-inventory system (Sarkar et al., 2010). Yang (2014) discussed the optimum profit under shortages and without shortages for stock-dependent demand. An order quantity model was formulated by Önal et al. (2016) where demand depended on the stock level, and capacity constraint also took place. They a use MINLP solver to solve this model. Under the consideration of partial backorder, a stock-level dependent inventory model was presented by Shaikh et al. (2019). Recently Mandal et al. (2021) calculate the importance of stock and advertise dependent demand for an inventory model.

The selling price of the product is another critical parameter to decide the demand for the product. Preservation technology and the selling price of the product had great importance on the deterioration product Shastri et al. (2014). The demand for deteriorating products is very much disturbed by the freshness of the product and the price of the product (Banerjee and Agrawal, 2017). The effect of the selling price in an inventory model was discussed by Dey et al. (2019b), where they also considered the effect of carbon emission to control the stainability. A two-echelon SCM for multi-item was developed by Pervin et al. (2019) where demand varies with the level of the stock and price of the product. An inventory model with stock-level dependent demand was evaluated by Cárdenas-Barrón et al. (2020). To control the inventory for the deteriorating items preservation is very essential stated by Das et al. (2021b). In this model, they considered that the demand varied with the price of the product and the backlog was partial. Sometimes advertisements claim a certain attention to determine the demand for the product (Dey et al., 2021a). An inventory model for return product was considered by Pando et al. (2021); Feng et al. (2022), where demand varies with the stock level and selling price of the product. Those models provide a close-loop solution for the decision variables.

4.2.2   INVENTORY MODEL WITH TRADE-CREDIT

In present days, customers want their desire product as early as possible, however, it is always not possible for the customer to pay full amount for the product at a single time. To overcome this problem, the company provides some delay period, in which no extra charge will be debited from the customers Khanra et al. (2011). By provide delay period both the company and customers are gainers, as customers get their desire product without paying full amount, and simultaneously the company can reduce their holding cost Sarkar (2012). Soni (2013) proposed the replenishment strategy for the deteriorating items in the light of delay in payment and demand subordinate on stock and price. Khanra et al. (2013) formed an inventory system which considered credit financing policy and demanded subordinates on time. Defective products with variable lead time in an EOQ model were calculated by Sarkar et al. (2014) They also discussed how lead time and delay period impacted system cost. Sarkar et al. (2015) showed how changeable decay rate and lead time had an impact on an inventory system having stationary lifetime products. Mahata and De (2017) formulated an inventory system considering trade-credit policy in a partial manner for a supply chain. Mashud et al. (2020) developed an inventory system to find out the influence of trade-credit, preservation technology, and advertisement policy. Sarkar et al. (2018) studied a sustainable SCM in the light of multi-delay-in-period trade-credit policy and variability in carbon emission. Shin et al. (2018) discussed the effect of trade-credit and human-error for a two-echelon SCM. Tiwari et al. (2019) proposed an order policy under the multi-trade-credit policies and carbon emissions for deteriorating products. An advanced inventory system of deteriorating products including various price discount policies was established by Saren et al. (2020). Sarkar et al. (2020b) implemented a replenishment strategy and trade-credit for the deteriorating products. Sepehri et al. (2021b) discussed pricing policy and permissible delay payments in an inventory system. Ahmed et al. (2021) studied an imperfect production system with a multi-period delay-in payment policy under partial backlogging. Recently different inventory or SCM was developed with trade-credit financing (Shaikh et al., 2021; Jani et al., 2021; Singh et al., 2021; Mashud et al., 2021; Das et al., 2021a). Basically, trade-credit financing is much more effective for deteriorating items. Thus, an in-depth analysis regarding deteriorate products is performed in the next segment.

4.2.3 MODEL FOR DETERIORATING PRODUCTS

In this study, we focus on deteriorating items, which may be any type of food products or products related to medical science. Deterioration of those types of products is really a headache for the industry. To optimize the loss, the company uses a different preservation technology or strategy, and "trade-credit or delay-in-payments" was one of them (Sarkar et al., 2017; Mishra et al., 2021; Sepehri et al., 2021a). Sarkar (2012) developed an inventory system with deteriorating items including the fact that the deterioration of products varies with time, also in this context the concept of delayed payment is imposed. Sett et al. (2012) proposed the impact of two warehouses for a time-dependent deteriorating item. A two-echelon SCM for deteriorating items was proposed by Sarkar (2013), where deterioration is probabilistic. An inventory system along with variable demand and variable deterioration was established by Sarkar and Sarkar (2013). The optimal credit period for deteriorating products was calculated by Wu et al. (2014), where they were also concerned about product's expiry. Taleizadeh and Nematollahi (2014) proposed an inventory system with deteriorating product, under the consideration of backorder. A SCM for deteriorating items along with the consideration of revenue sharing and investment contracts was developed by Zhang et al. (2015). An inventory system including decay products was studied by Shah and Cárdenas-Barrón (2015) where they considered the effect of discount policies for the credit period. In the same direction Kumar Sett et al. (2016) developed an optimal replenishment policy "with fixed life-time" items. Kumar Sett et al. (2016) model was extended by Sarkar and Saren (2017) along with different ordering and transfer policies. A sustainable SCM for deteriorating items under revenue sharing and promotional cost sharing was discussed by Bai et al. (2017). Jaggi et al. (2017) discussed the credit financing for a deteriorating item in a production model along with shortages and demand variability. Tiwari et al. (2018) developed Jaggi et al. (2017) model by considering SCM scenario and partial backorder. Application of the residue of deteriorating products through some remanufacturing facilities was proposed by Iqbal and Sarkar (2019). The decision for the selling price along with preservation technology for deteriorating items was studied by Li et al. (2019). A three-echelon SCM for deteriorating items was considered by Maihami et al. (2019), where they considered a probabilistic environment. The effect of carbon emission for a deteriorating product production system under two-stage inspection was performed by Shaw et al. (2020). A different solution approach for deteriorating items was invented by Calıskan (2020), where he used an analytic approach instated of derivative. Calıskan (2021) extend his own previous model by considering planned backorders. Duary et al. (2021) introduced an order quantity model "with product deterioration along with delay-in-payments" and a price discount strategy. Recently, Nematollahi et al. (2022) developed a supply chain model under the consideration of

Table 4.1

Novelty of Present Study Compare to Previous Study

Existing Study	Dependancy of Demand	Type of Model	Strategy	Deterioration Rate
Sarkar et al. (2010)	SD Dependent	Inv.	NA	NA
Soni (2013)	SD	Inv.	DIP	Yes
Khanra et al. (2013)	TD	Inv.	TC	NA
Taleizadeh and Nematollahi (2014)	Constant	Inv.	Backorder	Yes
Tiwari et al. (2018)	Constant	SCM	NA	Yes
Shaikh et al. (2019)	SD	Inv.	PD	Yes
Tiwari et al. (2019)	Constant	Inv.	MTD	Yes
Maihami et al. (2019)	SP	SCM	NA	Yes
Dey et al. (2019b)	SP	II	CR	NA
Mashud et al. (2020)	SD	Inv.	Backlogging	Yes
Mandal et al. (2021)	AD	Inv	TC	Yes
This paper	SP & SD	Inv.	TC	Constant & SD

NA, Not applicable; SD, Stock-dependent; II, Integrated inventory; CR, Cost reduction; TD, Time-dependent; Inv., Inventory; TC, Trade-credit; MTD, Multi-trade-credit; PD, Price discount; DIP, Delay in payment; AD, Advertisement-dependent; SP, Selling price; SCM, Supply chain management.

the shelf-life of the product. They also focused on the safety stock. In a similar direction, Tiwari et al. (2022) proposed a retailing strategy for an imperfect production model for deteriorating items under the consideration of two-level trade credit.

Different inventory systems with product deterioration were developed under separate scenarios, however, an inventory system considering stock level, variability in demand with price, and time under a variable deterioration rate is very rare. The gaps in literatures are summarized in the Table 4.1

4.3 DESCRIPTION OF PROBLEM, SYMBOLS, AND ASSUMPTIONS

Details analysis of the problems and the necessity of the model is discussed in this segment.

4.3.1 PROBLEM DEFINITION

The products which are used in our daily life, most of the products are deteriorating products. Thus, deciding on the deteriorating product at the exact time is really a very tough and essential decision for any industry. Most shop owners try to optimize their cost/profit by less loss. The main problem with the deteriorating product is that there is a fixed lifetime of those products and after that, those product is not usable for the purpose, for which it made and the lifetime of those products is very less. Industry managers have to make the perfect decision at the desired time; otherwise, the company faces a huge loss. Thus, exact time calculation is one of the most significant tasks for managers.

The current model was formulated for the deteriorating item, where demand varies with stock level and the selling price of the product. The product's rate of deterioration varies with inventory level, which makes the model more realistic. The large production rate helps to overcome the shortages situations. This study aims to find the exact time for production and the optimum selling price of the product, which leads to an optimum profit for the system. Moreover, to satisfy the customer and earn more profit, a delay period was provided. The effect of trade credit in an inventory model for deteriorating items, where demand is variable also established through this study.

4.3.2 SYMBOLS

Used symbols to illustrate the model are describe as follows:

Decision	Variables
Υ	Cycle length (month)
T_1	Time taken for inventory deplete to zero (month)
p	Per unit selling price ($/unit)
Parameters	
$D_m(\psi(t),t,p)$	Function representing rate of demand (unit)
$\psi(t)$	Level of inventory at any time $t \geq 0$ (unit)
Ω	Rate at which inventory deteriorated $0 < \Omega < 1$
m	Time allowed for delaying to stale the account (month)
R_1	Interest that obtains for investing a dollar per year
R_2	Interest that is payable in investing a dollar in stock ($/month)
	$R_1 \leq R_2$
c_h	Cost for holding unit inventory per cycle ($/unit/month)
ζ	Cost to produce a single unit ($/unit)
ϕ_0	Cost to set up the system in every cycle ($/setup)
a	Rate in which demand rate increase (unit/month)
b	Initial rate of demand (unit/month)
α, β, x	Shape parameters associated with demand function
γ	Scaling parameter associated with demand function
Q_0	Level of inventory at time $t = 0$
Ine_i	Total interest earned for case i, i=1,2,3
Inp_1	Total interest payable for case 1
avg_i	Average profit for case i ($ per cycle), i=1,2,3

4.3.3 ASSUMPTIONS

Every inventory model developed is with some assumptions, as it is near about impossible to formulate any inventory without proper assumptions. To decorate the current research, the following assumptions are considered, which are taken from previous studies and some are considered new for this model

1. The present study is conducted for a single deteriorate item, which is used in our daily life (for example, food products, dairy products, medicines, etc.). The demand vary with price of the product (Dey et al., 2019b), stock level and time (Mandal et al., 2021)
2. No exceptional installment to the provider amid putting an arrange that's $m <$ Υ intrigued is to be changed after graduation (Sarkar et al., 2020b).
3. Deterioration depends on the overall on-hand stock. On the off chance that stock is expansive, at that point deterioration rate moreover expanded (Soni, 2013).
4. Placing an order and get the ordered item on hand, takes some time, which is known as lead time, which is treated as negligible for developing the current study. Negligibility of the lead time converges to the system that replenishment performs just after placing the order.
5. To overcome the shortage situation, it is considered that the production rate is greater than the demand rate.
6. One other basic assumption for this study is that the planning period is infinite length and the ending inventory levels are zero.

4.4　MODEL FORMULATION

Time, selling price, and stock-level dependency demand rate for deteriorating items are considered as $D_m(\psi(t), t, p) = \alpha \psi(t) + \beta(at+b) + xp^{-\gamma}$ Consideration of $x = 0$ indicates independency of selling price. The basic differential equation is

$$\frac{d\psi(t)}{dt} = -\alpha \Omega \psi(t) - \beta(at+b) - xp^{-\gamma}, \ 0 \le t \le \Upsilon \tag{4.1}$$

with $\psi(0) = Q_0$, the initial inventory and $\psi(\Upsilon) = 0$
　　Within $0 \le t \le \Upsilon$, the value of $\psi(t)$ obtained as

$$
\psi(t) = \frac{e^{-t\alpha\Omega} p^{-\gamma} \left(-(e^{t\alpha\Omega} - e^{\Upsilon\alpha\Omega}) \alpha (x + bp^{-\gamma}\beta)\Omega - ap^{\gamma}\beta(e^{t\alpha\Omega}(-1 + t\alpha\Omega)) \right)}{\alpha^2 \Omega^2}
$$
$$
+ \frac{e^{-t\alpha\Omega} p^{-\gamma} e^{\Upsilon\alpha\Omega}(1 - \Upsilon\alpha\Omega)}{\alpha^2 \Omega^2} \tag{4.2}
$$

Clearly, the demand of an item will be more if a delay period is allowed by the distributor or wholesaler for the retailer or customers. Hence in this model we consider a delay period of m. Now three different cases may arise depending on delay period:

4.4.1　CASE 1

Let $m \le T_1 \le \Upsilon$

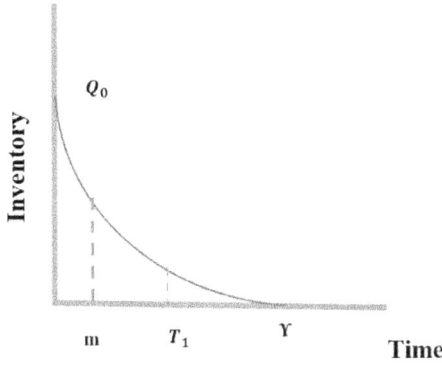

Figure 4.1 Level of inventory when $m \leq T_1 < \Upsilon$.

The inventory level for first case is graphically presented in Figure 4.1. The earned interest is calculated as:

$$
\begin{aligned}
Ine_1 &= pR_1 \int_0^m t D_m(t) dt \\
&= pR_1 \int_0^m t\{\alpha\psi(t) + \beta(at+b) + xp^{-\Upsilon}\} dt \\
&= pR_1 \left[\frac{1}{6\alpha^3\Omega^4} p^{-\Upsilon}(3\alpha(x+bp^\Upsilon\beta)\Omega(2e^{\Upsilon\alpha\Omega} + m^2\alpha^2(\Omega-1)\Omega^2 \right. \\
&\quad - 2e^{(\Upsilon-m)\alpha\Omega}(1+m\alpha\Omega)) + ap^\Upsilon\beta(6e^{\Upsilon\alpha\Omega}(\Upsilon\alpha\Omega-1) \\
&\quad - 6e^{(\Upsilon-m)\alpha\Omega}(1+m\alpha\Omega)(\Upsilon\alpha\Omega-1) \\
&\quad \left. + m^2\alpha^2(3+2m\alpha(\Omega-1)\Omega)\Omega^2)) \right]
\end{aligned}
\tag{4.3}
$$

After the delay period during the time interval $[m, T_1]$, for the unsold stock the customer/retailer will have to pay some interest, if R_2 annual rate and Inp_1 be the total payable interest then

$$
\begin{aligned}
Inp_1 &= pR_2 \int_m^{T_1} \psi(t) dt \\
&= pR_2 \left(\frac{1}{2\alpha^3\Omega^3} p^{-\Upsilon}(2\alpha(x+bp^\Upsilon\beta)\Omega(e^{(\Upsilon-m)\alpha\Omega} - e^{(\Upsilon-T_1)\alpha\Omega} \right. \\
&\quad + (m-T_1)\alpha\Omega) + ap^\Upsilon\beta(e^{(\Upsilon-T_1)\alpha\Omega}(2-2\Upsilon\alpha\Omega) + 2e^{(\Upsilon-m)\alpha\Omega}(-1+\Upsilon\alpha\Omega) \\
&\quad \left. + (m-T_1)\alpha\Omega(-2+(m+T_1)\alpha\Omega)))) \right)
\end{aligned}
\tag{4.4}
$$

If avg_1 be the average profit, then we have

$$avg_1 = \frac{1}{\Upsilon}\left[(p-\zeta)\int_0^\Upsilon D_m(\psi(t),t,p)dt - \phi_0 - c_h\int_0^\Upsilon \psi(t)dt\right.$$

$$\left. - \Omega\zeta\int_{T_1}^\Upsilon \psi(t)dt - Inp_1 + Ine_1\right]$$

$$= \frac{1}{\Upsilon}\left[\frac{1}{6\alpha^3\Omega^4}p^{-\gamma}\left(p(3\alpha(x+bp^\gamma\beta)\Omega(2e^{\Upsilon\alpha\Omega}(R_1+\alpha\Omega)+2e^{(\Upsilon-T_1)\alpha\Omega}R_2\Omega\right.\right.$$

$$-2e^{(\Upsilon-m)\alpha\Omega}(R_1+R_2\Omega+mR_1\alpha\Omega)+\alpha\Omega(-2+(-2mR_2+2R_2T_1$$

$$+m^2R_1\alpha(-1+\Omega)+2\Upsilon\alpha(-1+\Omega))\Omega))+ap^\gamma\beta(6e^{(\Upsilon-T_1)\alpha\Omega}R_2\Omega(-1+\Upsilon\alpha\Omega)$$

$$+6e^{\Upsilon\alpha\Omega}(R_1+\alpha\Omega)(-1+\Upsilon\alpha\Omega)-6e^{(\Upsilon-m)\alpha\Omega}(R_1+R_2\Omega+mR_1\alpha\Omega)(-1+\Upsilon\alpha\Omega)$$

$$+\alpha\Omega(6+\Omega(6mR_2+2m^3R_1\alpha^2(-1+\Omega)\Omega+3\Upsilon^2\alpha^2(-1+\Omega)\Omega$$

$$+3m^2\alpha(R_1-R_2\Omega)+3R_2T_1(-2+T_1\alpha\Omega)))))-6p^\gamma\alpha^3\Omega^4\phi_0$$

$$+3\Omega\left((2\alpha(x+bp^\gamma\beta)\Omega\left(1-e^{\Upsilon\alpha\Omega}+T\alpha\Omega\right)\right.$$

$$-ap^\gamma\beta(2-T^2\alpha^2\Omega^2+2e^{\Upsilon\alpha\Omega}(-1+\Upsilon\alpha\Omega)))c_h$$

$$+\left(-2\alpha(x+bp^\gamma\beta)\Omega(-\Omega+e^{(\Upsilon-T_1)\alpha\Omega}\Omega+\alpha(-1+e^{\Upsilon\alpha\Omega}-\Upsilon\alpha\Omega\right.$$

$$+(T_1+\Upsilon(-1+\alpha))\Omega^2))-ap^\gamma\beta\left(2\Omega+2e^{(\Upsilon-T_1)\alpha\Omega}\Omega(-1+\Upsilon\alpha\Omega)\right.$$

$$+\alpha\left(2+2e^{\Upsilon\alpha\Omega}(-1+\Upsilon\alpha\Omega)\right.$$

$$\left.\left.\left.\left.\left.\left.+\Omega^2\left(-2T_1+\alpha\ (T_1^2+\Upsilon^2\alpha)\Omega-\Upsilon^2\alpha(\alpha+\Omega)\right)\right)\right)\right)\zeta\right)\right)\right] \tag{4.5}$$

4.4.2 CASE 2

Let $T_1 < m < \Upsilon$

The inventory position for the second case is presented in Figure 4.2. Then by Mandal et al. (2021), interest earned for second case is

$$Ine_2 = pR_1\int_0^{T_1} tD_m(t)dt + pR_1(m-T_1)\int_0^{T_1} D_m(t)dt$$

$$= \frac{1}{6\alpha^3\Omega^4}p^{1-\gamma}R_1\left(-3\alpha(x+bp^\gamma\beta)\Omega\left(-(2m-T_1)T_1\alpha^2(-1+\Omega)\Omega^2\right.\right.$$

$$\left.\left.+2e^{(\Upsilon-T_1)\alpha\Omega}(1+m\alpha\Omega)-2e^{\Upsilon\alpha\Omega}(1+(m-T_1)\alpha\Omega)\right)\right.$$

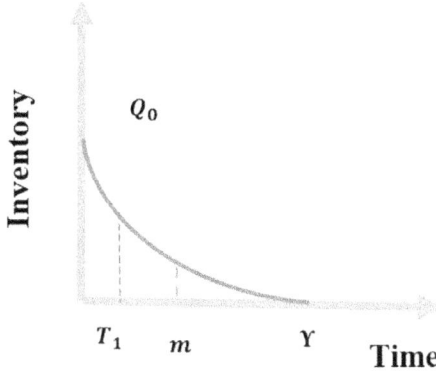

Figure 4.2 Level of inventory for $T_1 < m < \Upsilon$.

$$+ap^{\gamma}\beta\left(-6e^{(\Upsilon-T_1)\alpha\Omega}(1+m\alpha\Omega)(-1+\Upsilon\alpha\Omega)\right.$$

$$+6e^{\Upsilon\alpha\Omega}(-1+\Upsilon\alpha\Omega)(1+(m-T_1)\alpha\Omega)$$

$$\left.+T_1\alpha^2\Omega^2\left(T_1(-3-T_1\alpha(-1+\Omega)\Omega)+3m\left(2+T_1\alpha(-1+\Omega)\Omega\right)\right)\right)\right) \quad (4.6)$$

Clearly, in this case, all items are sold, hence there is no payable interest. If avg_2 is the average profit, then we have

$$avg_2 = \frac{1}{\Upsilon}[(p-\zeta)\int_0^{\Upsilon} D_m(\psi(t),t,p)dt - \phi_0 - c_h\int_0^{\Upsilon}\psi(t)dt - \Omega\zeta\int_{T_1}^{\Upsilon}\psi(t)dt + Ine_2]$$

$$= \frac{1}{6\Upsilon}\left(-6\phi_0 + \frac{1}{\alpha^3\Omega^4}p^{-\gamma}\left(3\alpha(x+bp^{\gamma}\beta)\Omega(2e^{\Upsilon\alpha\Omega}(pR_1+(-\zeta+p+mpR_1\right.\right.$$

$$-pR_1T_1)\alpha\Omega)-2e^{(\Upsilon-T_1)\alpha\Omega}(\zeta\Omega^2+p(R_1+mR_1\alpha\Omega))+\Omega(p\alpha(-2+(2\Upsilon+R_1(2m$$

$$-T_1)T_1)\alpha(-1+\Omega)\Omega)+2\zeta(\alpha+\Omega+\Upsilon\alpha^2\Omega-(T_1+\Upsilon(-1+\alpha))\alpha\Omega^2)))$$

$$+ap^{\gamma}\beta\left(-6\zeta\alpha\Omega+6e^{\Upsilon\alpha\Omega}(-1+\Upsilon\alpha\Omega)(pR_1+(-\zeta+p+mpR_1-pR_1T_1)\alpha\Omega)\right.$$

$$-6e^{(\Upsilon-T_1)\alpha\Omega}(-1+\Upsilon\alpha\Omega)\left(\zeta\Omega^2+p(R_1+mR_1\alpha\Omega)\right)+\Omega\left(3\zeta\Omega\left(-2\right.\right.$$

$$+\alpha\Omega\left(2T_1-T_1^2\alpha\Omega+\Upsilon^2\alpha(\alpha+\Omega-\alpha\Omega)\right)\right)+p\alpha\left(6+\alpha\Omega\left(3\Upsilon^2\alpha(-1+\Omega)\Omega\right.\right.$$

$$\left.\left.\left.\left.\left.+R_1T_1^2\left(-3-T_1\alpha(-1+\Omega)\Omega\right)+3mR_1T_1\left(2+T_1\alpha(-1+\Omega)\Omega\right)\right)\right)\right)\right)\right)$$

$$+3\Omega\left(2\alpha\left(x+bp^\gamma\beta\right)\Omega(1-e^{\Upsilon\alpha\Omega}+\Upsilon\alpha\Omega)-ap^\gamma\beta(2-\Upsilon^2\alpha^2\Omega^2\right.$$
$$\left.\left.+2e^{\Upsilon\alpha\Omega}(-1+\Upsilon\alpha\Omega))\right)c_h\right)\right)$$

(4.7)

4.4.3 CASE 1

Let $m \geq \Upsilon$.

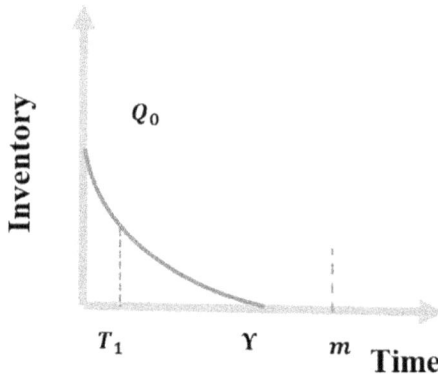

Figure 4.3 Level of inventory for $m \geq \Upsilon$.

The position of the inventory for $m \geq \Upsilon$ is represented in Figure 4.3. Thus, earned interest for this case is calculated as (See Mandal et al. (2021))

$$Ine_3 = pR_1 \int_0^{T_1} tD_m(t)dt + pR_1(m-T_1)\int_0^{T_1} D_m(t)dt$$
$$= \frac{1}{6\alpha^3\Omega^4}p^{1-\gamma}R_1\left(-3\alpha(x+bp^\gamma\beta)\Omega\left(-(2m-T_1)T_1\alpha^2(-1+\Omega)\Omega^2\right.\right.$$
$$\left.+2e^{(\Upsilon-T_1)\alpha\Omega}(1+m\alpha\Omega)-2e^{\Upsilon\alpha\Omega}(1+(m-T_1)\alpha\Omega)\right)$$
$$+ap^\gamma\beta\left(-6e^{(\Upsilon-T_1)\alpha\Omega}(1+m\alpha\Omega)(-1+\Upsilon\alpha\Omega)\right.$$
$$+6e^{\Upsilon\alpha\Omega}(-1+\Upsilon\alpha\Omega)(1+(m-T_1)\alpha\Omega)+T_1\alpha^2\Omega^2(T_1(-3$$
$$\left.\left.\left.-T_1\alpha(-1+\Omega)\Omega)+3m(2+T_1\alpha(-1+\Omega)\Omega)\right)\right)\right)$$

(4.8)

If avg_3 be the average profit, then we have

$$avg_3 = \frac{1}{\Upsilon}[(p-\zeta)\int_0^\Upsilon D_m(\psi(t),t,p)dt - \phi_0 - c_h \int_0^\Upsilon \psi(t)dt - \Omega\zeta \int_{T_1}^\Upsilon \psi(t)dt + Ine_3]$$

$$= \frac{1}{6\Upsilon}\left(-6\phi_0 + \frac{1}{\alpha^3\Omega^4}p^{-\gamma}\left(3\alpha(x+bp^\gamma\beta)\Omega(2e^{\Upsilon\alpha\Omega}(pR_1\right.\right.$$

$$+(-\zeta+p+mpR_1-pR_1T_1)\alpha\Omega)-2e^{(\Upsilon-T_1)\alpha\Omega}(\zeta\Omega^2+p(R_1+mR_1\alpha\Omega))$$

$$+\Omega(p\alpha(-2+(2\Upsilon+R_1(2m-T_1)T_1)\alpha(-1+\Omega)\Omega)$$

$$+2\zeta(\alpha+\Omega+\Upsilon\alpha^2\Omega-(T_1+\Upsilon(-1+\alpha))\alpha\Omega^2)))$$

$$+ap^\gamma\beta\left(-6\zeta\alpha\Omega+6e^{\Upsilon\alpha\Omega}(-1+\Upsilon\alpha\Omega)(pR_1+(-\zeta+p+mpR_1\right.$$

$$-pR_1T_1)\alpha\Omega)-6e^{(\Upsilon-T_1)\alpha\Omega}(-1+\Upsilon\alpha\Omega)\left(\zeta\Omega^2+p(R_1+mR_1\alpha\Omega)\right)$$

$$+\Omega\left(3\zeta\Omega(-2+\alpha\Omega(2T_1-T_1^2\alpha\Omega+\Upsilon^2\alpha(\alpha+\Omega-\alpha\Omega)))\right.$$

$$+p\alpha\left(6+\alpha\Omega\left(3\Upsilon^2\alpha(-1+\Omega)\Omega+R_1T_1^2(-3-T_1\alpha(-1+\Omega)\Omega)\right.\right.$$

$$\left.\left.\left.\left.+3mR_1T_1\left(2+T_1\alpha(-1+\Omega)\Omega\right)\right)\right)\right)\right)$$

$$+3\Omega\left(2\alpha(x+bp^\gamma\beta)\Omega(1-e^{\Upsilon\alpha\Omega}+\Upsilon\alpha\Omega)-ap^\gamma\beta(2-\Upsilon^2\alpha^2\Omega^2\right.$$

$$\left.\left.\left.+2e^{\Upsilon\alpha\Omega}(-1+\Upsilon\alpha\Omega))\right)c_h\right)\right) \tag{4.9}$$

4.5 SOLUTION METHODOLOGY

We note that Ine_3 and avg_3 are the same as Ine_2 and avg_2. Here during numerical calculation, we should take care of the fact that in **case 2**, $T_1 < m < \Upsilon$ and in **Case 3**, $m \geq \Upsilon$.

Our aim is to obtain the optimum average profit for all three cases. With the use of the following theorem, one can obtain the maximum value of average profit.

Theorem 4.1. *If a function $Z(T_1,\Upsilon,p)$ possesses a continuous second-order partial derivatives of T_1, Υ and p then $Z(T_1,\Upsilon,p)$ is maximum at $T_1 = T_1^*, \Upsilon = T^*, p = p^*$ if $d^2Z(T_1,T,p)$ is negative definite i.e., if $A < 0, B > 0$ and $C < 0$, where*

$$A = Z_{\Upsilon\Upsilon}, B = \begin{vmatrix} Z_{\Upsilon\Upsilon} & Z_{\Upsilon T_1} \\ Z_{T_1\Upsilon} & Z_{T_1 T_1} \end{vmatrix}, C = \begin{vmatrix} Z_{\Upsilon\Upsilon} & Z_{\Upsilon T_1} & Z_{\Upsilon p} \\ Z_{T_1\Upsilon} & Z_{T_1 T_1} & Z_{T_1 p} \\ Z_{p\Upsilon} & Z_{p T_1} & Z_{pp} \end{vmatrix}$$

where $Z_{\Upsilon\Upsilon} = \frac{\partial^2 Z}{\partial\Upsilon^2}, Z_{\Upsilon T_1} = \frac{\partial^2 Z}{\partial\Upsilon\partial T_1}, etc.

Proof. Similar as Mandal et al. (2021).   □

Base on the calculation, the optimum values are

$$Z(\Upsilon,T_1,p) = \begin{cases} avg_1(\Upsilon,T_1,p) & m \le T_1 < \Upsilon \\ avg_2(\Upsilon,T_1,p) & T_1 < m < \Upsilon \\ avg_3(\Upsilon,T_1,p) & m \ge \Upsilon \end{cases}$$

Due to the high complexity and high nonlinearity of the profit functions, the optimum values of the decision variables along with the optimum profit are obtained numerically, which are provided in the next segment. The concavity graphs are also provided to establish the concavity of the function.

4.6 NUMERICAL EXAMPLES

Due to the complexity of finding the maximum value of the average profit analytically, some numerical results and figures are provided below in Table 4.2, and in Figure 4.4, to apply this model practically. To calculate optimum result, the required value for the parameters was adopted from Mandal et al. (2021). The parametric values provided as: $b = 800$ units/month, $\beta = 0.6$ unit/month, $x = 0.7$ unit/month, $\phi_0 = \$ 90/$ order, $R_2 = \$0.12$/month, $a = 1.8$ units/month, $R_1 = \$0.10$/month, $c_h = \$1.2$ per month, $\alpha = 0.002$ units/month, $\gamma = 0.2$, $\zeta = \$1.8$/unit, $\Omega = 0.15$ and $m = 0.6, 0.9, 1.7$ month (for Case 1, Case 2, Case 3 resp.) week. Then by using the Equations (4.5), (4.7), and (4.9), one can find the best result, which is illustrated in Table 4.2. Table 4.2 proved that the optimal result is obtained for case 3 (Bold values) and which is \$2379.55, where the delay period is 0.82 months. For this case, the cycle time is 1.65 months. When the time for stale the account lies in between cycle time and time in which inventory reaches zero level the profit is \$2191.96 per cycle. In this case delay period is 0.8 per month. For case 1 the profit is \$1966.31. From the optimal result Table 4.2, it is clear that the profit of the system is optimum when $m \ge \Upsilon$.

Table 4.2
Illustration of Profit for Different Cases

	Cycle Length (month)	Time Length Inventory Reach Zero Level T_1 (month)	Selling Price ($)	Total Profit ($/cycle)
Case I: $m \le T_1 \le \Upsilon$	1.75	0.87	8.25	$1966.31
Case II: $T_1 < m < \Upsilon$	1.59	0.80	8.87	$2191.96
Case III: $m \ge \Upsilon$	**1.65**	**0.82**	**9.01**	**$2379.55**

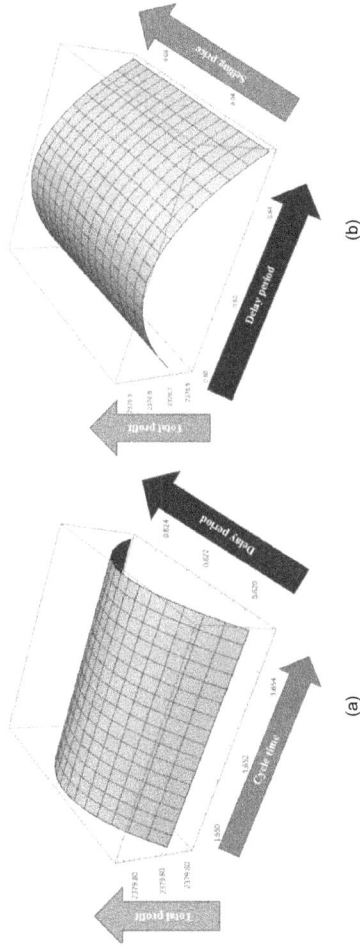

Figure 4.4 Graphical representation of concavity with respect to decision variables. (a) Graphical illustration of concavity with respect to the selling price and delay time. (b) Graphical illustration of concavity with respect to the time.

4.6.1 COMPARISON WITH EXISTING LITERATURE

Since, the numerical values are calculated with the help of the values of Mandal et al. (2021) model, thus in this section, a comparison is made along with the Mandal et al. (2021). In Mandal et al. (2021) model, they considered stock and advertisement-dependent demand pattern for a deteriorating item and their model provides the optimal profit when the delay period is 0.53 year and the profit is \$1840.41 per year, where the current model much more beneficial compared to Mandal et al. (2021). In this model, optimum profit is \$2379.55 for the entire cycle time when the delay period is greater than cycle time and it is quite obvious as more revenue can earn through interest. In the following Table 4.3 the comparison is made and a graphical representation is provided in Figure 4.5.

Table 4.3
Comparison with Previous Study

	Mandal et al. (2021)	This Study
Total Profit (\$/month)	1840.41	2379.55

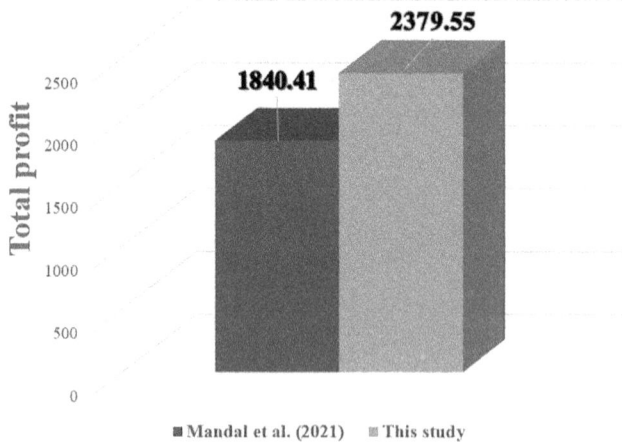

Figure 4.5 Comparison with Mandal et al. (2021).

4.7 SENSITIVITY ANALYSIS

This portion of the study recite how the increment and decrement in value of the parameters a, R_1, b, ϕ_0, α, β, x, ζ, γ, R_2, and c_h by 50% and 25% reflects upon the profit function respectively. The outcome in the profit function has been highlighted in Table 4.4 and graphically in Figure 4.6. As shown in Table 4.4 anyone can spontaneously understand the effect of changes in the parameter.

(i) Cost for holding items is exceptionally sensitive in nature. Increment in holding costs is pernicious for the manufacturing sectors. It is quite obvious that holding the item in the store is always very crucial for the deteriorating items. Thus, increase in unit holding cost is inversely proportional to the total system profit.

(ii) Initial demand for an ordering quantity model plays a vital role in optimizing the total profit of the system. Thus, a small change in initial demand has a high influence on the profit function. Since, the present model is developed for the deteriorating items, the initial demand highly influences the total profit.

Table 4.4
Sensitivity Analysis Table

Para-meters	Percentage Change (%)	Average Profit	Change in avg_3	Para-meters	Percentage Change (%)	Optimal Profit	Change in avg_3
b	50	3591.43	+50.93	β	50	3595.07	+51.08
	25	2985.49	+25.46		25	2987.31	+25.54
	−25	1773.61	−25.46		−25	1771.79	−25.54
	−50	1167.66	−50.93		−50	1164.02	−51.08
a	50	2383.19	+0.15	x	50	2381.07	+0.06
	25	2381.37	+0.08		25	2380.31	+0.03
	−25	2377.73	−0.08		−25	2378.79	−0.03
	−50	2375.91	−0.15		−50	2378.03	−0.06
α	50	2382.97	+0.14	γ	50	2378.95	−0.025
	25	2381.26	+0.07		25	2379.23	−0.013
	−25	2377.85	−0.07		−25	2379.9	+0.015
	−50	2376.1	−0.14		−50	2380.3	+0.031
R_1	50	2443.13	+2.67	ζ	50	2002.35	−15.85
	25	2411.35	+1.34		25	2190.95	−7.93
	−25	2347.77	−1.34		−25	2568.13	+7.93
	−50	2315.96	−2.67		−50	2756.74	+15.85
ϕ_0	50	2352.28	−1.15	c_h	50	2117.38	−11.02
	25	2365.91	−0.57		25	2248.46	−5.51
	−25	2393.18	+0.57		−25	2510.63	+5.51
	−50	2406.82	+1.15		−50	2641.71	+11.02

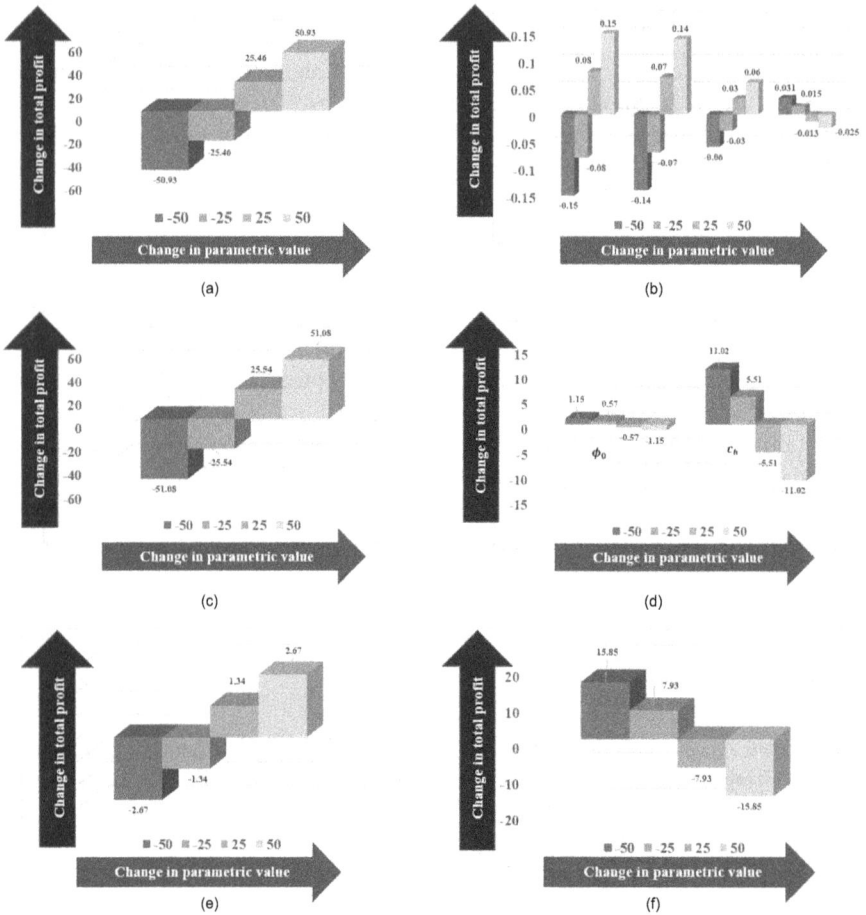

Figure 4.6 Graphical representation of the sensitivity of the key parameters. (a) Sensitivity of the parameter b on total profit, (b) sensitivity of the parameter β on total profit, (c) sensitivity of the parameter a, α, x, and γ on total profit, (d) sensitivity of setup cost and unit holding cost on total profit, (e) sensitivity of the parameter R_1 on total profit, and (f) sensitivity of the unit production cost on total profit.

(iii) As shown in the sensitivity Table 4.4, the basic setup cost is an authentic part of the manufacturing sector. Thus, change in setup cost is a little bit sensitive to the total profit.

(iv) The manufacturing cost ζ has a massive impression on the profit function. It is natural that if the manufacturing cost increases then total profit must be decreased and vice versa. Thus, unit manufacturing cost also plays a vital role in optimizing the profit of the system.

(v) The sensitivity table narrates the resultant that the earned interest rate R_1 has petty less impact on the profit.

(vi) Changes in the parameter β have an extensive dominance upon the profit of the system.

(vii) The scaling parameter γ associated with selling price has minor impact on the profit function.

(viii) The shape parameters a, α and x related to demand has low impact on profit function.

4.8 MANAGERIAL INSIGHTS

It demonstrates bargains through an inventory system having stock level and selling price subordinate demand design without considering shortages beneath trade-credit arrangement managers of the industry of deteriorating product can take several decisions for their industry. The Managerial insights for the study have been described a down.

(i) Optimization of profit is the primary motive for any stock demonstration. Display stock-level features have an awesome affect in the maximization of the benefit. The seller should follow the policy of making the item to be more strengthened in parallel with other existing items. The management for the treasury of items by the dealers should be that any client chooses their suitable items effortlessly. Regularly the request will increment on the off chance that the put-away item is higher than the request. Casually, the benefit will advance on the off chance that seller can offer for more items. In this demonstration, one can effortlessly discover that, when requested on the on-hand stock, the benefit is advanced.

(ii) Making a decision on the selling price of the item is very crucial to optimize the benefit that comes from the system. Generally, the items' selling price is inversely proportional to the demand, however, increase in selling price increases the sales revenue, on the other hand, a huge selling price is sometimes harmful for the demand, which is indirectly harmful for the industry. Thus, from the current study, managers of the industry can take proper decisions for the item's selling price.

(iii) In this competitive business environment, the delay period strategy is very much beneficial for several industries, but find out how much delay period is more beneficial, the industry managers can decide going through this research work. On the other hand, the measurement of the lifetime of a deteriorating product is another very crucial decision. Managers of those industries can take decisions on the lifetime of the product, when to provide more delay period and when to provide less delay period.

(iv) Making a decision of the demand is another major finding of this study. Managers of the industry can take decisions on the demand which can directly enhance the profit of the system.

(v) When to order is one of the most critical decisions for any inventory system which consists of deteriorating items. Managers of the shop can also take decisions on the time when to order through this study.

4.9   CONCLUSIONS

The present study focused on the retailing decision of an inventory system for deteriorating items, where the demand for the product depends on the stock level, time of deterioration, and the selling price of the product. Taking a decision when to place the order and how much will be the selling price which is beneficial for the deteriorating items are the two major and critical decisions that can be made through this current study. Parallelly, one can make the decision for the delay time. Initially, the differential equation helps to find out the level of inventory. In contrast, the second-order partial derivative helps calculate the value of the ruling variables and ensures the optimality of the profit function. Application of differential equation is too much essential for the inventory model, though some inventory model was solved analytically without the help of differential equation (Calıskan, 2020), however, to calculate the level of the inventory, a differential equation is beneficial.

Decisions on production and lead time are two major limitations of this study. Constant ordering and setup costs are other limitations of this study. The current research can be extended in several directions. One can consider multi-trade-credit policy Tiwari et al. (2019) to extend this study. Some intelligent technology such as autonomation (Dey et al., 2021b) can be utilized to identify the rate of deterioration. With the consideration of lead time and advanced technology to reduce lead time (Dey et al., 2019a), one can extend the current study. Instated of a single item, one can consider a multi-item assembled deteriorating product (Sarkar et al., 2020a; Dey et al., 2021c) in future research. By using the residue of the deteriorating item, one can produce a secondary item from those deteriorating products (Garai et al., 2021). This model can be explored by addressing different environment issues (Singh et al., 2021; Sarkar et al., 2021b, 2022b), (Bairagi et al., 2021). Waste in terms of deteriorating product management through secondary products is another very interesting research direction (Ullah et al., 2021; Sarkar and Sarkar, 2020; Habib et al., 2019). Instated of fixed production rate, variable smart production rate (Sarkar et al., 2022a) is another interesting research direction, which can be performed in the future.

REFERENCES

Ahmed, W., Moazzam, M., Sarkar, B., and Rehman, S. U. (2021). Synergic effect of reworking for imperfect quality items with the integration of multi-period delay-in-payment and partial backordering in global supply chains. *Engineering*, 7(2):260–271.

Bai, Q., Chen, M., and Xu, L. (2017). Revenue and promotional cost sharing contract versus two-part tariff contract in coordinating sustainable supply chain systems with deteriorating items. *International Journal of Production Economics*, 187:85–101.

Bairagi, N., Bhattacharya, S., Auger, P., and Sarkar, B. (2021). Bioeconomics fishery model in presence of infection: Sustainability and demand-price perspectives. *Applied Mathematics and Computation*, 405:126225.

Banerjee, S. and Agrawal, S. (2017). Inventory model for deteriorating items with freshness and price dependent demand: Optimal discounting and ordering policies. *Applied Mathematical Modelling*, 52:53–64.

Calıskan, C. (2020). A derivation of the optimal solution for exponentially deteriorating items without derivatives. *Computers & Industrial Engineering*, 148:106675.

Calıskan, C. (2021). A simple derivation of the optimal solution for the EOQ model for deteriorating items with planned backorders. *Applied Mathematical Modelling*, 89:1373–1381.

Cárdenas-Barrón, L. E., Chung, K.-J., and Trevino-Garza, G. (2014). Celebrating a century of the economic order quantity model in honor of ford whitman harris. *International Journal of Production Economics*, 155:1–7.

Cárdenas-Barrón, L. E., Shaikh, A. A., Tiwari, S., and Trevino-Garza, G. (2020). An EOQ inventory model with nonlinear stock-dependent holding cost, nonlinear stock-dependent demand and trade credit. *Computers & Industrial Engineering*, 139:105557.

Choi, S. K., Lim, K. E., and Lee, E. Y. (2008). A partial replenishment model for an inventory with constant demand. *Applied Mathematical Modelling*, 32(9):1790–1796.

Das, S., Khan, M. A.-A., Mahmoud, E. E., Abdel-Aty, A.-H., Abualnaja, K. M., and Shaikh, A. A. (2021a). A production inventory model with partial trade credit policy and reliability. *Alexandria Engineering Journal*, 60(1):1325–1338.

Das, S. C., Manna, A. K., Rahman, M. S., Shaikh, A. A., and Bhunia, A. K. (2021b). An inventory model for non-instantaneous deteriorating items with preservation technology and multiple credit periods-based trade credit financing via particle swarm optimization. *Soft Computing*, 25(7):5365–5384.

Dey, B. K., Bhuniya, S., and Sarkar, B. (2021a). Involvement of controllable lead time and variable demand for a smart manufacturing system under a supply chain management. *Expert Systems with Applications*, 184:115464.

Dey, B. K., Pareek, S., Tayyab, M., and Sarkar, B. (2021b). Autonomation policy to control work-in-process inventory in a smart production system. *International Journal of Production Research*, 59(4):1258–1280.

Dey, B. K., Sarkar, B., and Pareek, S. (2019a). A two-echelon supply chain management with setup time and cost reduction, quality improvement and variable production rate. *Mathematics*, 7(4):328.

Dey, B. K., Sarkar, B., Sarkar, M., and Pareek, S. (2019b). An integrated inventory model involving discrete setup cost reduction, variable safety factor, selling price dependent demand, and investment. *RAIRO-Operations Research*, 53(1):39–57.

Dey, B. K., Sarkar, B., and Seok, H. (2021c). Cost-effective smart autonomation policy for a hybrid manufacturing-remanufacturing. *Computers & Industrial Engineering*, 162:107758.

Duary, A., Das, S., Arif, M. G., Abualnaja, K. M., Khan, M. A.-A., Zakarya, M., and Shaikh, A. A. (2021). Advance and delay-in-payments with the price discount inventory model for deteriorating items under capacity constraint and partially backlogged shortages. *Alexandria Engineering Journal*, 61:1735–1745.

Feng, L., Wang, W.-C., Teng, J.-T., and Cárdenas-Barrón, L. E. (2022). Pricing and lot-sizing decision for fresh goods when demand depends on unit price, displaying stocks and product age under generalized payments. *European Journal of Operational Research*, 296(3):940–952.

Garai, A., Chowdhury, S., Sarkar, B., and Roy, T. K. (2021). Cost-effective subsidy policy for growers and biofuels-plants in closed-loop supply chain of herbs and herbal medicines: An interactive bi-objective optimization in t-environment. *Applied Soft Computing*, 100:106949.

Habib, M. S., Sarkar, B., Tayyab, M., Saleem, M. W., Hussain, A., Ullah, M., Omair, M., and Iqbal, M. W. (2019). Large-scale disaster waste management under uncertain environment. *Journal of Cleaner Production*, 212:200–222.

Iqbal, M. W. and Sarkar, B. (2019). Recycling of lifetime-dependent deteriorated products through different supply chains. *RAIRO-Operations Research*, 53(1):129–156.

Jaggi, C. K., Tiwari, S., and Goel, S. K. (2017). Credit financing in economic ordering policies for non-instantaneous deteriorating items with price dependent demand and two storage facilities. *Annals of Operations Research*, 248(1–2):253–280.

Jani, M. Y., Betheja, M. R., Chaudhari, U., and Sarkar, B. (2021). Optimal investment in preservation technology for variable demand under trade-credit and shortages. *Mathematics*, 9(11):1301.

Khanra, S., Ghosh, S. K., and Chaudhuri, K. (2011). An EOQ model for a deteriorating item with time-dependent quadratic demand under permissible delay in payment. *Applied Mathematics and Computation*, 218(1):1–9.

Khanra, S., Mandal, B., and Sarkar, B. (2013). An inventory model with time-dependent demand and shortages under trade credit policy. *Economic Modelling*, 35:349–355.

Kumar Sett, B., Sarkar, S., Sarkar, B., and Young Yun, W. (2016). Optimal replenishment policy with variable deterioration for fixed lifetime products. *Scientia Iranica*, 23(5):2318–2329.

Li, G., He, X., Zhou, J., and Wu, H. (2019). Pricing, replenishment and preservation technology investment decisions for non-instantaneous deteriorating items. *Omega*, 84:114–126.

Mahata, G. C. and De, S. K. (2017). Supply chain inventory model for deteriorating items with maximum lifetime and partial trade credit to credit-risk customers. *International Journal of Management Science and Engineering Management*, 12(1):21–32.

Maihami, R., Govindan, K., and Fattahi, M. (2019). The inventory and pricing decisions in a three-echelon supply chain of deteriorating items under probabilistic environment. *Transportation Research Part E: Logistics and Transportation Review*, 131:118–138.

Mandal, B., Dey, B. K., Khanra, S., and Sarkar, B. (2021). Advance sustainable inventory management through advertisement and trade-credit policy. *RAIRO-Operations Research*, 55(1):261–284.

Mashud, A. H. M., Hui-Ming, W., and Huang, C.-V. (2021). Preservation technology investment, trade credit and partial backordering model for a non-instantaneous deteriorating inventory. *RAIRO: Recherche Opérationnelle*, 55:51.

Mashud, A. H. M., Wee, H.-M., Sarkar, B., and Li, Y.-H. C. (2021). A sustainable inventory system with the advanced payment policy and trade-credit strategy for a two-warehouse inventory system. *Kybernetes*. 50(5): 1321–1348.

Mishra, U., Wu, J.-Z., and Sarkar, B. (2021). Optimum sustainable inventory management with backorder and deterioration under controllable carbon emissions. *Journal of Cleaner Production*, 279:123699.

Nematollahi, M., Hosseini-Motlagh, S.-M., Cárdenas-Barrón, L. E., and Tiwari, S. (2022). Coordinating visit interval and safety stock decisions in a two-level supply chain with shelf-life considerations. *Computers & Operations Research*, 139:105651.

Önal, M., Yenipazarli, A., and Kundakcioglu, O. E. (2016). A mathematical model for perishable products with price and displayed stock-dependent demand. *Computers & Industrial Engineering*, 102:246–258.

Pando, V., San-José, L. A., Sicilia, J., and Alcaide-Lopez-de Pablo, D. (2021). Maximization of the return on inventory management expense in a system with price and stock-dependent demand rate. *Computers & Operations Research*, 127:105134.

Pervin, M., Roy, S. K., and Weber, G. W. (2019). Multi-item deteriorating two-echelon inventory model with price and stock-dependent demand: A trade-credit policy. *Journal of Industrial & Management Optimization*, 15(3):1345.

Sana, S. S. and Chaudhuri, K. (2008). An inventory model for stock with advertising sensitive demand. *IMA Journal of Management Mathematics*, 19(1):51–62.

Saren, S., Sarkar, B., and Bachar, R. K. (2020). Application of various price discount policy for deteriorated products and delay-in-payments in an advanced inventory model. *Inventions*, 5(3):50.

Sarkar, B. (2012). An EOQ model with delay-in-payments and time varying deterioration rate. *Mathematical and Computer Modelling*, 55(3–4):367–377.

Sarkar, B. (2013). A production inventory model with probabilistic deterioration in two-echelon supply chain management. *Applied Mathematical Modelling*, 37(5):3138–3151.

Sarkar, B., Ahmed, W., and Kim, N. (2018). Joint effects of variable carbon emission cost and multi-delay-in-payments under single-setup-multiple-delivery policy in a global sustainable supply chain. *Journal of Cleaner Production*, 185:421–445.

Sarkar, B., Chaudhuri, K., and Sana, S. S. (2010). A stock-dependent inventory model in an imperfect production process. *International Journal of Procurement Management*, 3(4):361–378.

Sarkar, B., Dey, B. K., Pareek, S., and Sarkar, M. (2020a). A single-stage cleaner production system with random defective rate and remanufacturing. *Computers & Industrial Engineering*, 150:106861.

Sarkar, B., Dey, B. K., Sarkar, M., and AlArjani, A. (2021a). A sustainable online-to-offline (O2O) retailing strategy for a supply chain management under controllable lead time and variable demand. *Sustainability*, 13(4):1756.

Sarkar, B., Dey, B. K., Sarkar, M., Hur, S., Mandal, B., and Dhaka, V. (2020b). Optimal replenishment decision for retailers with variable demand for deteriorating products under a trade-credit policy. *RAIRO-Operations Research*, 54(6):1685–1701.

Sarkar, B., Gupta, H., Chaudhuri, K., and Goyal, S. K. (2014). An integrated inventory model with variable lead time, defective units and delay-in-payments. *Applied Mathematics and Computation*, 237:650–658.

Sarkar, B., Mandal, B., and Sarkar, S. (2017). Preservation of deteriorating seasonal products with stock-dependent consumption rate and shortages. *Journal of Industrial & Management Optimization*, 13(1):187.

Sarkar, B., Mridha, B., and Pareek, S. (2022a). A sustainable smart multi-type biofuel manufacturing with the optimum energy utilization under flexible production. *Journal of Cleaner Production*, 332:129869.

Sarkar, B., Mridha, B., Pareek, S., Sarkar, M., and Thangavelu, L. (2021b). A flexible biofuel and bioenergy production system with transportation disruption under a sustainable supply chain network. *Journal of Cleaner Production*, 317:128079.

Sarkar, B. and Saren, S. (2017). Ordering and transfer policy and variable deterioration for a warehouse model. *Hacettepe Journal of Mathematics and Statistics*, 46(5):985–1014.

Sarkar, B., Saren, S., and Cárdenas-Barrón, L. E. (2015). An inventory model with trade-credit policy and variable deterioration for fixed lifetime products. *Annals of Operations Research*, 229(1):677–702.

Sarkar, B. and Sarkar, S. (2013). Variable deterioration and demand—an inventory model. *Economic Modelling*, 31:548–556.

Sarkar, B., Ullah, M., and Sarkar, M. (2022b). Environmental and economic sustainability through innovative green products by remanufacturing. *Journal of Cleaner Production*, 332:129813.

Sarkar, M. and Sarkar, B. (2020). How does an industry reduce waste and consumed energy within a multi-stage smart sustainable biofuel production system? *Journal of Cleaner Production*, 262:121200.

Sepehri, A., Mishra, U., and Sarkar, B. (2021a). A sustainable production inventory model with imperfect quality under preservation technology and quality improvement investment. *Journal of Cleaner Production*, 310:127332.

Sepehri, A., Mishra, U., Tseng, M.-L., and Sarkar, B. (2021b). Joint pricing and inventory model for deteriorating items with maximum lifetime and controllable carbon emissions under permissible delay in payments. *Mathematics*, 9(5):470.

Sett, B. K., Sarkar, B., and Goswami, A. (2012). A two-warehouse inventory model with increasing demand and time varying deterioration. *Scientia Iranica*, 19(6):1969–1977.

Shah, N. H. and Cárdenas-Barrón, L. E. (2015). Retailer's decision for ordering and credit policies for deteriorating items when a supplier offers order-linked credit period or cash discount. *Applied Mathematics and Computation*, 259:569–578.

Shaikh, A. A., Das, S. C., Bhunia, A. K., and Sarkar, B. (2021). Decision support system for customers during availability of trade credit financing with different pricing situations. *RAIRO: Recherche Opérationnelle*, 55:1043.

Shaikh, A. A., Khan, M. A.-A., Panda, G. C., and Konstantaras, I. (2019). Price discount facility in an EOQ model for deteriorating items with stock-dependent demand and partial backlogging. *International Transactions in Operational Research*, 26(4):1365–1395.

Shastri, A., Singh, S., Yadav, D., and Gupta, S. (2014). Supply chain management for two-level trade credit financing with selling price dependent demand under the effect of preservation technology. *International Journal of Procurement Management*, 7(6):695–718.

Shaw, B. K., Sangal, I., and Sarkar, B. (2020). Joint effects of carbon emission, deterioration, and multi-stage inspection policy in an integrated inventory model. In: Shah, N. H. and Mittal, M. (eds.), *Optimization and Inventory Management*, pp. 195–208. Berlin/Heidelberg, Germany: Springer.

Shin, D., Mittal, M., and Sarkar, B. (2018). Effects of human errors and trade-credit financing in two-echelon supply chain models. *European Journal of Industrial Engineering*, 12(4):465–503.

Singh, S., Yadav, D., Sarkar, B., Sarkar, M., et al. (2021). Impact of energy and carbon emission of a supply chain management with two-level trade-credit policy. *Energies*, 14(6):1569.

Soni, H. N. (2013). Optimal replenishment policies for non-instantaneous deteriorating items with price and stock sensitive demand under permissible delay in payment. *International Journal of Production Economics*, 146(1):259–268.

Taleizadeh, A. A. and Nematollahi, M. (2014). An inventory control problem for deteriorating items with back-ordering and financial considerations. *Applied Mathematical Modelling*, 38(1):93–109.

Tiwari, S., Ahmed, W., and Sarkar, B. (2019). Sustainable ordering policies for non-instantaneous deteriorating items under carbon emission and multi-trade-credit-policies. *Journal of Cleaner Production*, 240:118183.

Tiwari, S., Cárdenas-Barrón, L. E., Malik, A. I., and Jaggi, C. K. (2022). Retailer's credit and inventory decisions for imperfect quality and deteriorating items under two-level trade credit. *Computers & Operations Research*, 138:105617.

Tiwari, S., Jaggi, C. K., Gupta, M., and Cárdenas-Barrón, L. E. (2018). Optimal pricing and lot-sizing policy for supply chain system with deteriorating items under limited storage capacity. *International Journal of Production Economics*, 200:278–290.

Ullah, M., Asghar, I., Zahid, M., Omair, M., Al Arjani, A., and Sarkar, B. (2021). Ramification of remanufacturing in a sustainable three-echelon closed-loop supply chain management for returnable products. *Journal of Cleaner Production*, 290:125609.

Wu, J., Ouyang, L.-Y., Cárdenas-Barrón, L. E., and Goyal, S. K. (2014). Optimal credit period and lot size for deteriorating items with expiration dates under two-level trade credit financing. *European Journal of Operational Research*, 237(3):898–908.

Yang, C.-T. (2014). An inventory model with both stock-dependent demand rate and stock-dependent holding cost rate. *International Journal of Production Economics*, 155:214–221.

Zhang, J., Liu, G., Zhang, Q., and Bai, Z. (2015). Coordinating a supply chain for deteriorating items with a revenue sharing and cooperative investment contract. *Omega*, 56:37–49.

5 Existence Results for the Positive Solutions of a Third-Order Boundary Value Problem & Numerical Algorithm Based on Bernoulli Polynomials

Santanu Biswas
Jadavpur University
Adamas University

CONTENTS

5.1 INTRODUCTION

The multi-point boundary value problems of ordinary differential equations play a vital role in both theory and application, and as a consequence, have attracted a great deal of interest over the years. The application can be observed to model various phenomena in physics, biology, and engineering. However, in many situations, including the subjects just mentioned above, only positive solutions are meaningful. That is why people are particularly interested in positive solutions (Khan et al., 2019; Alves and Boudjeriou, 2020; Chammem et al., 2021).

DOI: 10.1201/9781003227847-5

In this chapter, we are concerned about the existence of multiple positive solutions to the nonlinear third-order boundary value problem (following the work (Li, 2006; Bai, 2008; Feng and Liu, 2005; Sun, 2009; Guo and Ge, 2004; Avery and Peterson, 2001))

$$z'''(v) + g(v, z(v), z'(v)) = 0,$$
$$z(0) = 0,$$
$$z'(0) = z'(1),$$
$$z''(0) = \alpha z'(\xi) \tag{5.1}$$

where $0 < v < 1, \xi \in [0,1), \alpha \in [0, \frac{1}{\xi}), g \in C([0,1] \times [0,\infty] \times \mathbb{R}, [0,\infty])$.

Here, we establish some simple criteria for the existence of positive solutions for problems (5.1) based on the known Guo-Krasnosel'skii & Avery-Peterson fixed-point theorem in a cone. Also, we solved equation (5.1) numerically by Bernoulli polynomials.

5.2　PRELIMINARIES

Consider the following boundary value problem:-

$$z'''(v) + g(v, z(v), z'(v)) = 0,$$
$$z(0) = 0,$$
$$z'(0) = z'(1),$$
$$z''(0) = \alpha z'(\xi) \tag{5.2}$$

where $0 < v < 1, \xi \in [0,1), \alpha \in [0, \frac{1}{\xi})$. Next, we will prove some important Lemmas to deduce original results.

Lemma 5.1. *Let $\alpha\xi \neq 1$ & $h(u) \in L^1[0,1]$, then equation (5.2) has a unique solution*

$$z(x) = \int_0^1 G(x, y)h(y)dy,$$

where, $G(\gamma, \delta) = \frac{1}{2}G_1(\gamma, \delta) + xG_2(\xi, \delta)$

$$G_1(\gamma, \delta) = \begin{cases} (2\gamma - \gamma^2 - \delta)\delta & \text{if } 0 \le \delta \le \gamma \le 1 \\ \gamma^2(1 - \delta) & \text{if } 0 \le \gamma \le \delta \le 1 \end{cases}$$

$$G_2(\gamma, \delta) = \begin{cases} -\delta + \gamma\delta + \frac{1}{\alpha}(1 - \delta) & \text{if } 0 \le \delta \le \gamma \le 1 \\ -\gamma + \gamma\delta + \frac{1}{\alpha}(1 - \delta) & \text{if } 0 \le \gamma \le \delta \le 1 \end{cases}$$

Proof. Applying the method of variation of parameter in $z'''(s) + h(s) = 0$, we obtain that $z(s) = -\frac{1}{2}\int_0^s (s - y)^2 h(y)dy + As^2 + Bs + C$, where $A, B, C \in \mathbb{R}$. From $z(0) = 0$, we get $C = 0$.

$z'(0) = z'(1) \Rightarrow A = \frac{1}{2}\int_0^1 (1-y)h(y)dy.$
Again, $z''(0) = \alpha z'(\xi)$, we obtain that

$$2A = \alpha[-\int_0^\xi (\xi - \rho)h(\rho)d\rho + 2A\xi + B],$$

$$B = (\frac{1}{\alpha} - \xi)\int_0^1 (1-\rho)h(\rho)d\rho + \int_0^\xi (\xi - \rho)h(\rho)d\rho.$$

So, combining the results, we get that

$$z(s) = -\frac{1}{2}\int_0^s (s-\rho)^2 h(\rho)d\rho + \frac{s^2}{2}\int_0^1 (1-\rho)h(\rho)d\rho + Bs$$

$$= \frac{1}{2}\int_0^1 G_1(s,\rho)h(\rho)d\rho + s\int_0^1 G_2(\xi,\rho)h(\rho)d\rho$$

$$= \int_0^1 G(s,\rho)h(\rho)d\rho$$

$\square$

Lemma 5.2. *Let $\alpha\xi \neq 1$ & $h(u) \in L^1[0,1]$, then the boundary value problem (5.2) has a unique solution*

$$z'(\gamma) = \int_0^1 G'(\gamma,\rho)h(\rho)d\rho,$$

where, $G'(\gamma,\delta) = \frac{1}{2}G_3(\gamma,\delta) + G_2(\xi,\delta)$

$$G_3(\gamma,\delta) = \frac{\partial}{\partial t}G_1(\gamma,\delta)\begin{cases} 2(1-\gamma)\delta & \text{if } 0 \leq \delta \leq \gamma \leq 1 \\ 2(1-\delta)\gamma & \text{if } 0 \leq \gamma \leq \delta \leq 1 \end{cases}$$

Lemma 5.3. *For all $(\xi,s) \in [0,1] \times [0,1]$, we see that*

$$\frac{1}{\alpha}(1-s) - s \leq G_2(\xi,s) \leq \frac{1}{\alpha}(1-s).$$

Proof. We see that,
$1 \geq (1-\xi) \geq 0 \Rightarrow -1 \leq -(1-\xi) \leq 0,$ &
$1 \geq (1-s) \geq 0 \Rightarrow -1 \leq -(1-s) \leq 0.$
$G_2(u,v)$ can be defined as,

$$G_2(u,v) = \begin{cases} -v + uv + \frac{1}{\alpha}(1-v) & \text{if } 0 \leq v \leq u \leq 1 \\ -u + uv + \frac{1}{\alpha}(1-v) & \text{if } 0 \leq u \leq v \leq 1 \end{cases}$$

From the above results, we get that,

$$\frac{1}{\alpha}(1-s) - s \leq G_2(\xi,s) \leq \frac{1}{\alpha}(1-s).$$

$\square$

Lemma 5.4. *For all* $(u,v) \in [0,1] \times [0,1]$, *we get*

$$\gamma_0 G_1(1,v) \leq G_1(u,v) \leq G_1(1,v),$$

where $\gamma_0 = \tau^2$ *and* τ *satisfies* $\int_\tau^1 (1-s)sds > 0$.

Proof. Please see (Torres, 2013). $\qquad\qquad\qquad\qquad\qquad\qquad\qquad\qquad\qquad$ □

Lemma 5.5. *For all* $(u,v) \in [0,1] \times [0,1]$, *we get*

$$0 \leq G_3(u,v) \leq 2v(1-v).$$

Lemma 5.6. *Let* $h(u) \in \mathbf{C}^+[0,1]$. *The unique solution* $z(u)$ *of equation (5.2) is non-negative &*

$$\min_{\tau \leq v \leq 1} z(v) \geq \gamma_0 \max_{0 \leq v \leq 1} |z(v)|.$$

Proof. Clearly $z(v)$ is non negative. For $v \in [0,1]$, from Lemma (5.1) and (5.4), we get that

$$
\begin{aligned}
z(v) &= \int_0^1 G(v,\rho)h(\rho)d\rho \\
&= \int_0^1 (\frac{1}{2}G_1(v,\rho) + vG_2(\xi,\rho))h(\rho)d\rho \\
&\leq \frac{1}{2}\int_0^1 \rho(1-\rho)h(\rho)d\rho + v\int_0^1 G_2(\xi,\rho))h(\rho)d\rho.
\end{aligned}
$$

So,

$$\max_{0 \leq v \leq 1} |z(v)| \leq \frac{1}{2}\int_0^1 \rho(1-\rho)h(\rho)d\rho + v\int_0^1 G_2(\xi,\rho))h(\rho)d\rho.$$

For any $v \in [\tau,1]$, we have,

$$
\begin{aligned}
z(v) &= \int_0^1 G(v,\rho)h(\rho)d\rho \\
&= \int_0^1 (\frac{1}{2}G_1(v,\rho) + vG_2(\xi,\rho))h(\rho)d\rho \\
&\geq \frac{1}{2}\gamma_0 \int_0^1 \rho(1-\rho)h(\rho)d\rho + \gamma_0 \int_0^1 G_2(\xi,\rho))h(\rho)d\rho \\
&= \gamma_0[\frac{1}{2}\int_0^1 \rho(1-\rho)h(\rho)d\rho + \int_0^1 G_2(\xi,\rho))h(\rho)d\rho] \qquad\qquad (5.3)
\end{aligned}
$$

So, we get that,

$$z(v) \geq \gamma_0 \max_{0 \leq v \leq 1} |z(v)|.$$

Hence,

$$\min_{\tau \leq v \leq 1} z(v) \geq \gamma_0 \max_{0 \leq v \leq 1} |z(v)|.$$

$\qquad\qquad\qquad\qquad\qquad\qquad\qquad\qquad\qquad\qquad\qquad\qquad\qquad\qquad\qquad$ □

Lemma 5.7. *Let $h(v) \in \mathbf{C}^{+}[0,1]$, then the solution $z(v)$ of equation (5.2) satisfies the following property:*

$$\gamma_1(s) \max_{0 \leq v \leq 1} |z'(v)| \geq \max_{0 \leq v \leq 1} |z(v)|,$$

where $\gamma_1(\rho) = 1 + \frac{\frac{1}{2}G_1(1,\rho)}{G_2(\xi,\rho)}$.

Proof. We see that,

$$\frac{G(v,\rho)}{G'(v,\rho)} \leq \frac{G(1,\rho)}{G'(v,\rho)}$$

$$= \frac{G(1,\rho)}{\frac{1}{2}G_3(v,\rho) + G_2(\xi,\rho)}$$

$$\leq \frac{\frac{1}{2}G_1(1,\rho) + G_2(\xi,\rho)}{G_2(\xi,\rho)}$$

$$= 1 + \frac{\frac{1}{2}G_1(1,\rho)}{G_2(\xi,\rho)} \qquad (5.4)$$

So,

$$G(v,\rho) \leq \gamma_1(\rho)G'(v,\rho).$$

Hence,

$$\gamma_1(\rho) \max_{0 \leq v \leq 1} |z'(v)| \geq \max_{0 \leq v \leq 1} |z(v)|.$$

$\square$

5.3 EXISTENCE OF SOLUTIONS

In this section, we will discuss the existence of a solution for equation (5.1).

Consider the space

$$X = \{z : z(v) \in \mathbf{C}[0,1] \cap \mathbf{C}^1[0,1]\}.$$

On X, we define the norm

$$\|z\| = \sup\{\max_{0 \leq v \leq 1} |z(v)|, \max_{0 \leq v \leq 1} |z'(v)|\}, \quad z \in X.$$

Lemma 5.8. $(X, \|\cdot\|)$ *is a Banach space.*

Denote by P the cone on X,

$$P = \{z \in X : z(v) \geq 0; \min_{\tau \leq v \leq 1} z(v) \geq \gamma_0 \max_{0 \leq v \leq 1} |z(v)|; \gamma_1 \max_{0 \leq v \leq 1} |z'(v)| \geq \max_{0 \leq v \leq 1} |z(v)|\},$$

where $\sup \gamma_1(s) = \gamma_1$.

For any $z(v) \in X$ define,

$$Tz(v) = \int_0^1 G(v,\rho)g(\rho, z(\rho), z'(\rho))d\rho.$$

Lemma 5.9. *Assume that $f(v,z(v),z'(v))$ is continuous on $[0,1] \times [0,\infty) \times R$. Then $T : P \to P$ is completely continuous.*

Define functionals as,
$\delta(z) = \sup_{0 \le v \le 1} |z(v)|, \beta(z) = \sup_{0 \le v \le 1} |z'(v)|.$
Then we see that, $\delta(\xi w) = |\xi|\delta(z), \beta(\xi z) = |\xi|\beta(z), z \in X, \xi \in R.$
Also, $\delta(z_1) \le \delta(z_2)$, for $z_1 \le z_2, z_1, z_2 \in P.$
Let us define the constants M_1, M_2 & M_3 as,

$$M_1 = \int_0^1 \frac{\rho(1-\rho)}{2} d\rho + \int_0^1 \frac{(1-\rho)}{\alpha} d\rho$$

$$M_2 = \int_\tau^1 \gamma_0 G_1(1,\rho) d\rho$$

$$M_3 = \int_0^1 \rho(1-\rho) d\rho + \int_0^1 \frac{(1-\rho)}{\alpha}$$

To apply the fixed-point theorem in P, we need to modify g as follows:

$$\hat{g}(\sigma,\rho,\lambda) = \begin{cases} g(\sigma,\rho,\lambda) & \text{if } (\sigma,\rho,\lambda) \in [0,1] \times [0,a] \times R \\ g(\sigma,a,\lambda) & \text{if } (\sigma,\rho,\lambda) \in [0,1] \times [a,\infty) \times R. \end{cases}$$

$$g^*(\sigma,\rho,\lambda) = \begin{cases} \hat{g}(\sigma,\rho,\lambda) & \text{if } (\sigma,\rho,\lambda) \in [0,1] \times [0,\infty) \times [-M,M] \\ \hat{g}(\sigma,\rho,-M) & \text{if } (\sigma,\rho,\lambda) \in [0,1] \times [0,\infty) \times (-\infty,-M] \\ \hat{g}(\sigma,\rho,M) & \text{if } (\sigma,\rho,\lambda) \in [0,1] \times [0,\infty) \times [M,\infty) \end{cases}$$

Then, $g^* \in C([0,1] \times [0,\infty) \times R, [0,\infty)).$
Define,

$$Tz(v) = \int_0^1 G(v,\rho) g^*(\rho,z(\rho),z'(\rho)) d\rho.$$

Theorem 5.1. *Assume that $g(u,z(u),z'(u))$ is continuous on $[0,1] \times [0,\infty) \times R$ and there are $M > a > \gamma_0 a > c > 0$ such that $g(\sigma,\rho,\lambda)$ satisfies the conditions:-*

- $g(\sigma,\rho,\lambda) < \frac{c}{M_1}$ *for* $(\sigma,\rho,\lambda) \in [0,1] \times [0,c] \times [-M,M]$
- $g(\sigma,\rho,\lambda) \ge \frac{a}{M_2}$ *for* $(\sigma,\rho,\lambda) \in [0,1] \times [\gamma_0 a,a] \times [-M,M]$
- $g(\sigma,\rho,\lambda) < \frac{M}{M_3}$ *for* $(\sigma,\rho,\lambda) \in [0,1] \times [0,a] \times [-M,M]$

Then the boundary value problem (5.1) has at least one positive solution $z(u)$ satisfying

$$c < \delta(z) < b, |z'(v)| < M.$$

Proof. Take,
$$\Omega_1 = \{z \in X : |z(v)| < c, |z'(v)| < L\}.$$
$$\Omega_2 = \{z \in X : |z(v)| < b, |z'(v)| < L\}.$$

two boundary open sets in X, and D_1, D_2 can be defined as
$D_1 = \{z \in X : \delta(z) = c\}, D_2 = \{z \in X : \delta(z) = b\}$.
We have proved $T : P \to P$ is completely continuous, and there exists a non negative function δ such that $\delta(z + \lambda p) \geq \delta(z)$ for all $z \in p, \lambda \geq 0$ where $p \in (\Omega_2 \cap P) \setminus \{0\}$.

For $\delta(z) = c, z \in D_1 \cap P$, we have that

$$
\begin{aligned}
\delta(Tz) &= \max_{0 \leq v \leq 1} \left| \int_0^1 G(v, \rho) g^*(\rho, z(\rho), z'(\rho)) d\rho \right| \\
&\leq \max_{0 \leq v \leq 1} \frac{c}{M_1} \left| \int_0^1 \frac{\rho(1-\rho)}{2} d\rho + \int_0^1 \frac{v(1-\rho)}{\alpha} d\rho \right| \\
&\leq \frac{c}{M_1} \left| \int_0^1 \frac{\rho(1-\rho)}{2} d\rho + \int_0^1 \frac{(1-\rho)}{\alpha} d\rho \right| \\
&= c
\end{aligned}
$$

For $\delta(z) = a, z \in D_2 \cap P$, we get that

$$
\begin{aligned}
\delta(Tz) &= \max_{0 \leq v \leq 1} \left| \int_0^1 G(v, \rho) g^*(\rho, z(\rho), z'(\rho)) d\rho \right| \\
&> \max_{0 \leq v \leq 1} \left| \int_\tau^1 G(v, \rho) \frac{a}{M_2} d\rho \right| \\
&\geq \frac{a}{M_2} \left| \int_\tau^1 \gamma_0 G_1(1, \rho) d\rho \right| \\
&= a
\end{aligned}
$$

Now, for $z \in P$,

$$
\begin{aligned}
\beta(Tz) &= \max_{0 \leq v \leq 1} |(\acute{T}z)(v)| \\
&= \max_{0 \leq v \leq 1} \left| \int_0^1 G'(v, \rho) g^*(\rho, z(\rho), z'(\rho)) d\rho \right| \\
&\leq \max_{0 \leq v \leq 1} \frac{L}{M_3} \left| \int_0^1 \rho(1-\rho) d\rho + \int_0^1 \frac{(1-\rho)}{\alpha} d\rho \right| \\
&= M
\end{aligned}
$$

Hence, $\beta(Tz) < M$. So, $T(z(v)) = z(v)$ in $(\Omega_2 \setminus \overline{\Omega_1}) \cap P$ by applying the Guo - Krasnosel'skii fixed-point theorem (Guo and Ge, 2004). □

Finally, we shall now state and prove the existence of three positive solutions for equation (5.1). Let the continuous nonnegative concave functional δ_1, the continuous nonnegative convex functionals γ, θ and the continuous nonnegative functional ψ be defined on the cone P by

$$
\psi(z) = \theta(z) = \max_{0 \leq v \leq 1} |z(v)|, \delta_1(z) = \min_{\tau \leq v \leq 1} |z(v)|, \gamma(z) = \max_{0 \leq v \leq 1} |z'(v)|.
$$

By Lemma (5.6) and (5.7) the functionals defined above satisfy the following conditions:

$$\gamma_1 \gamma(z) \geq \psi(z), \gamma_1 \gamma(z) \geq \theta(z), \delta_1(z) \geq \gamma_0 \theta(z), \psi(z) \geq \delta_1(z).$$

$$||z|| \leq \gamma_1 \gamma(z).$$

Theorem 5.2. *Assume that there exists constants $0 < A < B < D$ such that $A < B < \frac{M_2 D}{M_3}, C = \frac{B}{\gamma_0}$ and $g(\sigma, \rho, \lambda)$ satisfies*

- $g(\sigma, \rho, \lambda) \leq \frac{D}{M_3}$ *for* $(\sigma, \rho, \lambda) \in [0,1] \times [0, \gamma_1 D] \times [-D, D]$
- $g(\sigma, \rho, \lambda) > \frac{B}{M_2}$ *for* $(\sigma, \rho, \lambda) \in [0,1] \times [B,C] \times [-D, D]$
- $g(\sigma, \rho, \lambda) < \frac{A}{M_1}$ *for* $(\sigma, \rho, \lambda) \in [0,1] \times [0,A] \times [-D, D]$

then equation (5.1) has three positive solutions z_1, z_2, z_3 satisfying

- $\max_{0 \leq u \leq 1} |z_i'(u)| \leq D, i=1,2,3;$
- $B < \min_{\tau \leq u \leq 1} |z_1(u)|$
- $B > \min_{\tau \leq u \leq 1} |z_2(u)|$
- $A < \max_{0 \leq u \leq 1} |z_2(u)|$
- $A > \max_{0 \leq u \leq 1} |z_3(u)|.$

Proof. For any $z(u) \in X$ define

$$Tz(v) = \int_0^1 G(v, \rho) f(\rho, z(\rho), z'(\rho)) d\rho.$$

For $z \in \overline{P(\gamma, D)}$, we have $\gamma(z) = \max_{0 \leq v \leq 1} |z'(v)| < D$. From equation 5.2, we obtain $g(\sigma, \rho, \lambda) \leq \frac{D}{M_3}$. Thus

$$
\begin{aligned}
\gamma(Tz) &= \max_{0 \leq v \leq 1} |(\acute{T}z)(v)| \\
&= \max_{0 \leq v \leq 1} |\int_0^1 G'(v, \rho) g(\rho, z(\rho), z'(\rho)) d\rho| \\
&\leq \max_{0 \leq v \leq 1} \frac{D}{M_3} |\int_0^1 \rho(1-\rho) d\rho + \int_0^1 \frac{(1-\rho)}{\alpha} d\rho| \\
&= D
\end{aligned}
$$

Hence, $T : \overline{P(\gamma, D)} \to \overline{P(\gamma, D)}$.
Now, $z(v) = \frac{B}{\gamma_0} \in P(\gamma, \theta, \delta_1, B, C, D)$ and $\delta_1(\frac{B}{\gamma_0}) > B$ implies that

$$\{z \in P(\gamma, \theta, \delta_1, B, C, D) | \delta_1(z) > B\} \neq \phi.$$

For $v \in P(\gamma, \theta, \delta_1, B, C, D)$, we have $B \leq z(v) \leq \frac{B}{\gamma_0}$ and $|z'(v)| < D$ for $0 \leq v \leq 1$. Then,

$$g(\sigma, \rho, \lambda) > \frac{B}{M_2}.$$

So,

$$
\begin{aligned}
\delta_1(Tz) &= \min_{\tau \leq v \leq 1} |Tz(v)| \\
&= \min_{\tau \leq v \leq 1} \left| \int_0^1 G(v, \rho) g(\rho, z(\rho), z'(\rho)) d\rho \right| \\
&\geq \min_{\tau \leq v \leq 1} \left| \int_\tau^1 G(v, \rho) \frac{B}{M_2} d\rho \right| \\
&\geq \frac{B}{M_2} \left| \int_\tau^1 \gamma_0 G_1(1, \rho) d\rho \right| \\
&= B
\end{aligned}
$$

which implies that $\delta_1(Tz) > B$, for all $z \in P(\gamma, \theta, \delta_1, B, C, D)$.

Secondly, for all $z \in P(\gamma, \theta, \delta_1, B, C, D)$ with $\theta(Tz) > \frac{B}{\gamma_0}$,

$$
\begin{aligned}
\delta_1(Tz) &\geq \gamma_0 \theta(Tz) \\
&> \gamma_0 \frac{B}{\gamma_0} \\
&= B
\end{aligned}
$$

Suppose that, $z \in R(\gamma, \psi, A, D)$, $\psi(z) = A$. Then by assumptions

$$
\begin{aligned}
\psi(Tz) &= \max_{0 \leq v \leq 1} |(Tz)(v)| \\
&= \max_{0 \leq v \leq 1} \left| \int_0^1 G(v, \rho) g(\rho, z(\rho), z'(\rho)) d\rho \right| \\
&\leq \max_{0 \leq v \leq 1} \frac{A}{M_1} \left| \int_0^1 \frac{\rho(1-\rho)}{2} d\rho + v \int_0^1 \frac{(1-\rho)}{\alpha} d\rho \right| \\
&\leq \frac{A}{M_1} \left| \int_0^1 \frac{\rho(1-\rho)}{2} d\rho + \int_0^1 \frac{(1-\rho)}{\alpha} d\rho \right| \\
&= A
\end{aligned}
$$

Thus, all conditions of Avery & Peterson theorem (Avery and Peterson, 2001) is satisfied. Hence problem (5.1) has at least three positive solutions satisfying equation 5.2. $\qquad \square$

5.4 NUMERICAL METHODS BASED ON BERNOULLI POLYNOMIALS

In this section, we would develop the numerical scheme for solving equation (5.1) using Bernoulli polynomials. Bernoulli polynomials play an important role in mathematics & physics. Bernoulli polynomials $\beta_i(t)$ can be defined as (Magnus et al., 2013),

$$\sum_{i=0}^{n} \binom{n+1}{i} \beta_i(t) = (n+1)t^n.$$

The first few Bernoulli polynomials are

- $\beta_0(x) = 1$
- $\beta_1(x) = t - \frac{1}{2}$
- $\beta_2(x) = t^2 - t + \frac{1}{6}$
- $\beta_3(x) = t^3 - \frac{3}{2}t^2 + \frac{1}{2}t$
- $\beta_4(x) = t^4 - 2t^3 + t^2 - \frac{1}{30}.$

Let us assume that, the solution for equation (5.1) is approximated by the first five Bernoulli polynomials. Hence,

$$y(t) = \sum_{n=0}^{4} x_n \beta_n(t) = \beta(t)X,$$

where $\beta(t) = [\beta_0(t), \beta_1(t), \beta_2(t), \beta_3(t), \beta_4(t)]$
& $X^T = [x_0, x_1, x_2, x_3, x_4].$

The derivative of $y(t)$ can be represented as

$$y'''(t) = \beta'''(t)X.$$

Now, we know that

$$\frac{d}{dt}\beta_i(t) = n\beta_{i-1}(t) \ \ i \geq 1.$$

So, we can write that (Kadkhoda, 2020),

$$(\beta'(t))^T = M\beta^T(t) \Rightarrow \beta'(t) = \beta(t)M^T,$$

where $M = \begin{pmatrix} 0 & 0 & 0 & 0 & 0 \\ 1 & 0 & 0 & 0 & 0 \\ 0 & 2 & 0 & 0 & 0 \\ 0 & 0 & 3 & 0 & 0 \\ 0 & 0 & 0 & 4 & 0 \end{pmatrix}$

Accordingly, the third derivative of $B(t)$ can be given as

$$\beta'''(t) = \beta(t)(M^T)^3.$$

From equation (5.1), we get that,

$$\beta(t)(M^T)^3 X + a(t)f(t, \beta(t)X) = 0.$$

After collocating at points $t = 0, 1$, we get

$$G(M^T)^3 X = H \tag{5.5}$$

where, $G = \begin{pmatrix} \beta_0(0) & \beta_1(0) & \beta_2(0) & \beta_3(0) & \beta_4(0) \\ \beta_0(1) & \beta_1(1) & \beta_2(1) & \beta_3(1) & \beta_4(1) \end{pmatrix}$,

& $H = \begin{pmatrix} a(0)f(0,\beta(0)X) \\ a(1)f(1,\beta(1)X) \end{pmatrix}$.

Now, the boundary conditions (if $\alpha = 1, \xi = 0$) are given by

$$y(0) = 0 \Rightarrow \beta(0)X = 0$$
$$y'(0) = y'(1) \Rightarrow (\beta(0)M^T - \beta(1)M^T)X = 0$$
$$y''(0) = y'(0) \Rightarrow (\beta(0)(M^T)^2 - \beta(0)M^T)X = 0$$

$$(5.6)$$

So, the solutions for equation (5.1) can be given by equation (5.5) & the above boundary conditions (5.6).

5.5 NUMERICAL RESULTS

In this section, we will explain the above results with examples.

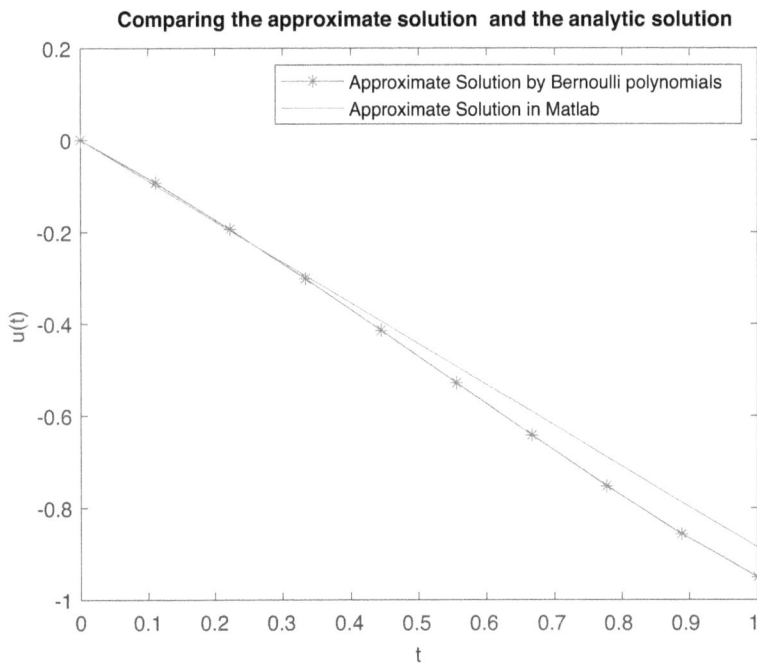

Figure 5.1 Comparing the approximate solution and the analytic solution.

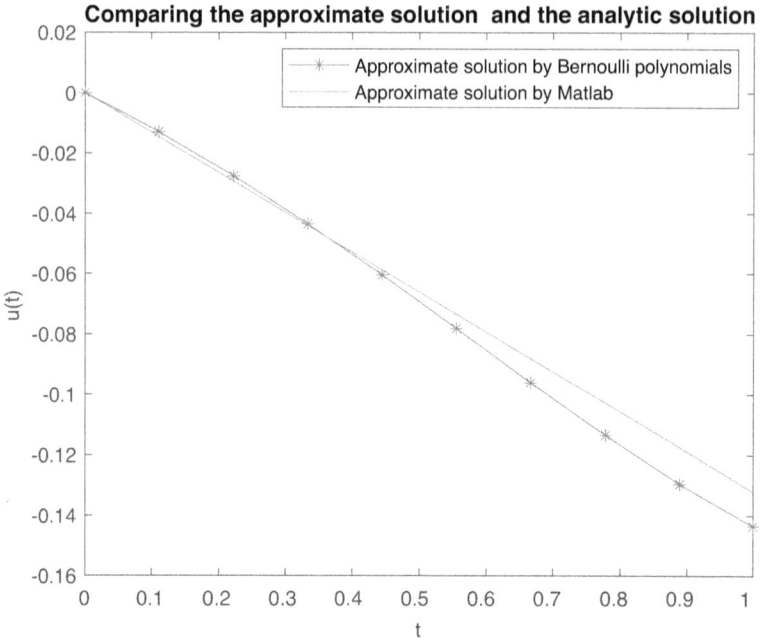

Figure 5.2 Comparing the approximate solution and the analytic solution.

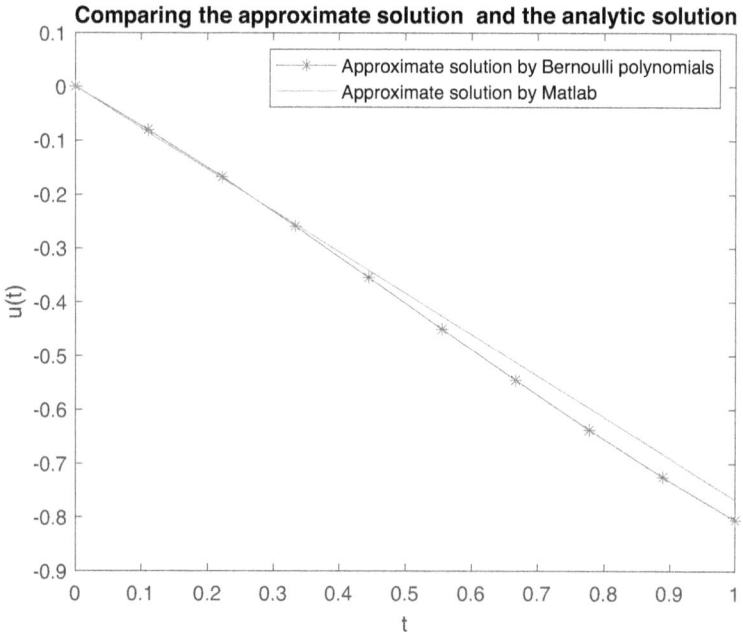

Figure 5.3 Comparing the approximate solution and the analytic solution.

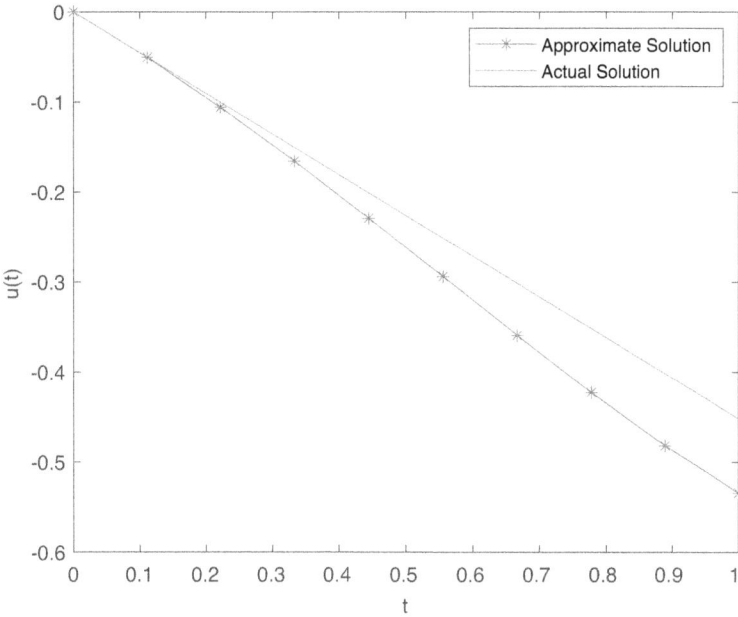

Figure 5.4 Comparing the approximate solution and the analytic solution.

Example 5.1

$$z'''(v) + g(v, z(v), z'(v)) = 0$$
$$z(0) = 0,$$
$$z'(0) = z'(1),$$
$$z''(0) = \alpha z'(\xi)$$

$$g(t, u, v) = \frac{t}{100} + u^3 + \frac{|v|^{\frac{1}{4}}}{1000} \tag{5.7}$$

where, $0 < v < 1, \alpha = 0.5, \tau = 0.5, \gamma_0 = 0.25$.
So, $M_1 = 1.08, M_2 = 0.04, M_3 = 1.17$.
Choosing $c = 0.50, b = 50, M = 10^8$,
We have

- $g(\sigma, \rho, \lambda) \leq 0.24 < 0.46 = \frac{c}{M_1}$ for $(\sigma, \rho, \lambda) \in [0, 1] \times [0, 0.5] \times [-10^8, 10^8]$
- $g(\sigma, \rho, \lambda) \geq 1953.22 > 1199.04 = \frac{b}{M_2}$ for $(\sigma, \rho, \lambda) \in [0, 1] \times [12.50, 50]$
  $\times [-10^8, 10^8]$
- $g(\sigma, \rho, \lambda) \leq 125000.11 < 85.71 \times 10^6 < \frac{L}{M_3}$ for $(\sigma, \rho, \lambda) \in [0, 1] \times [0, 50]$
  $\times [-10^8, 10^8]$.

then the boundary value problem (5.7) has at least one positive solution $z(v)$ satisfying

$$0.50 < \max_{0 \le v \le 1} |z(v)| < 50, |z'(v)| < 10^8.$$

Example 5.2

$$z'''(v) + g(v, z(v), z'(v)) = 0$$
$$z(0) = 0,$$
$$z'(0) = z'(1),$$
$$z''(0) = \alpha z'(\xi) \qquad\qquad (5.8)$$

$$g(t, u, v) = \begin{cases} \frac{u^4}{10(t+1)} + \sin(\frac{|v|}{10^{12}}) & \text{if } 0 \le u \le 10 \\ \frac{10000}{10(t+1)} + \sin(\frac{|v|}{10^{12}}) & \text{otherwise} \end{cases}$$

where, $\alpha = 0.5, \tau = 0.5, \gamma_0 = 0.25$.
So, $M_1 = 1.08, M_2 = 0.04, M_3 = 1.17$.
Choosing $A = 2, B = 10, D = 10^{12}$,

- $g(\sigma, \rho, \lambda) \le 1000.84 < 85.71 \times 10^{10} = \frac{d}{M_3}$ for $(\sigma, \rho, \lambda) \in [0, 1] \times [0, 10^{12}]$ $\times [-10^{12}, 10^{12}]$
- $g(\sigma, \rho, \lambda) \ge 500 > 239.81 = \frac{b}{M_2}$ for $(\sigma, \rho, \lambda) \in [0, 1] \times [10, 40] \times [-10^{12}, 10^{12}]$
- $g(\sigma, \rho, \lambda) \le 1.64 < 1.84 = \frac{a}{M_1}$ for $(\sigma, \rho, \lambda) \in [0, 1] \times [0, 2] \times [-10^{12}, 10^{12}]$.

then the boundary value problem (5.8) has three positive solutions u_1, u_2, u_3 satisfying

- $\max_{0 \le v \le 1} |z_i'(v)| \le 10^{12}, i = 1, 2, 3;$
- $2 < \min_{0.50 \le v \le 1} |z_1(v)|$
- $2 > \min_{0.50 \le v \le 1} |z_2(v)|$
- $10 < \max_{0 \le v \le 1} |z_2(v)|$
- $10 > \max_{0 \le v \le 1} |z_3(v)|$.

Example 5.3
Equation (5.1) can be written as if $\xi = 0, \alpha = 1, g(t, z(t), z'(t)) = t + e^t + z$.

$$z'''(v) + g(v, z(v), z'(v)) = 0$$
$$z(0) = 0,$$
$$z'(0) = z'(1),$$
$$z''(0) = \alpha z'(\xi)$$
$$g(t, u, v) = t + e^t + z \qquad\qquad (5.9)$$

The approximate solution $u(t)$ can be approximated as

$$z(t) = B(t)X,$$

where

$$X^T = [-0.4733, -.9514, 0, 0.3139, 0.0736].$$

Figure (5.1) compares the approximate solution in Matlab Software & the approximated solution by Bernoulli polynomials for equation (5.9).

Example 5.4

Equation (5.1) can be written as if $\xi = 0, \alpha = 1, g(t, z(t), z'(t)) = t + z'$.

$$z'''(v) + g(v, z(v), z'(v)) = 0$$
$$z(0) = 0,$$
$$z'(0) = z'(1),$$
$$z''(0) = \alpha z'(\xi)$$
$$g(t, z(t), z'(t)) = t + z' \qquad (5.10)$$

The approximate solution $u(t)$ can be approximated as

$$z(t) = \beta(t)X,$$

where $\beta(t) = [\beta_0(t), \beta_1(t), \beta_2(t), \beta_3(t), \beta_4(t)]$ & $X^T = [-0.0704, -0.1435, 0, 0.0648, 0.0417]$.

Figure (5.2) compares the approximate solution in Matlab Software & the approximated solution by Bernoulli polynomials for equation (5.10).

Example 5.5

Equation (5.1) can be written as if $\xi = 0, \alpha = 1, g(t, z(t), z'(t)) = t + 2 + z + z'$.

$$z'''(v) + g(v, z(v), z'(v)) = 0$$
$$z(0) = 0,$$
$$z'(0) = z'(1),$$
$$z''(0) = \alpha z'(\xi)$$
$$g(t, z(t), z'(t)) = t + 2 + z + z' \qquad (5.11)$$

The approximate solution $u(t)$ can be approximated as

$$z(t) = \beta(t)X,$$

where $\beta(t) = [\beta_0(t), \beta_1(t), \beta_2(t), \beta_3(t), \beta_4(t)]$ & $X^T = [-0.4026, -0.8057, 0, 0.2348, 0.0021]$.

Figure (5.3) compares the approximate solution in Matlab Software & the approximated solution by Bernoulli polynomials for equation (5.11).

Example 5.6
Equation (5.1) can be written as if $\xi = 0, \alpha = 1, g(t,z(t),z'(t)) = t + exp(t) + 2z + z'$.

$$z'''(v) + g(v,z(v),z'(v)) = 0$$
$$z(0) = 0,$$
$$z'(0) = z'(1),$$
$$z''(0) = \alpha z'(\xi)$$
$$g(t,z(t),z'(t)) = t + exp(t) + 2z + z' \tag{5.12}$$

The approximate solution $u(t)$ can be approximated as

$$z(t) = \beta(t)X,$$

where $\beta(t) = [\beta_0(t), \beta_1(t), \beta_2(t), \beta_3(t), \beta_4(t)]$ & $X^T = [-0.4026, -0.8057, 0, 0.2348, 0.0021]$.

Figure (5.4) compares the approximate solution in Matlab Software & the approximated solution by Bernoulli polynomials for equation (5.12).

5.6   CONCLUSION

In this chapter, we have discussed the existence criterion of multiple positive solutions in third-order boundary value problem. the application of Bernoulli polynomials for solving equation (5.1) is also analyzed here. The numerical scheme & the existence results may be helpful for Eco-epidemiology & food chain models (Biswas et al., 2015a,b, 2016a,b, 2017; Saifuddin et al., 2016a,b, 2017; Pal et al., 2015; Ghosh et al., 2016a,b; Biswas, 2017a,b; Elmojtaba et al., 2016, 2017; Sardar et al., 2016; Biswas, 2022).

5.7   ACKNOWLEDGEMENTS

Research of Santanu Biswas is supported by Dr. D. S. Kothari Postdoctoral Fellowship under the University Grants Commission scheme (Ref. No. F.4-2/2006 (BSR)/MA/19-20/0057). Special thanks to Prof. Sudeshna Banerjee for her valuable suggestions to improve the quality of the article.

REFERENCES

Alves, C. O. and Boudjeriou, T. (2020). Existence of solution for a class of non-variational Kirchhoff type problem via dynamical methods. *Nonlinear Analysis*, 197:111851.

Avery, R. I. and Peterson, A. C. (2001). Three positive fixed points of nonlinear operators on ordered Banach spaces. *Computers & Mathematics with Applications*, 42(3–5):313–322.

Bai, Z. (2008). Existence of solutions for some third-order boundary value problems. *Electronic Journal of Differential Equations*, 25:1–6.

Biswas, S. (2017a). Mathematical modeling of visceral leishmaniasis and control strategies. *Chaos, Solitons & Fractals*, 104:546–556.

Biswas, S. (2017b). Optimal predator control policy and weak Allee effect in a delayed preypredator system. *Nonlinear Dynamics*, 90(4):2929–2957.

Biswas, S. (2022). Forecasting and comparative analysis of Covid-19 cases in India and US. *The European Physical Journal Special Topics*, 231(18–20): 3537–3544.

Biswas, S., Saifuddin, M., Sasmal, K. S., Samanta, S., Pal, N., Ababneh, F., and Chattopadhyaya, J. (2016a). A delayed prey-predator system with prey subject to the strong Allee effect and disease. *Nonlinear Dynamics*, 84:1569–1594.

Biswas, S., Sasmal, K. S., Samanta, S., Saifuddin, M., Khan, Q. J. A., Alquranc, M., and Chattopadhyaya, J. (2015a). A delayed eco-epidemiological system with infected prey and predator subject to the weak Allee effect. *Mathematical Biosciences*, 263:198–208.

Biswas, S., Sasmal, S. K., Saifuddin, M., and Chattopadhyay, J. (2015b). On existence of multiple periodic solutions for Lotka-Volterra's predator-prey model with Allee effects . *Nonlinear Studies*, 22(2):189–199.

Biswas, S., Sasmal, S. K., Samanta, S., Saifuddin, M., Pal, N., and Chattopadhyay, J. (2016b). Complex dynamics in a delayed eco-epidemiological model with Allee effects and harvesting. *Nonlinear Dynamics*, 87(3):1553–1573.

Biswas, S., Subramanian, A., Elmojtaba, I., Chattopadhyay, J., and Sarkar, R. (2017). Optimal combinations of control strategies and cost-effective analysis for visceral leishmaniasis disease transmission. *PLoS One*, 12(2):e0172465.

Chammem, R., Ghanmi A., and Sahbani., A. (2021). Existence of solution for a singular fractional laplacian problem with variable exponents and indefinite weights. *Complex Variables and Elliptic Equations*, 66(8):1320–1332.

Elmojtaba, I. M., Biswas, S., and Chattopadhyay, J. (2016). Global dynamics and sensitivity analysis of a vector-host-reservoir model. *Sultan Qaboos University Journal for Science*, 21(2):120–138.

Elmojtaba, I. M., Biswas, S., and Chattopadhyay, J. (2017). Global analysis and optimal control of a periodic visceral leishmaniasis model. *Mathematics*, 5(4):80.

Feng, Y. and Liu, S. (2005). Solvability of a third-order two-point boundary value problem. *Applied Mathematics Letters*, 18(9):1034–1040.

Ghosh, K., Samanta, S., Biswas, S., Rana, S., Elmojtaba, I., Kesh, D., and Chattopadhyay, J. (2016a). Stability and bifurcation analysis of an eco-epidemiological model with multiple delays. *Nonlinear Studies*, 23(2):167–208.

Ghosh, K., Sardar, T., Biswas, S., Samanta, S., and Chattopadhyay, J. (2016b). An eco-epidemiological model with periodic transmission. *Nonlinear Studies*, 23(3):345–363.

Guo, Y. and Ge, W. (2004). Positive solutions for three-point boundary value problems with dependence on the first order derivative. *Journal of Mathematical Analysis and Applications*, 290:291–301.

Kadkhoda, N. (2020). A numerical approach for solving variable order differential equations using bernstein polynomials. *Alexandria Engineering Journal*, 59(5):3041–3047.

Khan, H., Li, Y., Khan, A., and Khan, A. (2019). Existence of solution for a fractional-order Lotka-Volterra reaction-diffusion model with Mittag-Leffler kernel. *Mathematical Methods in the Applied Sciences*, 42(9):3377–3387.

Li, S. (2006). Positive solutions of nonlinear singular third-order two-point boundary value problem. *Journal of Mathematical Analysis and Applications*, 323(1): 413–425.

Magnus, W., Oberhettinger, F., and Soni, R. P. (2013). *Formulas and Theorems for the Special Functions of Mathematical Physics*. Berlin, Germany: Springer Science & Business Media.

Pal, N., Samanta, S., Biswas, S., Alquran, M., Al-Khaled, K., and Chattopadhyay, J. (2015). Stability and bifurcation analysis of a three-species food chain model with delay. *International Journal of Bifurcation and Chaos*, 25(9):1550123.

Saifuddin, M., Biswas, S., Samanta, S., Sarkar, S., and Chattopadhyay, J. (2016a). Complex dynamics of an eco-epidemiological model with different competition coefficients and weak Allee in the predator. *Chaos, Solitons & Fractals*, 91: 270–285.

Saifuddin, M., Samanta, S., Biswas, S., and Chattopadhyay, J. (2017). An eco-epidemiological model with different competition coefficients and strong Allee in the prey. *International Journal of Bifurcation and Chaos*, 27(08):1730027.

Saifuddin, M., Sasmal, K. S., Biswas, S., Sarkar, S., Alquranc, M., and Chattopadhyaya, J. (2016b). Effect of emergent carrying capacity in an eco-epidemiological system. *Mathematical Methods in Applied Sciences*, 39, 806–823.

Sardar, T., Biswas, S., and Chattopadhyay, J. (2016). Global analysis of a periodic epidemic model on cholera in presence of bacteriophage. *Mathematical Methods in the Applied Sciences*, 39(14):4181–4195.

Sun, Y. (2009). Positive solutions for third-order three-point nonhomogeneous boundary value problems. *Applied Mathematics Letters*, 22:45–51.

Torres, F. J. (2013). Positive solutions for a third-order three-point boundary value problem. *Electronic Journal of Differential Equations*, 147:1–11.

6 An Interred, Inclined Long Strike-Slip Fault in a Striped Elastic/Viscoelastic Medium

Bula Mondal
J.B.V. Girls' High School

Asish Karmakar
Udairampur Pallisree Sikshayatan (H.S.)

CONTENTS

6.1 INTRODUCTION

Among many supernatural things one of the significant phenomena is an earthquake. The existence of life in highly seismically active countries is a big question. Therefore, to discover the answer to this question or prediction of an earthquake is the core objective of the study of this chapter. For such a prediction we have to understand the mechanism of stress accumulation in seismically active regions. So the rate of stress accumulation/release due to tectonic forces and the effect of fault movement on the nature of stress accumulation patterns during aseismic period may give us more insightful learning into the earthquake mechanism.

In this chapter, we introduced a theoretical model of multi-layered lithosphere-asthenosphere structure, in which the first layer is elastic and the second layer is the viscoelastic, overlying on a viscoelastic half-space. Here layers are supposed to be soldered in contact. The material of viscoelastic layer and half-space are of Maxwell type with different rigidity and viscosity. One of the basic causes to take the viscoelastic model is related to the plate tectonic theory of the Earth. The main postulate of this theory is that the lithosphere is roughly brittle elastic in nature and the material of the asthenosphere is imperfectly elastic. So, the lithospheric-asthenospheric surface of the earth can be chosen as a multi-layered half-space model. A good number of papers were done by the following authors such as (Maruyama, 1966; Rybicki, 1971, 1973; Mukhopadhyay, 1984, 1986; Sen and Debnath, 2012; Debnath and Sen, 2013a,b,c, 2014, 2015a,b; Mondal and Sen, 2016; Karmakar and Sen, 2016; Mondal et al., 2018b,a,c, 2019; Manna and Sen, 2019; Manna et al., 2019; Mondal and Debnath, 2021) etc. The outcome of the study of this chapter is discussed detail in equation 6.5. According to the numerical results and discussion, we are very much hopeful that the further study of this chapter may also be applicable in aseismic period of seismically active regions. Such studies will help the seismologists to formulate an effective program for earthquake prediction.

6.2   FORMULATION

Here we developed the main theory of the model based on lithosphere-asthenosphere structure. Also, we take an interred, long strike-slip fault, the slope of the fault with the horizon is θ_1 and the fault location is at the second layer. The long fault is taken so that the length of the fault is much more greater than its width l. The ends of the fault are taken to be horizontal. $(h_1 + r_1)$ is the depth of the upper end of the fault from the top surface. The depth of the upper end of the fault from the top surface is r_1.

The (y_1, y_2, y_3) is a Cartesian co-ordinate system with the plane free surface as the plane $y_3 = 0$ have been considered with y_3-axis taking downwards in the medium and y_1-axis is directed to the strike of the fault on the top surface of the earth. The plane $y_3 = h_1$ and $y_3 = h_2$ respectively represent the surface of separation of the first and second layers and second layer and half-space. With this choice of co-ordinate axes elastic layer occupies the region $(0 \leq y_3 \leq h_1, |y_2| < \infty)$, the viscoelastic layer occupies the region $(h_1 \leq y_3 \leq h_2, |y_2| < \infty)$ and the viscoelastic half-space occupies the region $(y_3 \geq h_2, |y_2| < \infty)$. Here we impose another Cartesian co-ordinate system (y_1', y_2', y_3') with the upper end of the fault F can be written as $F : (y_2' = 0, 0 \leq y_3' \leq l)$. The above two co-ordinate systems are connected by the following relations

$$\left.\begin{aligned}
y_1 &= y_1', \\
y_2 &= y_2' \sin \theta_1 + y_3' \cos \theta_1 \\
y_3 &= (h_1 + r_1) - y_2' \cos \theta_1 + y_3' \sin \theta_1
\end{aligned}\right\} \tag{6.1}$$

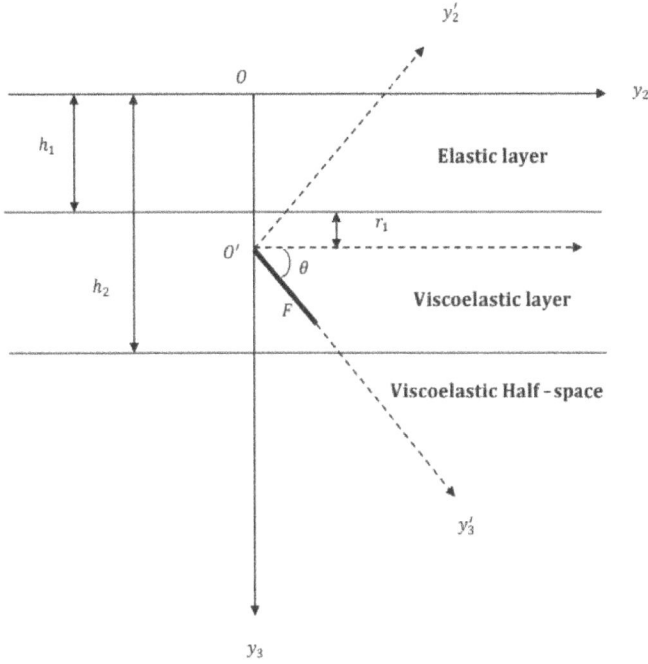

Figure 6.1 Sectinal representation of the model by $y_1 = 0$.

The plane $y_1 = 0$ gives the sectional figure of the model is given by Figure 6.1.

Since the fault is very long, the displacements, stresses, and strains are not depend on y_1 but functions of y_2, y_3, t. As explained in the paper done by Mondal and Sen (2017) the displacements, stresses, and strain components associated with strike-slip movement of the faults are $(u_1, \tau_{12}, \tau_{13}, e_{12}, e_{13})$ for the elastic layer, $(u'_1, \tau'_{12}, \tau'_{13}, e'_{12}, e'_{13})$ for viscoelastic layer and $(u''_1, \tau''_{12}, \tau''_{13}, e''_{12}, e''_{13})$ for viscoelastic half-space respectively. We assume μ_1 to be the rigidity of the elastic layer, μ_2 is effective rigidity and η_2 is the effective viscosity of the second layer and μ_3 is effective rigidity and η_3 is effective viscosity of viscoelastic half-space.

6.2.1 CONSTITUTIVE EQUATIONS

6.2.1.1 Stress-Strain Relations

$$\left.\begin{array}{l} \tau_{12} = \mu_1 \dfrac{\partial u_1}{\partial y_2} \\[4mm] \tau_{13} = \mu_1 \dfrac{\partial u_1}{\partial y_3} \end{array}\right\} \tag{6.2}$$

for elastic layer $(0 \le y_3 \le h_1, |y_2| < \infty), t \ge 0$

where the prefix constant of rigidity for the first layer is μ_1.

$$\left.\begin{array}{l}\left(\dfrac{1}{\eta_2}+\dfrac{1}{\mu_2}\dfrac{\partial}{\partial t}\right)\tau'_{12}=\dfrac{\partial^2 u'_1}{\partial t\partial y_2}\\[4mm]\left(\dfrac{1}{\eta_2}+\dfrac{1}{\mu_2}\dfrac{\partial}{\partial t}\right)\tau'_{13}=\dfrac{\partial^2 u'_1}{\partial t\partial y_3}\end{array}\right\}$$ (6.3)

for the viscoelastic layer $(h_1 \leq y_3 \leq h_2,\ -\infty < y_2 < \infty),\ t \geq 0$

where μ_2 is the effective rigidity and η_2 is the effective viscosity of the viscoelastic layer which are assumed to be constants.

$$\left.\begin{array}{l}\left(\dfrac{1}{\eta_3}+\dfrac{1}{\mu_3}\dfrac{\partial}{\partial t}\right)\tau''_{12}=\dfrac{\partial^2 u''_1}{\partial t\partial y_2}\\[4mm]\left(\dfrac{1}{\eta_3}+\dfrac{1}{\mu_3}\dfrac{\partial}{\partial t}\right)\tau''_{13}=\dfrac{\partial^2 u''_1}{\partial t\partial y_3}\end{array}\right\}$$ (6.4)

for the viscoelastic half-space $(y_3 \geq h_2,\ -\infty < y_2 < \infty),\ t \geq 0$

where μ_3 is the effective rigidity and η_3 is the effective viscosity of the viscoelastic half-space which are assumed to be constants. The time t is count from an appropriate state when there exists no seismic disturbance.

6.2.1.2   Equation of Motion for Stress

Between two seismic events i.e. in aseismic period we are neglecting the very small inertial terms with respect to other terms. Therefore, the relevant stresses fulfill the following relations:

$$\frac{\partial \tau_{12}}{\partial y_2}+\frac{\partial \tau_{13}}{\partial y_3}=0$$
$$\text{for } 0 \leq y_3 \leq h_1\,|y_2| < \infty$$ (6.5)

$$\frac{\partial \tau'_{12}}{\partial y_2}+\frac{\partial \tau'_{13}}{\partial y_3}=0$$
$$\text{for } h_1 \leq y_3 \leq h_2\,|y_2| < \infty$$ (6.6)

$$\frac{\partial \tau''_{12}}{\partial y_2}+\frac{\partial \tau''_{13}}{\partial y_3}=0$$
$$\text{for } y_3 \geq h_2\,|y_2| < \infty$$ (6.7)

From equations (6.2) and (6.5) we get

$$\nabla^2 u_1 = 0$$
$$\text{for } 0 \leq y_3 \leq h_1,\ |y_2| < \infty$$ (6.8)

and from equations (6.3) and (6.6) we get

$$\frac{\partial}{\partial t}(\nabla^2 u_1') = 0$$

which is satisfied if

$$\nabla^2 u_1' = 0$$
$$\text{for } h_1 \leq y_3 \leq h_2, \ |y_2| < \infty \tag{6.9}$$

and from equations (6.4) and (6.7) we get

$$\frac{\partial}{\partial t}(\nabla^2 u_1'') = 0$$

which is satisfied if

$$\nabla^2 u_1'' = 0$$
$$\text{for } y_3 \geq h_2, \ |y_2| < \infty \tag{6.10}$$

6.2.2 BOUNDARY CONDITIONS

Since the free surface is stress free and mediums are in welded contact then

$$\left.\begin{aligned}
\tau_{13} &= 0 \text{ at } y_3 = 0 \\
u_1 &= u_1' \text{ at } y_3 = h_1 \\
\tau_{13} &= \tau_{13}' \text{ at } y_3 = h_1 \\
u_1' &= u_1'' \text{ at } y_3 = h_2 \\
\tau_{13}' &= \tau_{13}'' \text{ at } y_3 = h_2 \\
\tau_{13}'' &\to 0 \text{ as } y_3 \to \infty \\
&\text{for } |y_2| < \infty
\end{aligned}\right\} \tag{6.11}$$

6.2.3 PRIMODIAL CONDITIONS

We count time t from an appropriate moment when the system is seismically at rest. The values of u_1, u_1', ..., e_{12}'' at time $t = 0$ are $(u_1)_0$, $(u_1')_0$, ... $(e_{12}'')_0$ and they obeys all the above equations.

6.2.4 CONDITIONS IN THE RANGE $|Y_2| \to \infty$

A time-dependent strain occurs far away from the fault which is maintained/ effected by the tectonic forces. So

$$\left.\begin{array}{l}
e_{12} \rightarrow (e_{12})_{0\infty} + g(t) \\
\text{as } |y_2| \rightarrow \infty,\ t \geq 0 \\
\text{for } 0 \leq y_3 \leq h_1 \\
\\
e'_{12} \rightarrow (e'_{12})_{0\infty} + g(t) \\
\text{as } |y_2| \rightarrow \infty,\ t \geq 0 \\
\text{for } h_1 \leq y_3 \leq h_2 \\
\\
e''_{12} \rightarrow (e''_{12})_{0\infty} + g(t) \\
\text{as } |y_2| \rightarrow \infty,\ t \geq 0 \\
\text{for } y_3 \geq h_2
\end{array}\right\} \tag{6.12}$$

where

$$(e_{12})_{0\infty} = \lim_{|y_2| \to \infty} (e_{12})_0$$
$$(e'_{12})_{0\infty} = \lim_{|y_2| \to \infty} (e'_{12})_0$$

Since the layers are soldered in contact, so same $g(t)$ is taken throughout the medium. Also taking an increasing function $g(t)$ of time with $g = 0$ at $t = 0$.

6.3 COMPONENTS OF FORCES IN THE ABSENCE OF FAULT MOVEMENT

We consider the condition when the model is totally in quasi-static, aseismic period before any sudden movement occurs in the model. The method of finding the solutions for displacements, stresses, and strains in the absence of any fault movement is stated in Appendix. These solutions are done by Mondal and Sen (2017) for the first layer, second layer, and viscoelastic half-space, respectively. The strike-slip movement due to the shear stress $\tau'_{1'2'}$ is given by

$$\left.\begin{array}{l}
\tau'_{1'2'} = \tau'_{12}(y_2, y_3, t) \sin \theta_1 - \tau'_{13}(y_2, y_3, t) \cos \theta_1 \\
\\
= (\tau'_{1'2'})_0 \exp\left\{-\dfrac{\mu_2 t}{\eta_2}\right\} + \mu_2 \sin \theta_1 \displaystyle\int_0^t g_1(\tau) \exp\left\{-\dfrac{\mu_2}{\eta_2}(t - \tau)\right\} d\tau
\end{array}\right\} \tag{6.13}$$

where $(\tau'_{1'2'})_0 = (\tau'_{12})_0 \sin \theta_1 - (\tau'_{13})_0 \cos \theta_1$.

From the solution (6.13) we can say that if there occurs any cumulated shear strain at a long distance from the fault for tectonic forces, then we suppose that the cumulation of shear stress $\tau'_{1'2'}$ in the viscoelastic layer. The movement across the fault will set in if this cumulated stress overcomes the entire cohesive and inertial forces across the fault after a suitable time. The local rheological behavior and the characteristics of the fault F are assumed to be such that the fault undergoes a sudden slipping movement generating an earthquake and becomes locked when the cumulated stress got released. So the above solutions are no longer valid and we need some modifications.

6.4 COMPONENTS OF FORCES IN ASEISMIC PERIOD FOLLOWED BY SUDDEN MOVEMENT ACROSS THE FAULT

In the case of sudden fault movement, the cumulated stress will be free after a few years. Here we omit the short span of the seismic period when the inertial terms are too large. We simply consider the aseismic period and the constitutive equations (6.2)–(6.12) are valid for $t \geq 0$ i.e. in the second phase of aseismic period. This type of sudden movement was discussed by Mondal and Sen (2017).

To solve the above BVP (Boundary Value Problem) we follow the paper done by Mondal and Sen (2017).

To find the solutions of the components of forces after the sudden movement across the fault, we apply a similar technique as explained by Mondal and Sen (2017) which is discussed in detail in the subsection 6.5.2 of Appendix and finally we obtain the complete result, which is given below :

1. **For elastic layer**$(0 \leq y_3 \leq h_1, |y_2| < \infty)$

$$
\left.
\begin{aligned}
u_1(y_2,y_3,t) &= (u_1)_p + y_2 g(t) + \frac{U}{\pi} \psi_1'(y_2,y_3,t) \\
\tau_{12}(y_2,y_3,t) &= (\tau_{12})_p + \mu_1 g(t) + \frac{\mu_1 U}{\pi} \psi_2'(y_2,y_3,t) \\
\tau_{13}(y_2,y_3,t) &= (\tau_{13})_p + \frac{\mu_1 U}{\pi} \psi_3'(y_2,y_3,t) \\
e_{12} &= (e_{12})_p + g(t) + \frac{U}{\pi} \psi_2'(y_2,y_3,t)
\end{aligned}
\right\}
\tag{6.14}
$$

2. **For viscoelastic layer**$(h_1 \leq y_3 \leq h_2, |y_2| < \infty)$

$$
\left.
\begin{aligned}
u_1'(y_2,y_3,t) &= (u_1')_p + y_2 g(t) + \frac{U}{2\pi} \phi_1'(y_2,y_3,t) \\
\tau_{12}'(y_2,y_3,t) &= (\tau_{12}')_p exp\left(-\frac{\mu_2 t}{\eta_2}\right) + \mu_2 \int_0^t g_1(\tau) exp\left\{-\frac{\mu_2(t-\tau)}{\eta_2}\right\} d\tau \\
&\quad + \frac{\mu_2 U}{2\pi} \phi_2'(y_2,y_3,t) \\
\tau_{13}'(y_2,y_3,t) &= (\tau_{13}')_p exp(-\frac{\mu_2 t}{\eta_2}) + \frac{\mu_2 U}{2\pi} \phi_3'(y_2,y_3,t) \\
\tau_{1'2'}'(y_2,y_3,t) &= (\tau_{1'2'}')_p exp\left(-\frac{\mu_2 t}{\eta_2}\right) \\
&\quad + \mu_2 \sin\theta_1 \int_0^t g_1(\tau) exp\left\{-\frac{\mu_2(t-\tau)}{\eta_2}\right\} d\tau \\
&\quad + \frac{\mu_2 U}{2\pi}(\phi_2' \sin\theta_1 - \phi_3' \cos\theta_1)
\end{aligned}
\right\}
\tag{6.15}
$$

3. **For viscoelastic half-space**$(y_3 \geq h_2, |y_2| < \infty)$

$$
\left.
\begin{aligned}
u_1''(y_2, y_3, t) &= (u_1'')_p + y_2 g(t) - \frac{U}{\mu_3 \pi} \chi_1'(y_2, y_3, t) \\
\tau_{12}''(y_2, y_3, t) &= (\tau_{12}'')_p exp\left(-\frac{\mu_3 t}{\eta_3}\right) \\
&\quad + \mu_3 \int_0^t g_1(\tau) exp\left\{-\frac{\mu_3(t-\tau)}{\eta_3}\right\} d\tau - \frac{U}{\pi} \chi_2'(y_2, y_3, t) \\
\tau_{13}''(y_2, y_3, t) &= (\tau_{13}'')_p exp\left(-\frac{\mu_3 t}{\eta_3}\right) - \frac{U}{\pi} \chi_3'(y_2, y_3, t)
\end{aligned}
\right\} \quad (6.16)
$$

where $\psi_1'(y_2, y_3, t)$, $\psi_2'(y_2, y_3, t)$, $\psi_3'(y_2, y_3, t)$, $\phi_1'(y_2, y_3, t)$, $\phi_2'(y_2, y_3, t)$, $\phi_3'(y_2, y_3, t)$ and $\chi_1'(y_2, y_3, t)$, $\chi_2'(y_2, y_3, t)$, $\chi_3'(y_2, y_3, t)$ are given by Appendix.

The above solutions (6.14)–(6.16) are valid at any point of the respective medium not lying on the fault F (i.e. $y_2' \neq 0$) and for $t \geq 0$ during the aseismic period of the model after the sudden movement across the fault and will remain valid as long as this aseismic state continues. For locked faults analytical investigations show that the components of forces will remain finite everywhere in the problem including the tip of the fault if $f(y_3')$ satisfies the same sufficient conditions given by Mondal and Sen (2016).

6.5 NUMERICAL RESULTS AND DISCUSSION

To learn the surface displacements, accumulation, and/or release of surface shear strain and the shear stress close to the crack led to cause strike-slip movement we choose $f(y_3') = \dfrac{y_3'^2 (y_3' - 1)^2}{\left(\frac{1}{2}\right)^4}$, such that displacements, stresses, and strains remain finite.

Now we compute the following quantities with the same values of parameters as explained in the paper by Mondal and Sen (2017).

1. Excess shear strain after one year of re-establishment of the aseismic period.

$$
E_{12} = [e_{12} - (e_{12})_p - g(t)]_{y_3=0, \, t=1 \text{ year}}
$$
$$
= \left[\frac{U}{\pi} \psi_2'(y_2, y_3, t)\right]_{y_3=0, \, t=1 \text{ year}}
$$

2. Shear stress Change in the elastic layer after one year of fault movement [i.e. after one year when the aseismic state re-established after sudden movement of fault]

$$
T_{12} = [\tau_{12} - (\tau_{12})_p - \mu_1 g(t)]_{t=1 \text{ year}}
$$
$$
= \left[\frac{\mu_1 U}{\pi} \psi_2'(y_2, y_3, t)\right]_{t=1 \text{ year}}
$$

3. Change in shear stress component along the fault in the second layer one year after the fault movement is

$$
T'_{1'2'} = \left[\tau'_{1'2'} - (\tau'_{1'2'})_p \exp\left(-\frac{\mu_2 t}{\eta_2}\right) - \mu_2 \sin\theta_1 \int_0^t g_1(\tau)\exp\left\{-\frac{\mu_2(t-\tau)}{\eta_2}\right\} d\tau \right]_{t=1 \text{ year}}
$$

$$
= \left[\frac{\mu_2 U}{2\pi} (\phi'_2 \sin\theta_1 - \phi'_3 \cos\theta_1) \right]_{t=1 \text{ year}}
$$

4. Change in shear stress in the second layer after one year of fault movement is

$$
T'_{12} = \left[\tau'_{12} - (\tau'_{12})_p \exp\left(-\frac{\mu_2 t}{\eta_2}\right) - \mu_2 \sin\theta_1 \int_0^t g_1(\tau)\exp\left\{-\frac{\mu_2(t-\tau)}{\eta_2}\right\} d\tau \right]_{t=1 \text{ year}}
$$

$$
= \left[\frac{\mu_2 U}{2\pi} \phi'_2(y_2, y_3, t) \right]_{t=1 \text{ year}}
$$

Figure 6.2 shows the excess shear strain at the top surface of the earth. The order of this shear strain is near about 10^{-7} which is almost coincide with the recorded

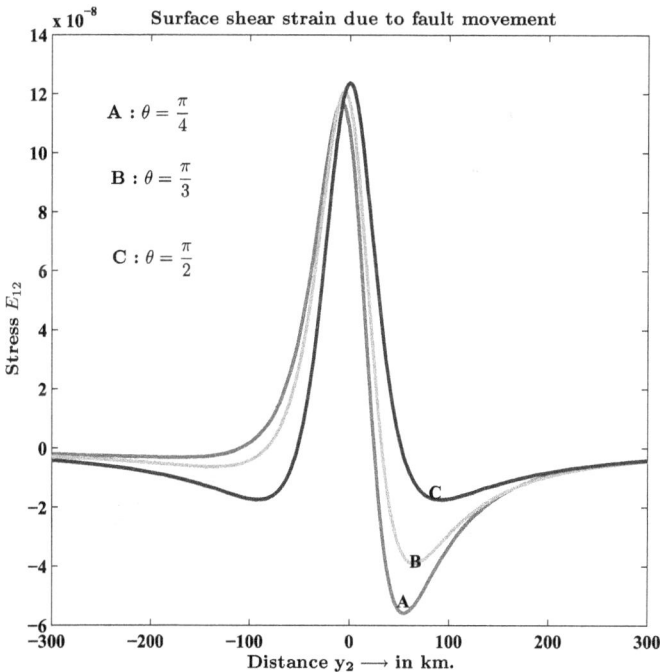

Figure 6.2 Redundant part of shear strain due to fault movement at the surface.

data. Also, when $\theta_1 \neq 90°$, the magnitude of this shear strain is highest and symmetric with respect to the crack $(y_2 = 0, \ y_3 = 0)$ otherwise it is asymmetric.

The contour representation of the variation of shear stress in the first layer due to the slip of the fault in the second layer shown in Figure 6.3 for $\theta_1 = \dfrac{\pi}{2}$ when the fault situated in the second layer.

The contour representation of the variation of shear stress in the second layer due to the slip of the fault in the second layer is shown in Figure 6.4 for $\theta_1 = \dfrac{\pi}{2}$ when the fault situated in the second layer.

The contour representation of the variation of shear stress in the second layer due to the slip of the fault in the second layer is shown in Figure 6.5 for $\theta_1 = \dfrac{\pi}{3}$ when the fault situated in the second layer.

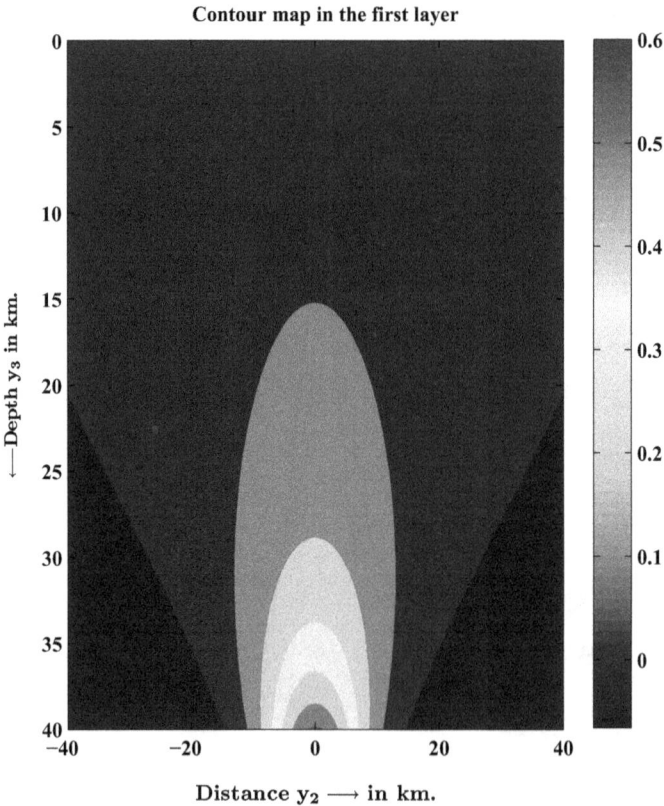

Figure 6.3   Contour representation of the shear stress in first layer.

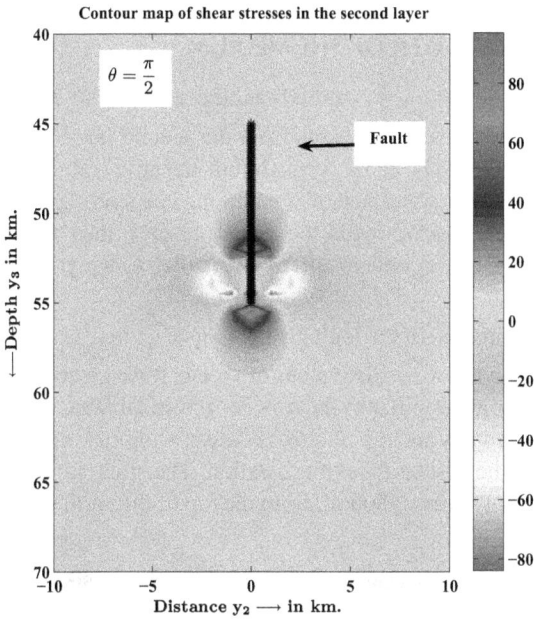

Figure 6.4 Contour representation of shear stress in second layer for $\theta_1 = \dfrac{\pi}{2}$.

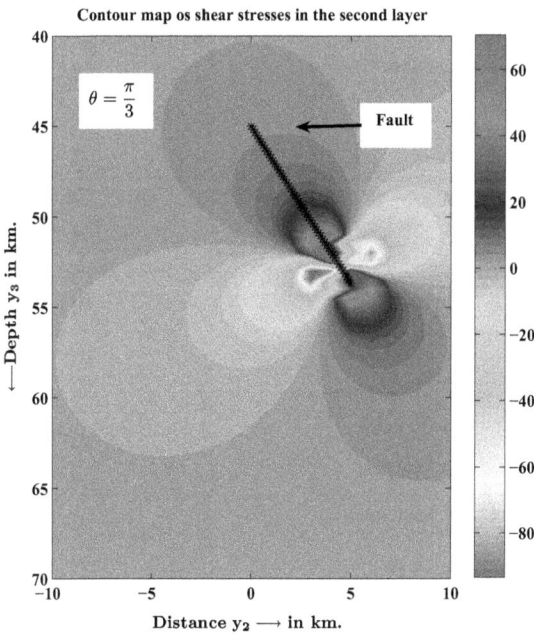

Figure 6.5 Contour representation of shear stress in second layer for $\theta_1 = \dfrac{\pi}{3}$.

6.5.1 CHANGE OF SHEAR STRESS T'_{12} (DUE TO FAULT SLIP) WITH DEEP IN THE SECOND LAYER OF THE MEDIUM

For a vertical fault with $\theta_1 = \dfrac{\pi}{2}$, we find that the shear stress T'_{12} is negative (stress release region) as we move downwards from the second layer at different distances from the fault. The stresses attain a maximum negative value at a deep of about 15 kilometers from the first surface of separation. The stress gradually dies out and goes to zero as we go downwards in the second layer. It may also be noted that the magnitudes of the stress become smaller and smaller as we go away from the fault (Figure 6.6).

However the inclination of the fault given by $\theta_1 = \dfrac{\pi}{3}$ there is a stress accumulation zone in the second layer for smaller values of y_2. But if we go away from the fault, the pattern of stress becomes similar to the case for a vertical fault. Most of the region in the second layer becomes an area of stress release, with maximum release at a point about 10 km. from the first surface of separation. The stress ultimately tends to zero if we go further downwards (≥ 80 km. from the top of the earth surface) (Figure 6.7).

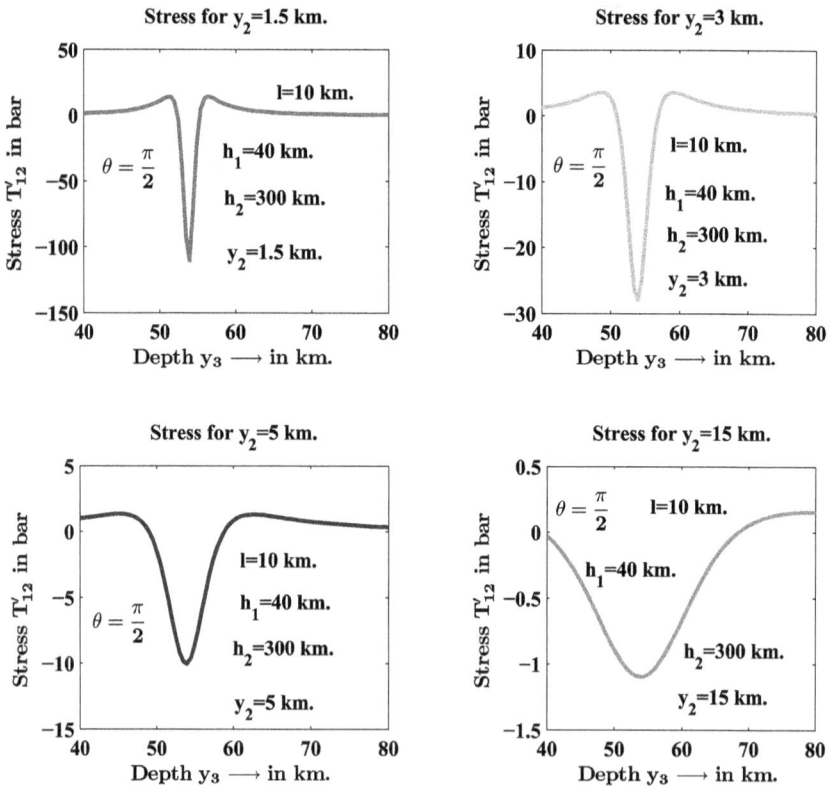

Figure 6.6 Shear stress in the second layer for different y_2 at $\theta_1 = \dfrac{\pi}{2}$.

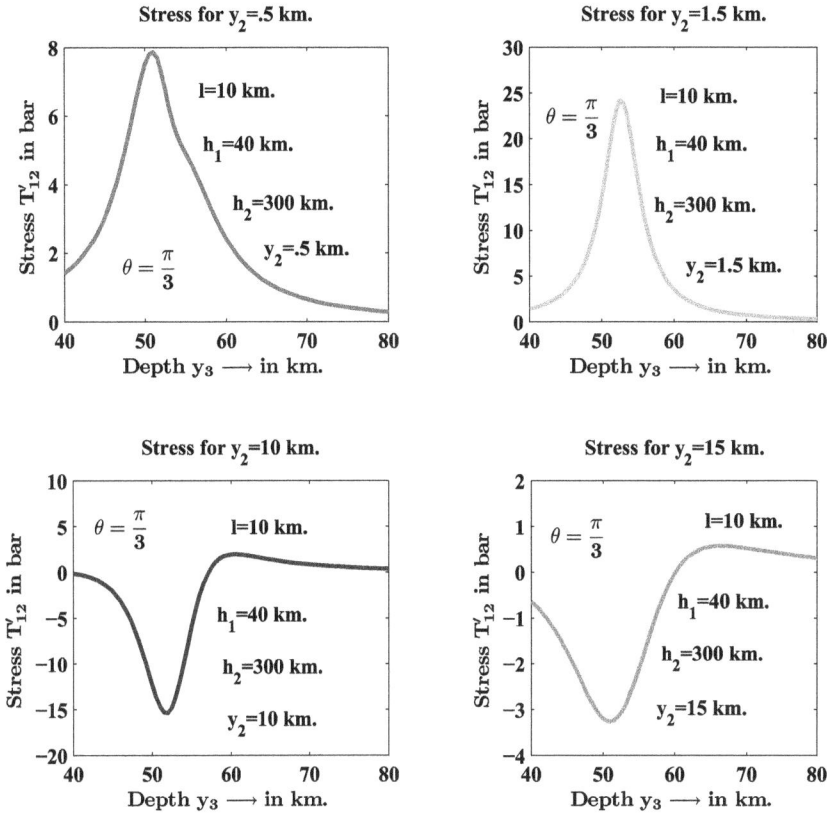

Figure 6.7 Shear stress in the second layer for different y_2 at $\theta_1 = \dfrac{\pi}{3}$.

6.5.2 SHEAR STRESS T'_{12} WITH DIFFERENT WIDTHS OF THE FAULT

We consider a vertical fault with different widths of 10, 15, 20, 25 km. Figure 6.8 shows that the stress accumulation pattern remains almost similar characteristically. Almost the entire second layer region becomes a region of stress release. The depth of the points of the maximum stress release increases with the increase in the width of the fault. The magnitude of the stress release also increases with the increasing l.

APPENDIX

Here we briefly discussed the calculations.

A.1 COMPONENTS OF FORCES IN THE ABSENCE OF ANY FAULT SLIP

Displacements, stresses, and strains in the absence of any fault movement can be obtained using similar methods as explained in the paper by Mondal and Sen (2017)

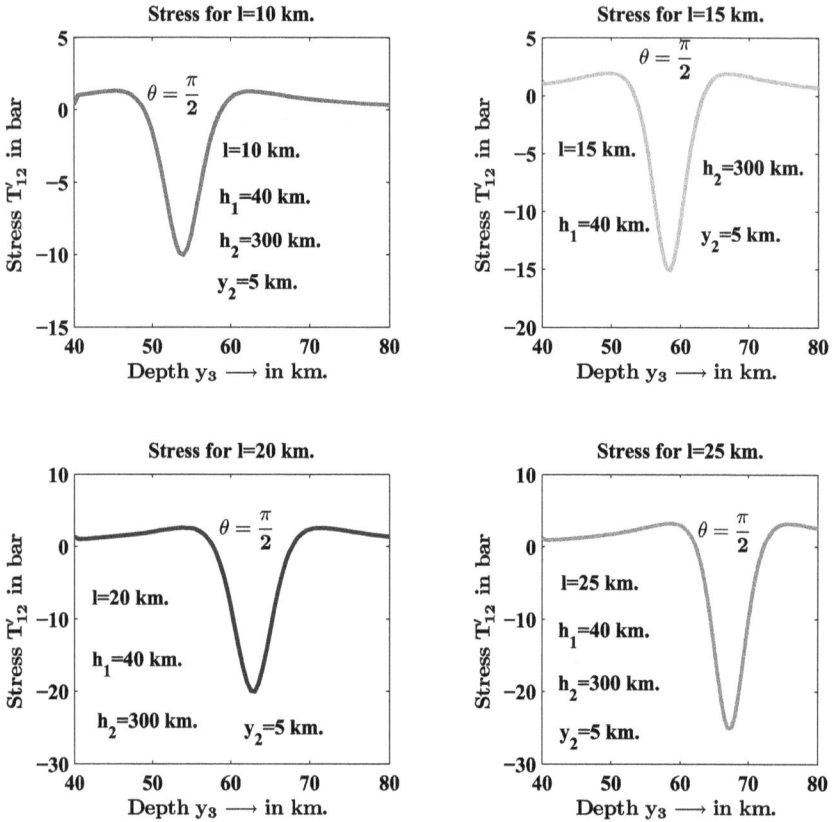

Figure 6.8 Stress pattern in the second layer for different fault length l at $\theta_1 = \dfrac{\pi}{2}$.

A.2 COMPONENTS OF FORCES AFTER THE RESTORATION OF ASEISMIC STATE FOLLOWING A SUDDEN STRIKE-SLIP MOVEMENT ACROSS THE FAULT

The displacements, stresses, and strains after the restoration of a new aseismic state followed by a sudden movement across the fault F can be obtained by following the same technique as explained in the paper by Mondal and Sen (2017). To find the components $(u_1)_2$, $(\tau_{12})_2$, ..., $(\tau_{13}'')_2$ we apply Laplace transforms and we get the boundary value problem, discussed in the paper of Mondal and Sen (2017). To solve this boundary value problem in the transformed domain we apply the modified Green's function technique developed by Maruyama (1966) and Rybicki (1973). Following them we get the solution for $(\bar{u}_1)_2$, $(\bar{u}_1')_2$, $(\bar{u}_1'')_2$ as given in the paper done by Mondal and Sen (2017).

Suppose $P(x_1', x_2', x_3')$ is any point on the fault F with respect to origin $O'(0,0,h_1 + r_1)$ and $P(x_1, x_2, x_3)$ is any point on F with respect to origin O and they are related by the relations

$$\left.\begin{array}{l} x_1 = x_1' \\ x_2 = x_2' \sin \theta_1 + x_3' \cos \theta_1 \\ x_3 = h_1 + r_1 - x_2' \cos \theta_1 + x_3' \sin \theta_1 \end{array}\right\} \tag{6.17}$$

On F : $x_2' = 0$ and $0 \le x_3' \le l$. So $x_2 = x_3' \cos \theta_1$ and $x_3 = h_1 + r_1 + x_3' \sin \theta_1$.

So from the paper done by Mondal and Sen (2017),we get

$$(\bar{u}_1)_2(Q_1,P) = \frac{U}{p} \int_0^l f(x_3') \left[G_{12(1)}^1(Q_1,P) \sin \theta_1 - G_{13(1)}^1(Q_1,P) \cos \theta_1 \right] dx_3'$$

$$\tag{6.18}$$

$$(\bar{u}_1')_2(Q_2,P) = \frac{U}{p} \int_0^l f(x_3') \left[G_{12(2)}^1(Q_2,P) \sin \theta_1 - G_{13(2)}^1(Q_2,P) \cos \theta_1 \right] dx_3'$$

$$\tag{6.19}$$

$$(\bar{u}_1'')_2(Q_3,P) = \frac{U}{p} \int_0^l f(x_3') \left[G_{12(3)}^1(Q_3,P) \sin \theta_1 - G_{13(3)}^1(Q_3,P) \cos \theta_1 \right] dx_3'$$

$$\tag{6.20}$$

where $Q_1(y_1,y_2,y_3)$, $Q_2(y_1,y_2,y_3)$, $Q_3(y_1,y_2,y_3)$ are the field points in the elastic layer, viscoelastic layer, and viscoelastic half-space respectively and $P(x_1,x_2,x_3)$ is any point on the fault F with respect to the origin O.

Where $G_{12(1)}^1$, $G_{12(2)}^1$, $G_{12(3)}^1$ can be obtained from the paper by Mondal and Sen (2017) respectively by replacing

$$\left.\begin{array}{l} d = x_3 + y_3 = h_1 + r_1 + x_3' \sin \theta_1 + y_3 \\ d_1 = x_2 - y_2 = x_3' \cos \theta_1 - y_2 \\ d_2 = y_3 - x_3 = y_3 - h_1 - r_1 - x_3' \sin \theta_1 \end{array}\right\} \tag{6.21}$$

Now we find the values of $G_{13(1)}^1$, $G_{13(2)}^1$ and $G_{13(2)}^1$ which are given by Rybicki (1973) as follows

$$\left.\begin{array}{l} G_{13(2)}^1 = \int_0^\infty [C_2(\lambda)e^{-\lambda y_3} + D_2(\lambda)e^{\lambda y_3}] \cos[\lambda(x_2 - y_2)] d\lambda \\ \quad - \frac{1}{2\pi} \frac{(x_3 - y_3)}{(x_2 - y_2)^2 + (x_3 - y_3)^2} \end{array}\right\} \tag{6.22}$$

$$G_{13(1)}^1 = \int_0^\infty [C_1(\lambda)e^{-\lambda y_3} + D_1(\lambda)e^{\lambda y_3}] \cos[\lambda(x_2 - y_2)] d\lambda \tag{6.23}$$

$$G_{13(3)}^{1} = \int_0^\infty C_3(\lambda)e^{-\lambda y_3}\sin[\lambda(x_2-y_2)]d\lambda \tag{6.24}$$

where

$$C_1 = D_1 = -\frac{\bar{\gamma}_1}{\pi\Delta_2}\left[(\bar{\gamma}_2-1)e^{\lambda(2h_1+x_3)}+(\bar{\gamma}_2+1)e^{\lambda(2h_1+2h_2-x_3)}\right]$$

$$C_2 = -\frac{1}{2\pi\Delta_2}\left[(\bar{\gamma}_1+1)(\bar{\gamma}_2+1)e^{\lambda(2h_1+2h_2-x_3)}+(\bar{\gamma}_1-1)(\bar{\gamma}_2-1)e^{\lambda(4h_1+x_3)}\right.$$
$$\left.+(\bar{\gamma}_1-1)(\bar{\gamma}_2+1)e^{\lambda(4h_1+2h_2-x_3)}+(\bar{\gamma}_1+1)(\bar{\gamma}_2-1)e^{\lambda(2h_1+x_3)}\right]$$

$$D_2 = \frac{(\bar{\gamma}_2-1)}{2\pi\Delta_2}\left\{(\bar{\gamma}_1-1)\left[e^{\lambda(4h_1-x_3)}-e^{\lambda x_3}\right]\right.$$
$$\left.+(\bar{\gamma}_1+1)\left[e^{\lambda(2h_1-x_3)}-e^{\lambda(2h_1+x_3)}\right]\right\}$$

$$C_3 = \frac{1}{\pi\Delta_2}\left\{(\bar{\gamma}_1-1)\left[e^{\lambda(4h_1+2h_2-x_3)}-e^{\lambda(2h_2+x_3)}\right]\right.$$
$$\left.+(\bar{\gamma}_1+1)\left[e^{\lambda(2h_1+2h_2-x_3)}-e^{\lambda(2h_1+2h_2+x_3)}\right]\right\}$$

$$\left.\begin{aligned}\Delta_2 &= (\bar{\gamma}_2-1)e^{2\lambda h_1}\left[(\bar{\gamma}_1+1)+(\bar{\gamma}_1-1)e^{2\lambda h_1}\right]\\&+(\bar{\gamma}_2+1)e^{2\lambda h_2}\left[(\bar{\gamma}_1-1)+(\bar{\gamma}_1+1)e^{2\lambda h_1}\right]\end{aligned}\right\}$$

Now we apply the same technique explained in the paper of Mondal and Sen (2017), we get

$$\frac{(\bar{\gamma}_1+1)(\bar{\gamma}_2+1)}{\Delta_2} = \frac{e^{-2\lambda(h_1+h_2)}}{M}$$

$$\frac{(\bar{\gamma}_1-1)(\bar{\gamma}_2-1)}{\Delta_2} = \frac{a_1c_1e^{-2\lambda(h_1+h_2)}}{M}$$

$$\frac{(\bar{\gamma}_1-1)(\bar{\gamma}_2+1)}{\Delta_2} = \frac{c_1e^{-2\lambda(h_1+h_2)}}{M}$$

$$\frac{(\bar{\gamma}_1+1)(\bar{\gamma}_2-1)}{\Delta_2} = \frac{a_1e^{-2\lambda(h_1+h_2)}}{M}$$

where

$$\left.\begin{aligned}M &= \left(1+a_1e^{-2\lambda h_2}+a_1c_1e^{\lambda(2h_1-2h_2)}+c_1e^{-2\lambda h_1}\right)\\a_1 &= \frac{\bar{\gamma}_2-1}{\bar{\gamma}_2+1},\ c_1 = \frac{\bar{\gamma}_1-1}{\bar{\gamma}_1+1},\ \bar{\gamma}_2 = \frac{\bar{\mu}_3}{\bar{\mu}_2},\ \bar{\gamma}_1 = \frac{\bar{\mu}_1}{\mu_1}\end{aligned}\right\}$$

As stated in the paper by Mondal and Sen (2016), we get

$$M^{-1} = 1 - a_1e^{-2\lambda h_2} - a_1c_1e^{\lambda(2h_1-2h_2)} - c_1e^{-2\lambda h_1}$$

$$
\begin{aligned}
C_2(\lambda)e^{-\lambda y_3} = -\frac{1}{2\pi}\Big[& e^{-\lambda(x_3+y_3)} - a_1 e^{-\lambda(x_3+y_3+2h_2)} \\
& -a_1 c_1 e^{-\lambda(x_3+y_3-2h_1+2h_2)} - c_1 e^{-\lambda(x_3+y_3+2h_1)} \\
& +a_1 c_1 e^{-\lambda(y_3-x_3+2h_2-2h_1)} - a_1^2 c_1 e^{-\lambda(4h_2-2h_1-x_3+y_3)} \\
& -a_1^2 c_1^2 e^{-\lambda(4h_2-4h_1-x_3+y_3)} - a_1 c_1^2 e^{-\lambda(2h_2-x_3+y_3)} \\
& +c_1 e^{-\lambda(x_3+y_3-2h_1)} - a_1 c_1 e^{-\lambda(2h_2-2h_1+x_3+y_3)} \\
& -a_1 c_1^2 e^{-\lambda(2h_2-4h_1+x_3+y_3)} - c_1^2 e^{-\lambda(y_3+x_3)} \\
& +a_1 e^{-\lambda(2h_2-x_3+y_3)} - a_1^2 e^{-\lambda(4h_2-x_3+y_3)} \\
& - a_1^2 c_1 e^{-\lambda(4h_2-2h_1-x_3+y_3)} - a_1 c_1 e^{-\lambda(2h_2+2h_1-x_3+y_3)}\Big]
\end{aligned}
$$

$$
\begin{aligned}
D_2(\lambda)e^{\lambda y_3} = \frac{1}{2\pi}\Big[& a_1 c_1 e^{-\lambda(2h_2-2h_1-y_3+x_3)} - a_1^2 c_1 e^{-\lambda(4h_2-2h_1-y_3+x_3)} \\
& -a_1^2 c_1^2 e^{-\lambda(4h_2-4h_1-y_3+x_3)} - a_1 c_1^2 e^{-\lambda(2h_2-y_3+x_3)} \\
& -a_1 c_1 e^{-\lambda(2h_1+2h_2-x_3-y_3)} + a_1^2 c_1 e^{-\lambda(4h_2+2h_1-x_3-y_3)} \\
& +a_1^2 c_1^2 e^{-\lambda(4h_2-x_3-y_3)} + a_1 c_1^2 e^{-\lambda(4h_1+2h_2-x_3-y_3)} \\
& +a_1 e^{-\lambda(2h_2+x_3-y_3)} - a_1^2 e^{-\lambda(4h_2+x_3-y_3)} \\
& -a_1^2 c_1 e^{-\lambda(4h_2-2h_1+x_3-y_3)} - a_1 c_1 e^{-\lambda(2h_2+2h_1+x_3-y_3)} \\
& -a_1 e^{-\lambda(2h_2-x_3-y_3)} + a_1^2 e^{-\lambda(4h_2-x_3-y_3)} \\
& + a_1^2 c_1 e^{-\lambda(4h_2-2h_1-x_3-y_3)} + a_1 c_1 e^{-\lambda(2h_2+2h_1-x_3-y_3)}\Big]
\end{aligned}
$$

Putting the values of $C_2(\lambda)e^{-\lambda y_3}$ and $D_2(\lambda)e^{\lambda y_3}$ in equation (6.22) we get

$$
\begin{aligned}
G^1_{13(2)} = \frac{1}{2\pi} \int_0^\infty \Bigg[&-e^{-\lambda(x_3+y_3)} + a_1\Big\{ e^{-\lambda(x_3+y_3+2h_2)} \\
&-e^{-\lambda(-x_3+y_3+2h_2)} + e^{-\lambda(x_3-y_3+2h_2)} - e^{-\lambda(-x_3-y_3+2h_2)} \Big\} \\
&+c_1\Big\{ e^{-\lambda(x_3+y_3+2h_1)} - e^{-\lambda(x_3+y_3-2h_1)} \Big\} \\
&+a_1c_1\Big\{ 2e^{-\lambda(x_3+y_3+2h_2-2h_1)} - e^{-\lambda(-x_3+y_3+2h_2-2h_1)} \\
&+e^{-\lambda(-x_3+y_3+2h_2+2h_1)} + e^{-\lambda(x_3-y_3+2h_2-2h_1)} \\
&-e^{-\lambda(x_3-y_3+2h_2+2h_1)} \Big\} + a_1^2\Big\{ e^{-\lambda(-x_3+y_3+4h_2)} \\
&-e^{-\lambda(x_3-y_3+4h_2)} + e^{-\lambda(-x_3-y_3+4h_2)} \Big\} + c_1^2 e^{-\lambda(x_3+y_3)} \\
&+a_1^2c_1\Big\{ 2e^{-\lambda(-x_3+y_3+4h_2-2h_1)} - 2e^{-\lambda(x_3-y_3+4h_2-2h_1)} \\
&+e^{-\lambda(-x_3-y_3+4h_2+2h_1)} + e^{-\lambda(-x_3-y_3+4h_2-2h_1)} \Big\} \\
&+a_1c_1^2\Big\{ e^{-\lambda(-x_3+y_3+2h_2)} + e^{-\lambda(x_3+y_3+2h_2-4h_1)} \\
&-e^{-\lambda(x_3-y_3+2h_2)} + e^{-\lambda(-x_3-y_3+2h_2+4h_1)} \Big\} \\
&+a_1^2c_1^2\Big\{ e^{-\lambda(-x_3+y_3+4h_2-2h_1)} - e^{-\lambda(x_3-y_3+4h_2-4h_1)} \\
&+e^{-\lambda(-x_3-y_3+4h_2)} \Big\} \Bigg] \cos[\lambda(x_2-y_2)]\, d\lambda - \frac{1}{2\pi}\frac{(x_3-y_3)}{(x_2-y_2)^2+(x_3-y_3)^2}
\end{aligned}
$$

$$
\tag{6.25}
$$

On integrating we get,

$$
\begin{aligned}
G^1_{13(2)} = \frac{1}{2\pi} \Bigg[&-\frac{(x_3+y_3)}{(x_3+y_3)^2+(x_2-y_2)^2} + a_1 \Bigg\{ \frac{(2h_2+x_3+y_3)}{(2h_2+x_3+y_3)^2+(x_2-y_2)^2} \\
&-\frac{(2h_2-x_3+y_3)}{(2h_2-x_3+y_3)^2+(x_2-y_2)^2} + \frac{(2h_2+x_3-y_3)}{(2h_2+x_3-y_3)^2+(x_2-y_2)^2} \\
&-\frac{(2h_2-x_3-y_3)}{(2h_2-x_3-y_3)^2+(x_2-y_2)^2} \Bigg\} + c_1 \Bigg\{ \frac{(2h_1+x_3+y_3)}{(2h_1+x_3+y_3)^2+(x_2-y_2)^2} \\
&-\frac{(-2h_1+x_3+y_3)}{(2h_1-x_3-y_3)^2+(x_2-y_2)^2} \Bigg\} \\
&+a_1c_1 \Bigg\{ 2\frac{(2h_2-2h_1+x_3+y_3)}{(2h_2-2h_1+x_3+y_3)^2+(x_2-y_2)^2} \\
&-\frac{(2h_2-2h_1-x_3+y_3)}{(2h_2-2h_1-x_3+y_3)^2+(x_2-y_2)^2} \\
&+\frac{(2h_2+2h_1-x_3+y_3)}{(2h_2+2h_1-x_3+y_3)^2+(x_2-y_2)^2} \\
&+\frac{(2h_2-2h_1+x_3-y_3)}{(2h_2-2h_1+x_3-y_3)^2+(x_2-y_2)^2} \\
&-\frac{(2h_2+2h_1+x_3-y_3)}{(2h_2+2h_1+x_3-y_3)^2+(x_2-y_2)^2} \Bigg\} \\
&+a_1^2 \Bigg\{ \frac{(4h_2-x_3+y_3)}{(4h_2-x_3+y_3)^2+(x_2-y_2)^2} - \frac{(4h_2+x_3-y_3)}{(4h_2+x_3-y_3)^2+(x_2-y_2)^2} \\
&+\frac{(4h_2-x_3-y_3)}{(4h_2-x_3-y_3)^2+(x_2-y_2)^2} \Bigg\} + c_1^2\frac{(x_3+y_3)}{(x_3+y_3)^2+(x_2-y_2)^2} \\
&+a_1^2c_1 \Bigg\{ 2\frac{(4h_2-2h_1-x_3+y_3)}{(4h_2-2h_1-x_3+y_3)^2+(x_2-y_2)^2} \\
&-2\frac{(4h_2-2h_1+x_3-y_3)}{(4h_2-2h_1+x_3-y_3)^2+(x_2-y_2)^2} \\
&+\frac{(4h_2+2h_1-x_3-y_3)}{(4h_2+2h_1-x_3-y_3)^2+(x_2-y_2)^2} \\
&+\frac{(4h_2-2h_1-x_3-y_3)}{(4h_2-2h_1-x_3-y_3)^2+(x_2-y_2)^2} \Bigg\} \\
&+a_1c_1^2 \Bigg\{ \frac{(2h_2-x_3+y_3)}{(2h_2-x_3+y_3)^2+(x_2-y_2)^2} \\
&+\frac{(2h_2-4h_1+x_3+y_3)}{(2h_2-4h_1+x_3+y_3)^2+(x_2-y_2)^2} \\
&-\frac{(4h_1+2h_2+x_3-y_3)}{(4h_1+2h_2+x_3-y_3)^2+(x_2-y_2)^2} \\
&+\frac{(2h_2-x_3-y_3)}{(2h_2-x_3-y_3)^2+(x_2-y_2)^2} \Bigg\} \\
&+a_1^2c_1^2 \Bigg\{ \frac{(4h_2-2h_1-x_3+y_3)}{(4h_2-2h_1-x_3+y_3)^2+(x_2-y_2)^2} \\
&-\frac{(4h_2-4h_1+x_3-y_3)}{(4h_2-4h_1+x_3-y_3)^2+(x_2-y_2)^2} + \frac{(4h_2-x_3-y_3)}{(4h_2-x_3-y_3)^2+(x_2-y_2)^2} \\
&-\frac{(x_3-y_3)}{(x_3-y_3)^2+(x_2-y_2)^2} \Bigg\} \Bigg]
\end{aligned}
$$

$$(6.26)$$

Now putting $x_2 = x_3' \cos \theta_1$ and $x_3 = h_1 + r_1 + x_3' \sin \theta_1$, we get

$$
\begin{aligned}
G^1_{13(2)} = \frac{1}{2\pi} \Bigg[&-\frac{(h_1 + r_1 + x_3' \sin \theta_1 + y_3)}{(h_1 + r_1 + x_3' \sin \theta_1 + y_3)^2 + (x_3' \cos \theta_1 - y_2)^2} \\
&+ a_1 \Bigg\{ \frac{(2h_2 + h_1 + r_1 + x_3' \sin \theta_1 + y_3)}{(2h_2 + h_1 + r_1 + x_3' \sin \theta_1 + y_3)^2 + (x_3' \cos \theta_1 - y_2)^2} \\
&- \frac{(2h_2 - h_1 - r_1 - x_3' \sin \theta_1 + y_3)}{(2h_2 - h_1 - r_1 - x_3' \sin \theta_1 + y_3)^2 + (x_3' \cos \theta_1 - y_2)^2} \\
&+ \frac{(2h_2 + h_1 + r_1 + x_3' \sin \theta_1 - y_3)}{(2h_2 + h_1 + r_1 + x_3' \sin \theta_1 - y_3)^2 + (x_3' \cos \theta_1 - y_2)^2} \\
&- \frac{(2h_2 - h_1 - r_1 - x_3' \sin \theta_1 - y_3)}{(2h_2 - h_1 - r_1 - x_3' \sin \theta_1 - y_3)^2 + (x_3' \cos \theta_1 - y_2)^2} \Bigg\} \\
&+ c_1 \Bigg\{ \frac{(3h_1 + r_1 + x_3' \sin \theta_1 + y_3)}{(3h_1 + r_1 + x_3' \sin \theta_1 + y_3)^2 + (x_3' \cos \theta_1 - y_2)^2} \\
&- \frac{(-h_1 + r_1 + x_3' \sin \theta_1 + y_3)}{(-h_1 + r_1 + x_3' \sin \theta_1 + y_3)^2 + (x_3' \cos \theta_1 - y_2)^2} \Bigg\} \\
&+ a_1 c_1 \Bigg\{ 2\frac{(2h_2 - h_1 + r_1 + x_3' \sin \theta_1 + y_3)}{(2h_2 - h_1 + r_1 + x_3' \sin \theta_1 + y_3)^2 + (x_3' \cos \theta_1 - y_2)^2} \\
&- \frac{(2h_2 - 3h_1 - r_1 - x_3' \sin \theta_1 + y_3)}{(2h_2 - 3h_1 - r_1 - x_3' \sin \theta_1 + y_3)^2 + (x_3' \cos \theta_1 - y_2)^2} \\
&+ \frac{(2h_2 + h_1 - r_1 - x_3' \sin \theta_1 + y_3)}{(2h_2 + h_1 - r_1 - x_3' \sin \theta_1 + y_3)^2 + (x_3' \cos \theta_1 - y_2)^2} \\
&+ \frac{(2h_2 - h_1 + r_1 + x_3' \sin \theta_1 - y_3)}{(2h_2 - h_1 + r_1 + x_3' \sin \theta_1 - y_3)^2 + (x_3' \cos \theta_1 - y_2)^2} \\
&- \frac{(2h_2 + 3h_1 + r_1 + x_3' \sin \theta_1 - y_3)}{(2h_2 + 3h_1 + r_1 + x_3' \sin \theta_1 - y_3)^2 + (x_3' \cos \theta_1 - y_2)^2} \Bigg\} \\
&+ a_1^2 \Bigg\{ \frac{(4h_2 - h_1 - r_1 - x_3' \sin \theta_1 + y_3)}{(4h_2 - h_1 - r_1 - x_3' \sin \theta_1 + y_3)^2 + (x_3' \cos \theta_1 - y_2)^2} \\
&- \frac{(4h_2 + h_1 + r_1 + x_3' \sin \theta_1 - y_3)}{(4h_2 + h_1 + r_1 + x_3' \sin \theta_1 - y_3)^2 + (x_3' \cos \theta_1 - y_2)^2} \\
&+ \frac{(4h_2 - h_1 - r_1 - x_3' \sin \theta_1 - y_3)}{(4h_2 - h_1 - r_1 - x_3' \sin \theta_1 - y_3)^2 + (x_3' \cos \theta_1 - y_2)^2} \Bigg\}
\end{aligned}
$$

continued equation

$$+c_1^2 \frac{(h_1 + r_1 + x_3' \sin\theta_1 + y_3)}{(h_1 + r_1 + x_3' \sin\theta_1 + y_3)^2 + (x_3' \cos\theta_1 - y_2)^2}$$

$$+a_1^2 c_1 \left\{ 2 \frac{(4h_2 - 3h_1 - r_1 - x_3' \sin\theta_1 + y_3)}{(4h_2 - 3h_1 - r_1 - x_3' \sin\theta_1 + y_3)^2 + (x_3' \cos\theta_1 - y_2)^2} \right.$$

$$-2 \frac{(4h_2 - h_1 + r_1 + x_3' \sin\theta_1 - y_3)}{(4h_2 - h_1 + r_1 + x_3' \sin\theta_1 - y_3)^2 + (x_3' \cos\theta_1 - y_2)^2}$$

$$+ \frac{(4h_2 + h_1 - r_1 - x_3' \sin\theta_1 - y_3)}{(4h_2 + h_1 - r_1 - x_3' \sin\theta_1 - y_3)^2 + (x_3' \cos\theta_1 - y_2)^2}$$

$$\left. + \frac{(4h_2 - 3h_1 - r_1 - x_3' \sin\theta_1 - y_3)}{(4h_2 - 3h_1 - r_1 - x_3' \sin\theta_1 - y_3)^2 + (x_3' \cos\theta_1 - y_2)^2} \right\}$$

$$+a_1 c_1^2 \left\{ \frac{(2h_2 - h_1 - r_1 - x_3' \sin\theta_1 + y_3)}{(2h_2 - h_1 - r_1 - x_3' \sin\theta_1 + y_3)^2 + (x_3' \cos\theta_1 - y_2)^2} \right.$$

$$+ \frac{(2h_2 - 3h_1 + r_1 + x_3' \sin\theta_1 + y_3)}{(2h_2 - 3h_1 + r_1 + x_3' \sin\theta_1 + y_3)^2 + (x_3' \cos\theta_1 - y_2)^2}$$

$$- \frac{(5h_1 + 2h_2 + r_1 + x_3' \sin\theta_1 - y_3)}{(5h_1 + 2h_2 + r_1 + x_3' \sin\theta_1 - y_3)^2 + (x_3' \cos\theta_1 - y_2)^2}$$

$$\left. + \frac{(2h_2 - h_1 - r_1 - x_3' \sin\theta_1 - y_3)}{(2h_2 - h_1 - r_1 - x_3' \sin\theta_1 - y_3)^2 + (x_3' \cos\theta_1 - y_2)^2} \right\}$$

$$+a_1^2 c_1^2 \left\{ \frac{(4h_2 - 3h_1 - r_1 - x_3' \sin\theta_1 + y_3)}{(4h_2 - 3h_1 - r_1 - x_3' \sin\theta_1 + y_3)^2 + (x_3' \cos\theta_1 - y_2)^2} \right.$$

$$- \frac{(4h_2 - 3h_1 + r_1 + x_3' \sin\theta_1 - y_3)}{(4h_2 - 3h_1 + r_1 + x_3' \sin\theta_1 - y_3)^2 + (x_3' \cos\theta_1 - y_2)^2}$$

$$+ \frac{(4h_2 - h_1 - r_1 - x_3' \sin\theta_1 - y_3)}{(4h_2 - h_1 - r_1 - x_3' \sin\theta_1 - y_3)^2 + (x_3' \cos\theta_1 - y_2)^2}$$

$$\left. \left. - \frac{(h_1 + r_1 + x_3' \sin\theta_1 - y_3)}{(h_1 + r_1 + x_3' \sin\theta_1 - y_3)^2 + (x_3' \cos\theta_1 - y_2)^2} \right\} \right]$$

Putting the values of $G_{12(2)}^1$ and $G_{13(2)}^1$ in equation (6.19) we get,

$$(\bar{u}_1')_2(Q_2) = \frac{U}{2\pi} \bar{\phi}_1'(y_2, y_3, p) \tag{6.27}$$

where

$$
\bar{\phi}_1'(y_2, y_3, p) = \int_0^l f(x_3') \left[-\frac{(x_3' \cos\theta_1 - y_2)\sin\theta_1}{A_1} \right.
$$

$$
+ a_1 \left\{ \frac{(x_3' \cos\theta_1 - y_2)\sin\theta_1}{A_2} \right.
$$

$$
+ \frac{(x_3' \cos\theta_1 - y_2)\sin\theta_1}{A_3} + \frac{(x_3' \cos\theta_1 - y_2)\sin\theta_1}{A_4}
$$

$$
+ \left. \frac{(x_3' \cos\theta_1 - y_2)\sin\theta_1}{A_5} \right\}
$$

$$
+ c_1 \left\{ \frac{(x_3' \cos\theta_1 - y_2)\sin\theta_1}{A_6} - \frac{(x_3' \cos\theta_1 - y_2)\sin\theta_1}{A_7} \right\}
$$

$$
+ a_1 c_1 \left\{ 2\frac{(x_3' \cos\theta_1 - y_2)\sin\theta_1}{A_8} + \frac{(x_3' \cos\theta_1 - y_2)\sin\theta_1}{A_9} \right.
$$

$$
- \left. \frac{(x_3' \cos\theta_1 - y_2)\sin\theta_1}{A_{10}} + \frac{(x_3' \cos\theta_1 - y_2)\sin\theta_1}{A_{11}} - \frac{(x_3' \cos\theta_1 - y_2)\sin\theta_1}{A_{12}} \right\}
$$

$$
+ a_1^2 \left\{ -\frac{(x_3' \cos\theta_1 - y_2)\sin\theta_1}{A_{13}} - \frac{(x_3' \cos\theta_1 - y_2)\sin\theta_1}{A_{14}} \right.
$$

$$
- \left. \frac{(x_3' \cos\theta_1 - y_2)\sin\theta_1}{A_{15}} \right\} + c_1^2 \frac{(x_3' \cos\theta_1 - y_2)\sin\theta_1}{A_1}
$$

$$
+ a_1^2 c_1 \left\{ -2\frac{(x_3' \cos\theta_1 - y_2)\sin\theta_1}{A_{16}} - \frac{(x_3' \cos\theta_1 - y_2)\sin\theta_1}{A_{17}} \right.
$$

$$
- 2\left. \frac{(x_3' \cos\theta_1 - y_2)\sin\theta_1}{A_{18}} - \frac{(x_3' \cos\theta_1 - y_2)\sin\theta_1}{A_{19}} \right\}
$$

$$
+ a_1 c_1^2 \left\{ -\frac{(x_3' \cos\theta_1 - y_2)\sin\theta_1}{A_3} + \frac{(x_3' \cos\theta_1 - y_2)\sin\theta_1}{A_{20}} \right.
$$

$$
- \left. \frac{(x_3' \cos\theta_1 - y_2)\sin\theta_1}{A_5} - \frac{(x_3' \cos\theta_1 - y_2)\sin\theta_1}{A_{21}} \right\}
$$

$$
+ a_1^2 c_1^2 \left\{ -\frac{(x_3' \cos\theta_1 - y_2)\sin\theta_1}{A_{22}} - \frac{(x_3' \cos\theta_1 - y_2)\sin\theta_1}{A_{23}} \right.
$$

$$
- \frac{(x_3' \cos\theta_1 - y_2)\sin\theta_1}{A_{15}} - \frac{(x_3' \cos\theta_1 - y_2)\sin\theta_1}{A_{24}}
$$

$$
+ \left. \frac{(x_3' \cos\theta_1 - y_2)\sin\theta_1}{A_1} \right\} - a_1 \left\{ \frac{(2h_2 + h_1 + r_1 + x_3' \sin\theta_1 + y_3)\cos\theta_1}{A_2} \right.
$$

$$
- \frac{(2h_2 - h_1 - r_1 - x_3' \sin\theta_1 + y_3)\cos\theta_1}{A_3}
$$

$$
+ \frac{(2h_2 + h_1 + r_1 + x_3' \sin\theta_1 - y_3)\cos\theta_1}{A_5}
$$

$$
- \left. \frac{(2h_2 - h_1 - r_1 - x_3' \sin\theta_1 - y_3)\cos\theta_1}{A_4} \right\}
$$

$$
- c_1 \left\{ \frac{(3h_1 + r_1 + x_3' \sin\theta_1 + y_3)\cos\theta_1}{A_6} \right.
$$

$$
+ \left. \frac{(h_1 - r_1 - x_3' \sin\theta_1 - y_3)\cos\theta_1}{A_7} \right\}
$$

$$
\text{(6.28)}
$$

continued equation (6.28)

$$
\begin{aligned}
&+a_1 c_1 \left\{ 2\frac{(2h_2 - h_1 + r_1 + x_3' \sin\theta_1 + y_3)\cos\theta_1}{A_8} \right. \\
&\quad -\frac{(2h_2 - 3h_1 - r_1 - x_3' \sin\theta_1 + y_3)\cos\theta_1}{A_9} \\
&\quad +\frac{(2h_2 + h_1 - r_1 - x_3' \sin\theta_1 + y_3)\cos\theta_1}{A_{10}} + \frac{(2h_2 - h_1 + r_1 + x_3' \sin\theta_1 - y_3)\cos\theta_1}{A_{11}} \\
&\quad \left. -\frac{(2h_2 + 3h_1 + r_1 + x_3' \sin\theta_1 - y_3)\cos\theta_1}{A_{12}} \right\} \\
&-a_1^2 \left\{ \frac{(4h_2 - h_1 - r_1 - x_3' \sin\theta_1 + y_3)\cos\theta_1}{A_{13}} - \frac{(4h_2 + h_1 + r_1 + x_3' \sin\theta_1 - y_3)\cos\theta_1}{A_{14}} \right. \\
&\quad \left. +\frac{(4h_2 - h_1 - r_1 - x_3' \sin\theta_1 - y_3)\cos\theta_1}{A_{15}} \right\} - c_1^2 \frac{(h_1 + r_1 + x_3' \sin\theta_1 + y_3)\cos\theta_1}{A_1} \\
&-a_1^2 c_1 \left\{ 2\frac{(4h_2 - 3h_1 - r_1 - x_3' \sin\theta_1 + y_3)\cos\theta_1}{A_{18}} \right. \\
&\quad -2\frac{(4h_2 - h_1 + r_1 + x_3' \sin\theta_1 - y_3)\cos\theta_1}{A_{16}} + \frac{(4h_2 + h_1 - r_1 - x_3' \sin\theta_1 - y_3)\cos\theta_1}{A_{17}} \\
&\quad \left. +\frac{(4h_2 - 3h_1 - r_1 - x_3' \sin\theta_1 - y_3)\cos\theta_1}{A_{19}} \right\} \\
&-a_1 c_1^2 \left\{ \frac{(2h_2 - h_1 - r_1 - x_3' \sin\theta_1 + y_3)\cos\theta_1}{A_3} \right. \\
&\quad +\frac{(2h_2 - 3h_1 + r_1 + x_3' \sin\theta_1 + y_3)\cos\theta_1}{A_{20}} - \frac{(2h_2 + 5h_1 + r_1 + x_3' \sin\theta_1 - y_3)\cos\theta_1}{A_{25}} \\
&\quad \left. +\frac{(2h_2 - h_1 - r_1 - x_3' \sin\theta_1 - y_3)\cos\theta_1}{A_4} \right\} \\
&-a_1^2 c_1^2 \left\{ \frac{(4h_2 - 3h_1 - r_1 - x_3' \sin\theta_1 + y_3)\cos\theta_1}{A_{18}} \right. \\
&\quad -\frac{(4h_2 - 3h_1 + r_1 + x_3' \sin\theta_1 - y_3)\cos\theta_1}{A_{23}} + \frac{(4h_2 - h_1 - r_1 - x_3' \sin\theta_1 - y_3)\cos\theta_1}{A_{15}} \\
&\quad \left. +\frac{(h_1 + r_1 + x_3' \sin\theta_1 - y_3)\cos\theta_1}{A_{24}} \right] dx_3'
\end{aligned}
$$

Taking Laplace inverse transformation of equation (6.27) we get

$$
(u_1')_2(Q_2) = \frac{U}{2\pi}\phi_1'(y_2, y_3, t) \tag{6.29}
$$

where

$$\phi_1'(y_2, y_3, t) = L^{-1}\left\{ \bar\phi_1'(y_2, y_3, p) \right\} = \int_0^l f(x_3')\left[-\frac{(x_3' \cos\theta_1 - y_2)\sin\theta_1}{A_1} \right.$$

$$+ \left(\frac{a_3}{a_4} + \frac{a_2 a_5 - a_3 a_4}{a_4 a_5} e^{-\frac{a_5 t}{a_4}} \right)\left\{ \frac{(x_3' \cos\theta_1 - y_2)\sin\theta_1}{A_2} + \frac{(x_3' \cos\theta_1 - y_2)\sin\theta_1}{A_3} \right.$$

$$+ \frac{(x_3' \cos\theta_1 - y_2)\sin\theta_1}{A_4} + \frac{(x_3' \cos\theta_1 - y_2)\sin\theta_1}{A_5} \Bigg\}$$

$$+ \left\{ -1 + \left(1 + \frac{a_6}{a_8} \right)e^{-\frac{a_7}{a_8}t} \right\}\left\{ \frac{(x_3' \cos\theta_1 - y_2)\sin\theta_1}{A_6} + \frac{(x_3' \cos\theta_1 - y_2)\sin\theta_1}{A_7} \right\}$$

$$+ \left\{ K + \frac{T}{a_4}e^{-\frac{a_5}{a_4}t} + \frac{V}{a_8}e^{-\frac{a_7}{a_8}t} \right\}\left\{ 2\frac{(x_3' \cos\theta_1 - y_2)\sin\theta_1}{A_8} \right.$$

$$+ \frac{(x_3' \cos\theta_1 - y_2)\sin\theta_1}{A_9} - \frac{(x_3' \cos\theta_1 - y_2)\sin\theta_1}{A_{10}}$$

$$+ \frac{(x_3' \cos\theta_1 - y_2)\sin\theta_1}{A_{11}} - \frac{(x_3' \cos\theta_1 - y_2)\sin\theta_1}{A_{12}} \Bigg\}$$

$$+ \left\{ s_1 + \frac{s_2}{a_4}e^{-\frac{a_5}{a_4}t} + \frac{s_3 t}{a_4^2}e^{-\frac{a_5}{a_4}t} \right\}$$

$$\times \left\{ -\frac{(x_3' \cos\theta_1 - y_2)\sin\theta_1}{A_{13}} - \frac{(x_3' \cos\theta_1 - y_2)\sin\theta_1}{A_{14}} - \frac{(x_3' \cos\theta_1 - y_2)\sin\theta_1}{A_{15}} \right\}$$

$$+ \left\{ 1 + \frac{s_5}{a_8}e^{-\frac{a_7 t}{a_8}} + \frac{s_6 t}{a_8^2}e^{-\frac{a_7 t}{a_8}} \right\}\frac{(x_3' \cos\theta_1 - y_2)\sin\theta_1}{A_1}$$

$$+ \left\{ s_7 + \frac{s_8}{a_4}e^{-\frac{a_5}{a_4}t} + \frac{s_9 t}{a_4^2}e^{-\frac{a_5}{a_4}t} + \frac{s_{10}}{a_8}e^{-\frac{a_7}{a_8}t} \right\}$$

$$\times \left\{ -2\frac{(x_3' \cos\theta_1 - y_2)\sin\theta_1}{A_{16}} - \frac{(x_3' \cos\theta_1 - y_2)\sin\theta_1}{A_{17}} \right.$$

$$-2\frac{(x_3' \cos\theta_1 - y_2)\sin\theta_1}{A_{18}} - \frac{(x_3' \cos\theta_1 - y_2)\sin\theta_1}{A_{19}} \Bigg\}$$

$$+ \left\{ s_{11} + \frac{s_{12}}{a_4}e^{-\frac{a_5}{a_4}t} + \frac{s_{13}}{a_8}e^{-\frac{a_7}{a_8}t} + \frac{s_{14} t}{a_8^2}e^{-\frac{a_7}{a_8}t} \right\}$$

$$\times \left\{ -\frac{(x_3' \cos\theta_1 - y_2)\sin\theta_1}{A_3} + \frac{(x_3' \cos\theta_1 - y_2)\sin\theta_1}{A_{20}} \right.$$

$$- \frac{(x_3' \cos\theta_1 - y_2)\sin\theta_1}{A_5} - \frac{(x_3' \cos\theta_1 - y_2)\sin\theta_1}{A_{21}} \Bigg\}$$

$$+ \left\{ s_{15} + \frac{s_{16}}{a_4}e^{-\frac{a_5}{a_4}t} + \frac{s_{17} t}{a_4^2}e^{-\frac{a_5}{a_4}t} + \frac{s_{18}}{a_8}e^{-\frac{a_7}{a_8}t} + \frac{s_{19} t}{a_8^2}te^{-\frac{a_7}{a_8}t} \right\}$$

$$\times \left\{ -\frac{(x_3' \cos\theta_1 - y_2)\sin\theta_1}{A_{22}} - \frac{(x_3' \cos\theta_1 - y_2)\sin\theta_1}{A_{23}} - \frac{(x_3' \cos\theta_1 - y_2)\sin\theta_1}{A_{15}} \right\}$$

$$\tag{6.30}$$

continued equation (6.30)

$$
\begin{aligned}
&-\frac{(x_3'\cos\theta_1 - y_2)\sin\theta_1}{A_{24}} + \frac{(h_1 + r_1 + x_3'\sin\theta_1 + y_3)\cos\theta_1}{A_1} \\
&- \left(\frac{a_3}{a_4} + \frac{a_2 a_5 - a_3 a_4}{a_4 a_5} e^{-\frac{a_5 t}{a_4}}\right) \left\{ \frac{(2h_2 + h_1 + r_1 + x_3'\sin\theta_1 + y_3)\cos\theta_1}{A_2} \right. \\
&- \frac{(2h_2 - h_1 - r_1 - x_3'\sin\theta_1 + y_3)\cos\theta_1}{A_3} + \frac{(2h_2 + h_1 + r_1 + x_3'\sin\theta_1 - y_3)\cos\theta_1}{A_5} \\
&\left. - \frac{(2h_2 - h_1 - r_1 - x_3'\sin\theta_1 - y_3)\cos\theta_1}{A_4} \right\} - \left\{ -1 + (1 + \frac{a_6}{a_8})e^{-\frac{a_7 t}{a_8}} \right\} \\
&\times \left\{ \frac{(3h_1 + r_1 + x_3'\sin\theta_1 + y_3)\cos\theta_1}{A_6} \right\} + \frac{(h_1 - r_1 - x_3'\sin\theta_1 - y_3)\cos\theta_1}{A_7} \\
&- \left\{ K + \frac{T}{a_4}e^{-\frac{a_5}{a_4}t} + \frac{V}{a_8}e^{-\frac{a_7}{a_8}t} \right\} \left\{ 2\frac{(2h_2 - h_1 + r_1 + x_3'\sin\theta_1 + y_3)\cos\theta_1}{A_8} \right. \\
&- \frac{(2h_2 - 3h_1 - r_1 - x_3'\sin\theta_1 + y_3)\cos\theta_1}{A_9} + \frac{(2h_2 + h_1 - r_1 - x_3'\sin\theta_1 + y_3)\cos\theta_1}{A_{10}} \\
&\left. + \frac{(2h_2 - h_1 + r_1 + x_3'\sin\theta_1 - y_3)\cos\theta_1}{A_{11}} - \frac{(2h_2 + 3h_1 + r_1 + x_3'\sin\theta_1 - y_3)\cos\theta_1}{A_{12}} \right\} \\
&- \left\{ s_1 + \frac{s_2}{a_4}e^{-\frac{a_5}{a_4}t} + \frac{s_3 t}{a_4^2}e^{-\frac{a_5}{a_4}t} \right\} \left\{ \frac{(4h_2 - h_1 - r_1 - x_3'\sin\theta_1 + y_3)\cos\theta_1}{A_{13}} \right. \\
&\left. - \frac{(4h_2 + h_1 + r_1 - x_3'\sin\theta_1 - y_3)\cos\theta_1}{A_{14}} + \frac{(4h_2 - h_1 - r_1 - x_3'\sin\theta_1 - y_3)\cos\theta_1}{A_{15}} \right\} \\
&- \left\{ 1 + \frac{s_5}{a_8}e^{-\frac{a_7 t}{a_8}} + \frac{s_6 t}{a_8^2}e^{-\frac{a_7 t}{a_8}} \right\} \frac{(h_1 + r_1 + x_3'\sin\theta_1 + y_3)\cos\theta_1}{A_1} \\
&- \left\{ s_7 + \frac{s_8}{a_4}e^{-\frac{a_5}{a_4}t} + \frac{s_9 t}{a_4^2}e^{-\frac{a_5}{a_4}t} + \frac{s_{10}}{a_8}e^{-\frac{a_7}{a_8}t} \right\} \\
&\times \left\{ 2\frac{(4h_2 - 3h_1 - r_1 - x_3'\sin\theta_1 + y_3)\cos\theta_1}{A_{18}} - 2\frac{(4h_2 - h_1 + r_1 + x_3'\sin\theta_1 - y_3)\cos\theta_1}{A_{16}} \right. \\
&\left. + \frac{(4h_2 + h_1 - r_1 - x_3'\sin\theta_1 - y_3)\cos\theta_1}{A_{17}} + \frac{(4h_2 - 3h_1 - r_1 - x_3'\sin\theta_1 - y_3)\cos\theta_1}{A_{19}} \right\} \\
&- \left\{ s_{11} + \frac{s_{12}}{a_4}e^{-\frac{a_5}{a_4}t} + \frac{s_{13}}{a_8}e^{-\frac{a_7}{a_8}t} + \frac{s_{14} t}{a_8^2}e^{-\frac{a_7}{a_8}t} \right\} \\
&\times \left\{ \frac{(2h_2 - h_1 - r_1 - x_3'\sin\theta_1 + y_3)\cos\theta_1}{A_3} + \frac{(2h_2 - 3h_1 + r_1 + x_3'\sin\theta_1 + y_3)\cos\theta_1}{A_{20}} \right. \\
&\left. - \frac{(2h_2 + 5h_1 + r_1 + x_3'\sin\theta_1 - y_3)\cos\theta_1}{A_{25}} + \frac{(2h_2 - h_1 - r_1 - x_3'\sin\theta_1 - y_3)\cos\theta_1}{A_4} \right\} \\
&- \left\{ s_{15} + \frac{s_{16}}{a_4}e^{-\frac{a_5}{a_4}t} + \frac{s_{17} t}{a_4^2}e^{-\frac{a_5}{a_4}t} + \frac{s_{18}}{a_8}e^{-\frac{a_7}{a_8}t} + \frac{s_{19} t}{a_8^2}e^{-\frac{a_7}{a_8}t} \right\} \\
&\times \left\{ \frac{(4h_2 - 3h_1 - r_1 - x_3'\sin\theta_1 + y_3)\cos\theta_1}{A_{18}} - \frac{(4h_2 - 3h_1 + r_1 + x_3'\sin\theta_1 - y_3)\cos\theta_1}{A_{23}} \right. \\
&\left. + \frac{(4h_2 - h_1 - r_1 - x_3'\sin\theta_1 - y_3)\cos\theta_1}{A_{15}} \right\} + \left\{ \frac{(h_1 + r_1 + x_3'\sin\theta_1 - y_3)\cos\theta_1}{A_{24}} \right\} \Bigg] dx_3'
\end{aligned}
$$

The stress after the restoration of new aseismic state followed by a sudden movement across the fault F can be obtained by following the same technique as explained in the paper by Mondal and Sen (2017). To find the component $(\tau_{12})_2$ we apply Laplace transform and we get

$$\left.\begin{aligned}(\bar{\tau}'_{12})_2 &= \bar{\mu}_2 \frac{\partial(\bar{u}'_1)_2}{\partial y_2} \\ &= \frac{U}{2\pi}\bar{\phi}'_2(y_2,y_3,p)\end{aligned}\right\} \tag{6.31}$$

Taking Laplace inverse transformation of equation (6.31) we get

$$(\tau'_{12})_2 = \frac{U}{2\pi}\phi'_2(y_2,y_3,t) \tag{6.32}$$

Similarly,

$$\left.\begin{aligned}(\bar{\tau}'_{13})_2 &= \bar{\mu}_2 \frac{\partial(\bar{u}'_1)_2}{\partial y_3} \\ &= \frac{U}{2\pi}\bar{\phi}'_3(y_2,y_3,p)\end{aligned}\right\} \tag{6.33}$$

Taking Laplace inverse transformation of equation (6.33) we get

$$(\tau'_{13})_2 = \frac{\mu_2 U}{2\pi}\phi'_3(y_2,y_3,t) \tag{6.34}$$

In similar way we also derived the expressions $\psi'_1(y_2,y_3)$, $\psi'_2(y_2,y_3)$, $\psi_3('y_2,y_3)$ for elastic layer and $\chi'_1(y_2,y_3)$, $\chi'_2(y_2,y_3)$, $\chi'_3(y_2,y_3)$ for viscoelastic half-space, in which $s_{46},s_{46},\ldots,s_{94}$ are given in the paper of Mondal and Sen (2017) and $A_1,A_2,\ldots,A_{28}$ are given below.
where

$$\left.\begin{aligned}A_1 &= (h_1+r_1+x'_3\sin\theta_1+y_3)^2+(x'_3\cos\theta_1-y_2)^2 \\ A_2 &= (2h_2+h_1+r_1+x'_3\sin\theta_1+y_3)^2+(x'_3\cos\theta_1-y_2)^2 \\ A_3 &= (2h_2-h_1-r_1-x'_3\sin\theta_1+y_3)^2+(x'_3\cos\theta_1-y_2)^2 \\ A_4 &= (2h_2-h_1-r_1-x'_3\sin\theta_1-y_3)^2+(x'_3\cos\theta_1-y_2)^2 \\ A_5 &= (2h_2+h_1+r_1+x'_3\sin\theta_1-y_3)^2+(x'_3\cos\theta_1-y_2)^2 \\ A_6 &= (3h_1+r_1+x'_3\sin\theta_1+y_3)^2+(x'_3\cos\theta_1-y_2)^2 \\ A_7 &= (h_1-r_1-x'_3\sin\theta_1-y_3)^2+(x'_3\cos\theta_1-y_2)^2 \\ A_8 &= (2h_2-2h_1+h_1+r_1+x'_3\sin\theta_1+y_3)^2+(x'_3\cos\theta_1-y_2)^2 \\ A_9 &= (2h_2-3h_1-r_1-x'_3\sin\theta_1+y_3)^2+(x'_3\cos\theta_1-y_2)^2 \\ A_{10} &= (2h_2+h_1-r_1-x'_3\sin\theta_1+y_3)^2+(x'_3\cos\theta_1-y_2)^2 \\ A_{11} &= (2h_2-h_1+r_1+x'_3\sin\theta_1-y_3)^2+(x'_3\cos\theta_1-y_2)^2 \\ A_{12} &= (2h_2+3h_1+r_1+x'_3\sin\theta_1-y_3)^2+(x'_3\cos\theta_1-y_2)^2\end{aligned}\right\} \tag{6.35}$$

continued equation (6.35)

$$
\begin{aligned}
A_{13} &= (4h_2 - h_1 - r_1 - x_3' \sin\theta_1 + y_3)^2 + (x_3' \cos\theta_1 - y_2)^2 \\
A_{14} &= (4h_2 + h_1 + r_1 + x_3' \sin\theta_1 - y_3)^2 + (x_3' \cos\theta_1 - y_2)^2 \\
A_{15} &= (4h_2 - h_1 - r_1 - x_3' \sin\theta_1 - y_3)^2 + (x_3' \cos\theta_1 - y_2)^2 \\
A_{16} &= (4h_2 - h_1 + r_1 + x_3' \sin\theta_1 - y_3)^2 + (x_3' \cos\theta_1 - y_2)^2 \\
A_{17} &= (4h_2 + h_1 - r_1 - x_3' \sin\theta_1 - y_3)^2 + (x_3' \cos\theta_1 - y_2)^2 \\
A_{18} &= (4h_2 - 3h_1 - r_1 - x_3' \sin\theta_1 + y_3)^2 + (x_3' \cos\theta_1 - y_2)^2 \\
A_{19} &= (4h_2 - 3h_1 - r_1 - x_3' \sin\theta_1 - y_3)^2 + (x_3' \cos\theta_1 - y_2)^2 \\
A_{20} &= (2h_2 - 3h_1 + r_1 + x_3' \sin\theta_1 + y_3)^2 + (x_3' \cos\theta_1 - y_2)^2 \\
A_{21} &= (2h_2 + 3h_1 - r_1 - x_3' \sin\theta_1 - y_3)^2 + (x_3' \cos\theta_1 - y_2)^2 \\
A_{22} &= (4h_2 - 5h_1 - r_1 - x_3' \sin\theta_1 + y_3)^2 + (x_3' \cos\theta_1 - y_2)^2 \\
A_{23} &= (4h_2 - 3h_1 + r_1 + x_3' \sin\theta_1 - y_3)^2 + (x_3' \cos\theta_1 - y_2)^2 \\
A_{24} &= (h_1 + r_1 + x_3' \sin\theta_1 - y_3)^2 + (x_3' \cos\theta_1 - y_2)^2 \\
A_{25} &= (2h_2 + 5h_1 + r_1 + x_3' \sin\theta_1 - y_3)^2 + (x_3' \cos\theta_1 - y_2)^2 \\
A_{26} &= (3h_1 + r_1 + x_3' \sin\theta_1 - y_3)^2 + (x_3' \cos\theta_1 - y_2)^2 \\
A_{27} &= (2h_2 + h_1 - r_1 - x_3' \sin\theta_1 - y_3)^2 + (x_3' \cos\theta_1 - y_2)^2 \\
A_{28} &= (3h_1 - r_1 - x_3' \sin\theta_1 + y_3)^2 + (x_3' \cos\theta_1 - y_2)^2
\end{aligned}
$$

REFERENCES

Debnath, P. and Sen, S. (2014). Creeping movement across a long strike-slip fault in a half- space of linear viscoelastic material representing the lithosphere-asthenosphere system. *Frontiers in Science*, 4(2):21–28.

Debnath, P. and Sen, S. (2015a). A finite rectangular strike-slip fault in a linear viscoelastic half space creeping under tectonic forces. *International Journal of Current Research*, 7(07):18365–18373.

Debnath, P. and Sen, S. (2015b). A vertical creeping strike-slip fault in a viscoelastic half- space under the action of tectonic forces varying with time. *IOSR Journal of Mathematics (IOSR-JM)*, 11(3):105–114.

Debnath, S. and Sen, S. (2013a). Aseismic ground deformation in a viscoelastic layer overlying a viscoelastic half-space model of the lithosphere-asthenosphere system. *Geosciences*, 2(3):60–67.

Debnath, S. and Sen, S. (2013b). Pattern of stress-strain accumulation due to a long dip-slip fault movement in a viscoelastic layer over a viscoelastic half-space model of the lithosphere–asthenosphere system. *International Journal of Applied Mechanics and Engineering*, 18(3):653–670.

Debnath, S. and Sen, S. (2013c). Two interacting creeping vertical rectangular strike-slip faults in a viscoelastic half-space model of the lithosphere. *International Journal of Scientific and Engineering Research*, 4(6):1058–1071.

Karmakar, A. and Sen, S. (2016). A sudden movement across an inclined surface breaking strike-slip fault in an elastic layer overlying a viscoelastic layer and a viscoelastic half-space. *IOSR Journal of Applied Geology and Geophysics (IOSR-JAGG)*, 4(5):39–58.

Manna, K. and Sen, S. (2019). Stress-strain accumulation due to interactions of finite and long strike-slip faults in a viscoelastic half-space. *Bulletin of the Calcutta Mathematical Society*, 111(2):173–198.

Manna, K., Sen, S., and Ghosh, U. (2019). Interactions among finite rectangular faults in a viscoelastic half-space. *IOSR Journal of Applied Geology and Geophysics (IOSR-JAGG)*, 7(5):33–42.

Maruyama, T. (1966). On two dimensional dislocation in an infinite and semi-infinite medium. *Bulletin of Earthquake Research Institute*, 44:811–871.

Mondal, B. and Sen, S. (2016). Long vertical strike-slip fault in a multi-layered elastic media. *Geosciences*, 6(2):29–40.

Mondal, B. and Sen, S. (2017). Pattern of stress accumulation due to a sudden movement across a long vertical strike-slip fault in a three-layered elastic/ viscoelastic model. *IOSR-JAGG*, 5(I):34–39.

Mondal, D. and Debnath, P. (2021). An application of fractional calculus to geophysics: Effect of a strike-slip fault on displacement, stresses and strains in a fractional order maxwell type viscoelastic half-space. *International Journal of Applied Mathematics*, 34(5):873–888.

Mondal, S., Sen, S., and Debsarma, S. (2018a). A mathematical model to study the stress distribution due to a strike slip fault creeping with a reducing velocity. *Bulletin of the Calcutta Mathematical Society*, 110(4):265–280.

Mondal, S., Sen, S., and Debsarma, S. (2018b). A numerical approach for solution of aseismic ground deformation problem. *Journal of Geoscience and Geomatics*, 6(1):27–34.

Mondal, S., Sen, S., and Debsarma, S. (2018c). A numerical approach to determine the ground deformation due to creeping movement across a long dip-slip fault. *Bulletin of the Calcutta Mathematical Society*, 110(6):541–564.

Mondal, S., Sen, S., and Debsarma, S. (2019). A mathematical model fro analyzing the ground deformation due to a creeping movement across a strike-slip fault. *International Journal on Geomathematics*, 10:16.

Mukhopadhyay, A. E. A. (1984). On two interacting creeping vertical surface breaking strike-slip fault in the lithosphere. *Bulletin of the Indian Society of Earthquake Technology*, 21:163–191.

Mukhopadhyay, A. E. A. (1986). On two aseismically creeping and interacting buried vertical strike-slip faults in the lithosphere. *Bulletin of the Indian Society of Earthquake Technology*, 23:91–117.

Rybicki, K. (1971). The elastic residual field of a very long strike-slip fault in the presence of a discontinuity. *Bulletin of the Seismological Society of America*, 61:79–92.

Rybicki, K. (1973). Static deformation of a multi-layered half-space by very long strike slip fault. *Pure and Applied Geophysics*, 110:1955–1966.

Sen, S. and Debnath, S. (2012). A creeping vertical strike-slip fault of finite length in a viscoelastic half-space model of the lithosphere. *International Journal of Computing*, 2(3):687–697.

7 Combined Study on Time-Dependent Deterioration and Carbon Emission for Fixed Lifetime Substitutable/ Complementary Product in a Sustainable Supply Chain Management

Nilkamal Bar
Banasthali Vidyapith

Sharmila Saren
Government General Degree College

Isha Sangal
Banasthali Vidyapith

Biswajit Sarkar
Yonsei University
Saveetha University

CONTENTS

DOI: 10.1201/9781003227847-7

7.1   INTRODUCTION

In our day-to-day, we use so many products, some of which can be substituted by another product and some others are the complement of some other product. For example, generally, we start our day with brush and toothpaste, which are a complement to each other or may start with reading a newspaper say 'The Times of India', one can substitute it with another newspaper say 'The statesman'. That means products can be classified into two categories, (1) Substitutable product means if customers' liked product is not available in stock, they will buy a product which is similar to the original one, called its substitute (Fang et al., 2021). For example, if a customer likes to purchase cosmetics like type-A shampoo but for its unavailability, he/she will buy type-B shampoo. (2) Complementary product means when a customer buys a product he may be motivated to purchase another product which can be used together by the customer (Hemmati et al., 2018). For example, if a customer buys shampoo for hair care, he will buy a conditioner. Again, the demand for both types of products is directly related to their pricing. If we go through some examples, we can see that in the case of the substitutable product if the price of the perfume of type-A is unavailable in stock or its price is increased then demand for the perfume of type-B will be increased. Also, in the case of the complementary product if the demand for shampoo decreases due to its price increases or unavailability in stock then it will be the reason for the decrease in demand for conditioners. This fact is named after the cross-price elasticity of demand. Where for substitutable goods we call it positive cross-price elasticity of demand and for complementary goods, we call it negative cross-price elasticity of demand.

In the current business world, it is hard to get more profit working as an individual in the supply chain. Cooperation among the supply chain members plays an effective role in increasing the system's total profit. But the exchange of incomplete information and conflicting interest among the supply chain members make the supply chain disorganized (Halat et al., 2021). In the supply chain, some companies producing substitutable and complementary products collaborate with a common retailer to maximize their profit (Wang et al., 2021). Thus, the supply chain members need to follow suitable cooperation strategies to optimize their utility.

In the business field to incline customers' attention, retailers need to a stock variety of items as much as they can (Sing et al., 2017). Some researchers assume that products remain unchanged over time in their classical inventory management models (Moubed et al., 2021). But in a real situation, we are confronted by the fact that some products like fruits, liquid medicine, alcohol, and blood are volatile and difficult to be preserved by any manufacturing sector (Li et al., 2019) from perishing, declining, evaporating, and meeting the deadlines of consumption, and this fact is known as deterioration (Tiwari et al., 2021). Product deterioration occurs not only because of holding but also because of time consumed in long-distance transportation, and variability in demand and supply (Haung et al., 2018). In all these cases retailers, as well as the whole supply chain, have to confront high loss for this phenomenon. For example, every year, more than 1.3 billion tons of food are wasted or perished which is equivalent to one-third of food produced globally.[1]

The definition of sustainability from the point of view of economics is the specification of a set of actions to be taken by present persons that will not diminish the prospects of future persons to enjoy levels of consumption, wealth, utility, or welfare comparable to those enjoyed by present persons. Economic models of sustainability seek axiomatic guidance for the selection of rules regarding natural resource use (Geda et al., 2020). Though, from the definition, we can observe that there are many schedules to accomplish sustainability, in this study, we go through the process of abetting carbon emission and properly disposing of waste. By the time e-waste, e.g. TVs, fridges, mobile phones, computers, etc. become headaches for developing countries like India as in 2019–2020, the amount of e-waste have crossed one million tons.[2] Lack of proper disposal of this waste can be toxic, risky to the nearby environment, and become the origin of pollution (Yadav et al., 2021). The Union Ministry of

Environment, Forests, and Climate Change (MoEF&CC), India, published the new Solid Waste Management Rules (SWM), 2016.[3]

In this chapter, it is considered that both the manufacturers produce inventory in a single setup and supply in the account of the retailer by multi-shipment, which is named after the single-setup-multi-delivery (SSMD) policy. Following this policy, the number of transportation has increased thereby increasing the cost. But it has reduced holding expenses for the retailer, i.e., there is contraposition within holding inventory and the number of transportations. As the number of transportations increases, the emission due to it also increases. But Sarkar et al. (2016) investigated that with the adaptation of SSMD policy, it mitigated the amount of emission to the environment due to transportation and holding of products at the retailer stock.

Another issue is carbon emission as this is the most liable element for global warming. Activities related to members of the supply chain are responsible for carbon emissions. The present study considers two manufacturers that produce substitutable/complementary products and supply them to a common retailer following the SSMD policy. Carbon is emitted to the environment for the setup process, holding product, product deterioration, disposal of the product, and transportation. Among different policies, the carbon tax policy is one of the most powerful policies for carbon reduction (Yu et al., 2020; Liu et al., 2021; Luo et al., 2022). Thus, the supply chain makes more profit by proper utilization of carbon tax policy, waste management, and minimizing environmental pollution.

The research gap for the study is as follows:

1. In the literature, only a few researchers placed attention on both substitutable and complementary products with deterioration. But none of those considered the products' deterioration as a function of time for a fixed lifetime product.
2. In the literature, no researcher considered demand as a function exponentially depending on the selling price for deteriorating substitutable and complementary products with carbon emission.

The rest of this chapter is sorted out as follows: A brief description of the literature is provided in Section 7.2. Section 7.3, provides problem description, symbols used for the parameter, and basic assumptions of the model. Formulation of the model and methodology used to solve the model are described in Section 7.4. Section 7.5 contains a numerical example and Section 7.6 contains the Conclusions of the study and provides a future extension of the paper.

7.2 LITERATURE REVIEW

This section provides a brief study of the recent literature to understand the research gap which helps to search the knowledge for formulating the proposed model.

In the existing literature, many researchers agree that some products having no expiry date can be held indefinitely to deliver to the customers' demand. But fixed lifetime products like cosmetics are kept for use, sometimes those start to decay owing to changes in environmental conditions such as temperature and humidity change. An economic order quantity (EOQ) model for deteriorating items with constant deterioration and demand depends linearly on the selling price was formed by Shaikh and Cárdenas-Barrón (2020) and they optimize the total cost optimizing the delivery batch size by following SSMD policy. Palanivel and Uthayakumar (2017) formulated a production inventory model for the deteriorated product. They considered probabilistic deterioration rate and variable product cost in their model. Chang (2014) formulated a supplier-buyer model for the deteriorated item where deterioration follows a constant probability distribution. Maihami et al. (2019) constructed a three-player supply network with a non-constant probabilistic deterioration rate for each member participating in it. They optimize the total cost arguing price-dependent demand and the demand relate linearly with the price. Also, many other researchers like Rout et al. (2020) and Mishra et al. (2020) formulated different models taking deterioration as constant and deterministic/probabilistic. In reality, a product's decay rate will never be constant throughout its lifetime, but rather it varies with time. In this direction, Sarkar and Saren (2017) introduced a time-dependent variable deterioration in their study. They formulated an EOQ model in which demand function depends on inventory and maximizes the profit, optimizing cycle time and order quantity. An EOQ model for fixed lifetime deteriorated product was developed by Kaur et al. (2016), where deterioration was taken as a function of time and also as a non-increasing time-dependent demand that optimized the system cost. A two-player supply network Shah et al. (2016) and Giri et al. (2017) for a single product and Pattnaik and Gahan (2020) for multi-item was developed for time-varying deteriorated products with preservation investment. Liu et al. (2020) introduced a two-level supply network model evaluating time-dependent deterioration and price-sensitive demand for profit maximization. But they didn't bother about carbon emission and waste generated from the deteriorated product.

In supply network, demand for inventory is the key. In literature, most researchers like Rout et al. (2020), Ghiami et al. (2015), and Sarkar et al. (2021a) considered demand as a constant. But in reality, its changes depended on different criteria. In the case of the product types of two complementary products say mobile phone and SIM card, if the demand for mobile phone increases (for different purposes like less price with good features) then the demand for SIM card too increases, and vice versa. Also, in the case of two substitutable products say two different toothpastes manufactured by a company, say type-A and type-B, if type-A is unavailable in the market or the price becomes high then the demand for type-B will increase. Among different criteria, the selling price has a direct impact on demand. Bhuniya et al. (2019), Dey et al. (2019), Roy et al. (2018), Bhuniya et al. (2021), Basari and Heydari (2017), and Poormoaied and Atan (2020), and many more other researchers developed their model in different direction taking demand linearly dependent on the selling price. But for some highly deteriorated/perishable products like meat and vegetables, the demand decreases exponentially with selling value. That means a research diversion occurs and needs more literature to study. In this direction Li et al. (2017), Feng et al. (2017), Gan (2018), and Sebatjanea and Adetunji (2020) formulated their model with demand depending exponentially on selling price.

In the supply chain, the upstream player (like manufacturer) produced products and supplied them to the downstream player (like a retailer). The upstream player manufacture products in a single cycle and can deliver those by a single shipment or by multiple equal shipments. The former is called the SSSD (Single setup single delivery) policy while the latter is named as the SSMD policy. In SSSD policy transportation cost is less but the holding cost of the downstream player is high, while in SSMD policy the transportation cost is high but the downstream player's holding cost is low. In this direction, several researchers Wei et al. (2015), Bai et al. (2019), and Liu et al. (2020) formulated their model culturing SSSD policy, while some other literature exits like Sarkar et al. (2016, 2021a) and Rout et al. (2020).

The environmental issue is one of the pillars of a sustainable supply chain. Various activities conducted by the supply chain members were the causes for the emission of CO_2, that is, Greenhouse gas, which pollutes the environment and increases the temperature day by day. Thus, for mitigating the emission level researchers are working on it. In this direction Xu et al. (2016), formulated a two-player sustainable supply network considering carbon emission for production. Sarkar et al. (2018) explored a multi-objective sustainable supply network for production minimizing carbon emission. A sustainable supply chain for flexible bio-fuel and bio-energy manufacturing discipline was developed by Sarkar et al. (2021b) in which they considered carbon emission transportation and bio-fuel production. Sarkar et al. (2021a) established a three-player cooperative sustainable supply chain reducing defective products and carbon emission to transport inventory. Daryanto et al. (2019) explored a three-player sustainable supply network for deteriorated products. They considered carbon emission for product deterioration and transportation. Rout et al. (2020) formulated a cooperative sustainable supply chain management with an imperfect production system considering carbon emission for production setup and rework setup, and for production and rework, transportation.

The products available in the market may be complementary of other products, or may be substitutable for some others or they have no relation at all. However, for complementary and substitutable products, manufacturers have to adopt market strategy. In this direction, Ngendakuriyo and Taboubo (2016) explored a supply chain for complementary products. They optimize the system profit finding pricing strategies for complementary products. In the same direction, Sarkar and Lee (2017) developed an SCM for complimentary products and optimize the profit of SCM members under the Stackelberg game policy. Also, for pricing and warranty policies, Wei et al. (2015) and for pricing and servicing policies with cooperative/non-cooperative decision Wang et al. (2017), a continuous review policy with interrelated demand Poormoaied and Atan (2020), and pricing and channel power Giri et al. (2020) were studied for complementary items in two-player supply network consisting of two manufacturers and a retailer. A two-echelon closed-loop supply network was formulated by Jalali et al. (2020) for complementary product optimizing return rate and pricing. A two-level supply network having two producers and a single retailer for the substitutable product was formulated by Zhao et al. (2014) considering pricing decision and channel power, strategic inventory (SI)and pricing decision by Saha et al. (2021), green substitutable products considering sales effort and pricing by Basari and Heydari (2017), stock dependent demand and ending inventory level by Pan et al. (2018), and substitutable products with green product preference considering carbon emission by Bai et al. (2019). In the above-mentioned literature, no one discussed both complementary and substitutable products. Yadav et al. (2021) formulated a continuous review model for both types of products reducing waste for deteriorated products with a constant deterioration rate and Taleizadeh et al. (2019) developed an IM for the deteriorated complementary and substitutable products with constant deterioration and demand linearly depending on selling price.

The foregoing studies suggest that it is necessary to build up a new study for forming a sustainable supply chain for both substitutable and complementary products, which are deteriorating in nature. For this, a two-echelon supply chain with two manufacturers and a single retailer was studied. Manufacturers produce either complementary products or substitutable products with variable production rates and serve retailers abiding by SSMD policy. The retailer's end product deterioration is considered, and the deterioration is time-varying. To reduce the waste, deteriorated products are disposed of following the Government policy. Carbon emission for transportation, setup, holding product, product deterioration, and product disposal is considered. To reduce carbon emission carbon tax policy is considered for this chapter. For the novelty of the study, an author contribution table is provided in Table 11.1.

Table 7.1
Authors Contribution Table

Author(s)	Scope	Deterio-Ration	Price Dependent Demand	Positive CPE	Negative CPE	SSMD	Carbon Emission
Wei et al. (2015)	TMSR SCM	NA	Linearly	NA	AP	NA	NA
Ghiami et al. (2015)	SMMB SCM	Cons.	NA	NA	NA	NA	AP
Bai et al. (2019)	TMSR SCM	Cons.	Linearly	AP	NA	NA	AP
Taleizadeh et al. (2019)	IM	Cons.	Linearly	AP	AP	NA	NA
Rout et al. (2020)	SVSB SSCM	Cons.	NA	NA	NA	AP	AP
Liu et al. (2020)	SMSR SCM	TD	Linearly	AP	NA	NA	NA
Sebatjanea and Adetunji (2020)	Three-echelon	Cons.	Exp	NA	NA	NA	NA
Sarkar et al. (2021a)	Three-echelon SSCM	TD	NA	NA	NA	AP	AP
Yadav et al. (2021)	TMSR SSCM	Cons.	Linearly	AP	AP	AP	AP
This paper	TMSR SSCM	TD	Exp	AP	AP	AP	AP

CPE, Cross-price elasticity; AP, Applicable; Cons., Constant; NA, Not Applicable; SMSR, Single manufacturer single retailer; TMSR, Two manufacturer single retailer; SVSB, Single vendor single buyer; SMMB, Single manufacturer multi-buyer; Exp, Exponentially; TD, Time-dependent; IM, Inventory model.

7.3 PROBLEM DEFINITION, NOTATION, AND ASSUMPTIONS

This section contains problem illustration, notation being used to formulate mathematical model, and some assumptions to simplify the problem.

7.3.1 PROBLEM DEFINITION

A sustainable supply chain management is introduced in this chapter with two manufacturers who produce either a substitutable product or a complementary product with a single retailer, and the retailer receives two different products from the manufacturers. The retailer orders the products at the beginning of each cycle with cross-price elasticity of the demand. For a fixed lifetime of the products, a time-dependent deterioration rate is considered for this study. In all cycles, manufacturers produce products with a flexible production rate. Both the manufacturers produce a fixed lot size in every cycle and make deliveries to the common retailer in 'x', and 'y' number of batches following SSMD process. Adaptation of SSMD policy not only reduces the holding cost of the retailer but also reduces the rate of deterioration of the product. Now another problem arises because of the different activities of manufacturers and retailers like setup process, holding products, transportation, etc., resulting in carbon emission which harms the environment. For the abatement of carbon emission, carbon tax policy is considered for this study. Again, by following the Government policy for waste management, supply chain members are obliged to adopt some adequate waste management policy so that the retailer in this study adopted the disposal policy for the deteriorated products.

7.3.2 NOTATION

This section contains the notations that are used to developed the mathematical model.

7.3.3 ASSUMPTIONS

For simplification of the mathematical model some assumption are developed which are given as follows.

1. In our study a sustainable supply chain management is developed with two manufacturers (Ma_1, and Ma_2) and a single retailer who receives two different product with cross-price elasticity in demand.
2. For fixed lifetime of the products, a time-dependent deterioration is considered at the retailer end. Long time holding of huge product can causes more deterioration. To avoid this manufacturer apply the SSMD procedure which also reduce the holding cost of the retailer Sarkar et al. (2021a). Deterioration rate is defined as:

Decision	Variables
p_i	Production rate of item A_i of manufacturer Ma_i $i = 1,2$ (units/year)
T	Cycle length of retailer
x	Units of shipment that manufacturer Ma_1 deliver A_1 item for retailer in a every manufacturing period
y	Units of shipment that manufacturer Ma_2 deliver A_2 item for retailer in a every manufacturing period

Parameter	For Manufacturers Ma_1 and Ma_2
H_{A_1}	Ma_1 manufacturer's holding cost of the item A_1 ($/unit/year)
H'_{A_1}	Ma_1 manufacturer's carbon emission unit for holding per unit A_1 item (ton/unit)
H_{A_2}	Ma_2 manufacturer's holding cost of the item A_2 ($/unit/year)
H'_{A_2}	Ma_2 manufacturer's carbon emission unit for holding per unit A_2 item (ton/unit)
P_{c1}	Ma_1 manufacturer's manufacturing cost ($/year)
P_{c2}	Ma_2 manufacturer's manufacturing cost ($/year)
K_{A_1}	Ma_1 manufacturer's setup cost ($/setup)
K'_{A_1}	Ma_1 manufacturer emitting carbon for setup (ton/setup)
K_{A_2}	Ma_2 manufacturer's setup cost ($/setup)
K'_{A_2}	Ma_2 manufacturer emitting carbon for setup (ton/setup)
F_T	Constant transportation cost in every shipment ($/shipment)
f_i	Distance between manufacturer Ma_i, retailer (km)
T_c	Cost per each truck to transport product ($/truck)
τ	Capacity of truck (units/truck)
C_e	Carbon emission unit for traveling unit distance (ton/km)
α_1	Ma_1 manufacturer set selling price of item A_1
α_2	Ma_2 manufacturer set selling price of item A_2
u_i	Tool/die cost ($)($i = 1,2$)
v_i	Development cost ($)($i = 1,2$)

Parameter	For Retailer
S_{p_i}	Retailer's selling price of the item A_i ($/unit)
O_{r_1}	Expense to order item A_1 ($/order)
O_{r_2}	Expense to order item A_2 ($/order)
h'_{r_2}	Retailer emitting carbon for holding unit A_2 item (ton/unit)
d_{c_1}	Per unit deterioration value of item A_1 ($/unit)
d_{c_2}	Deterioration value per unit of item A_2 ($/unit)
D_{c_1}	Per unit cost to dispose A_1 item ($/unit)

$$r_1 = \frac{1}{1 + p_1 - t}$$

$$r_2 = \frac{1}{1 + p_2 - t},$$

3. According to the new rules implemented by the Union of Environment and Forest and Climate Change, India, for solid waste management in 2016, manufacturers are bound to invest dollars for disposing the deteriorate products.

4. Adaptation of the SSMD policy implies an increase in the number of transport products. For this, a fixed transportation cost and a variable transportation

Parameter	For Retailer
D'_{c_1}	Carbon emission for disposing per unit A_1 item (ton/unit)
D_{c_2}	Per unit cost to dispose A_2 item ($/unit)
D'_{c_2}	Carbon emission for disposing per unit A_2 item (ton/unit)
r_1	Deterioration rate at retailer of item A_1
r_2	Deterioration rate at retailer of item A_2
p_1	Maximum lifetime of item A_1
p_2	Maximum lifetime of item A_2
β_1	Demand of item A_1 (units/year)
β_2	Demand of item A_2 (units/year)
σ	Tax for per unit carbon emission ($/unit carbon)
E_i	Retailer's basic demand for product A_i (units)
γ_i, γ'_i	Price sensitivity coefficient ($i = 1, 2$)
a_i	Coefficient for positive cross-price elasticity ($i = 1, 2$)
b_i	Coefficient for negative cross-price elasticity ($i = 1, 2$)
t	Time in the interval $[0, T]$

(function of distance) are considered for this model and the manufacturers have to carry the whole transportation cost.

5. The increase in the number of transportation causes more carbon emissions in the environment. In addition to transportation, carbon is also emitted owing to set up for production, holding product, product deterioration, and product disposal. For healthy environment, a carbon tax policy is imposed for these activities of manufacturers and retailer.

6. Demand of the products are exponentially related to its retail price and linearly to cross-price elasticity. The demand function for "positive cross-price elasticity" can be expressed

$$\beta_1 = E_1 - \gamma_1 e^{-m_1 S_{p_1}} + \gamma_2 a_1 S_{p_2}$$
$$\beta_2 = E_2 - \gamma'_1 e^{-m_2 S_{p_2}} + \gamma'_2 a_2 S_{p_1}$$

Also the demand function for negative cross-price elasticity can be expressed as

$$\beta_1 = E_1 - \gamma_1 e^{-m_1 S_{p_1}} - \gamma_2 b_1 S_{p_2}$$
$$\beta_2 = E_2 - \gamma'_1 e^{-m_2 S_{p_2}} - \gamma'_2 b_2 S_{p_1}$$

The coefficient of cross-price elasticity lies in the interval $[0, 1]$.

7. Manufacturer Ma_1 provides x deliveries and manufacturer Ma_2 provides y deliveries to the retailer in their every production cycle, while the retailer has the cycle length $T = \frac{Q_1}{\beta_1} = \frac{Q_2}{\beta_2}$.

7.4   MATHEMATICAL EXPRESSION OF THE MODEL

This chapter introduces a model to determine the impact of time-dependent deterioration of a sustainable supply chain and estimate related costs. Two manufacturers are producing either substitutable or complementary products to accomplish the demand of a common retailer. We address manufacturer-I as 'He', and manufacturer-II as 'She'. Carbon emission is another concern that has been included in this chapter remembering its ominous effect on the environment. Nowadays, carbon tax policy has become a popular policy in many countries[4] who are aiming for zero carbon by 2050. Among the different activities of the members, carbon emission is given serious consideration, and carbon tax policy has been followed for emission abatement. The detailed derivation of different costs for setup, manufacturing, holding, ordering, deteriorated product, disposal of the deteriorated product, transportation of items of the members are in the following subsections.

7.4.1 MATHEMATICAL MODEL FOR RETAILER

The inventory volume of the retailer is illustrated in Figure 7.1 for two types of products A_1 and A_2, respectively. The cycle length of the retailer is T, and at the start of each period, the retailer has an inventory volume of Q_1, Q_2 for product A_1, A_2. Following the SSMD policy, the retailer receives 'x' and 'y' equal shipments from the manufacturer-I and manufacturer-II, respectively. Deterioration process take place for the products A_1, A_2 at a rate r_1, and r_2 for retaining physically at the retailer store, respectively. The retailer does not receive the next shipment from the manufacturer until the current batch depletes to zero level for customer demand and product decay. The inventory level of retailer is given in Figure 7.1.

The stock level for the product A_1 which depletes to zero level for customer demand and product decay is presented using the differential equation as follows:

$$\frac{dI_1(t)}{dt} = -\frac{1}{1+p_1-t}I_1 - \beta_1, \quad t \in [0,T]$$
$$\frac{dI_1(t)}{dt} + \frac{1}{1+p_1-t}I_1 = -\beta_1 \tag{7.1}$$

and given the initial conditions are, $I_1(0) = Q_1$ and $I_1(T) = 0$
Equation (7.1) is a linear differential equation having order one. Solving this equation

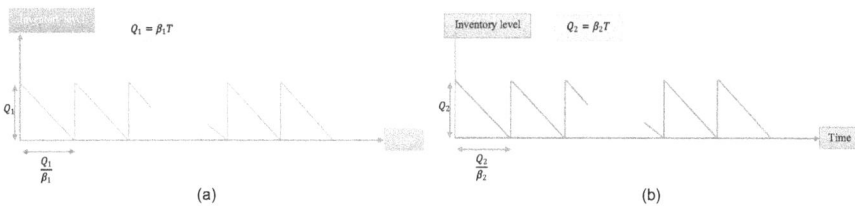

Figure 7.1 Retailer's inventory level. (a) Inventory of level first product (A_1) and (b) Inventory of level first product (A_2).

we obtain the inventory level of retailer for item A_1 as follows:

$$I_1 = \beta_1(1+p_1-t)\ln\left(\frac{1+p_1-t}{1+p_1-T}\right) \tag{7.2}$$

Using the condition $I_1(0) = Q_1$, we get the maximum inventory of retailer for item A_1 as follows:

$$Q_1 = \beta_1(1+p_1)\ln\left(\frac{1+p_1}{1+p_1-T}\right) \tag{7.3}$$

Again,

$$\int_0^T I_1(t)dt = \frac{\beta_1}{2}\left[(1+p_1)(\eta_1-T)+\frac{T^2}{2}\right] \tag{7.4}$$

$$\eta_1 = (1+p_1)\ln\left(\frac{1+p_1}{1+p_1-T}\right)$$

Now, the inventory level of the product A_2 received from manufacturer-II is obtained similarly as the product of A_1.

$$\frac{dI_2(t)}{dt} = -\frac{1}{1+p_2-t}I_2-\beta_2, \quad t\in[0,T]$$

$$\frac{dI_2(t)}{dt}+\frac{1}{1+p_2-t}I_2 = -\beta_2 \tag{7.5}$$

With the initial condition, $I_2(0) = Q_1$ and $I_2(T) = 0$
Equation (7.2) is a linear differential equation having order one solving this equation and using the initial condition we obtain the inventory level of retailer for product A_1 at any time t is as follows:

$$I_2 = \beta_2(1+p_2-t)\ln\left(\frac{1+p_2-t}{1+p_2-T}\right) \tag{7.6}$$

Given the condition that $I_2(0) = Q_2$, this provides us maximum inventory level of the retailer for the product A_2.

$$Q_2 = \beta_2(1+p_2)\ln\left(\frac{1+p_2}{1+p_2-T}\right) \tag{7.7}$$

$$\text{Again } \int_0^T I_2(t)dt = \frac{\beta_2}{2}\left[(1+p_2)(\eta_2-T)+\frac{T^2}{2}\right] \tag{7.8}$$

$$\eta_2 = (1+p_2)\ln\left(\frac{1+p_2}{1+p_2-T}\right)$$

7.4.1.1 Ordering Cost

At the beginning of each cycle, the retailer needs to order both types of products to the manufacturers depending on the customer demand. The retailer's period to sell out the products is T. Therefore, the ordering cost in each period of the retailer is found out as below.

$$OC_R = \frac{O_{r_1}}{T} + \frac{O_{r_2}}{T} \tag{7.9}$$

7.4.1.2 Holding Cost

After receiving the products from manufacturer-I and manufacturer-II, the retailer has to retain them at the store until it depletes to zero level with time owing to customer demand and deterioration. Activities associated with holding products can cause the emission of carbon in the environment. The retailer has to pay a carbon tax for carbon emission due to holding products. Thus, the holding cost for both the products along with carbon tax is determined as follows:

$$
\begin{aligned}
HC_R &= \frac{(h_{r_1} + \sigma h'_{r_1})}{T} \int_0^T I_1(t)dt + \frac{(h_{r_2} + \sigma h'_{r_2})}{T} \int_0^T I_2(t)dt \\
&= \frac{(h_{r_1} + \sigma h'_{r_1})\beta_1}{2T} \left[(1+p_1)(\eta_1 - T) + \frac{T^2}{2} \right] \\
&\quad + \frac{(h_{r_2} + \sigma h'_{r_2})\beta_2}{2T} \left[(1+p_2)(\eta_2 - T) + \frac{T^2}{2} \right]
\end{aligned} \tag{7.10}
$$

7.4.1.3 Deterioration Cost

When products are physically stored at the retailer house, products started to deteriorate at a rate r_1, and r_2 as time passes. But the deployment of preservation policy reduced the deterioration rate to $r_1 - M(\delta)$, and $r_2 - M(\delta)$, respectively. Per unit deterioration cost charged as \$ d_{c_1} and d_{c_2}, respectively for the products A_1 and A_2. Deteriorated products eject carbon to the environment and the retailer has to pay tax for carbon. Therefore, the deterioration cost including carbon tax is given as follows:

$$
\begin{aligned}
DC_R &= \frac{d_{c_1}\left(\int_0^T I_1(t) - \beta_1 T \right)}{T} + \frac{d_{c_2}\left(\int_0^T I_2(t) - \beta_2 T \right)}{T} \\
&= \frac{d_{c_1}\beta_1}{2T} \left[(1+p_1)(\eta_1 - T) + \frac{T^2}{2} - 2T \right] + \frac{d_{c_2}\beta_2}{2T} \left[(1+p_2)(\eta_2 - T) + \frac{T^2}{2} - 2T \right]
\end{aligned} \tag{7.11}
$$

7.4.1.4 Disposal Cost

The retailer maintains a proper disposal policy of the deteriorated products so that it does not amplify any negative impact on the environment. The cost which is needed for disposing of the deteriorated products with a carbon tax is derived as follows:

$$\text{DDC}_R = \frac{(D_{c_1} + \sigma D'_{c_1})\left(\int_0^T I_1(t) - \beta_1 T\right)}{T} + \frac{(D_{c_2} + \sigma D'_{c_2})\left(\int_0^T I_2(t) - \beta_2 T\right)}{T}$$

$$= \frac{(D_{c_1} + \sigma D'_{c_1})\beta_1}{2T}\left[(1 + p_1)(\eta_1 - T) + \frac{T^2}{2} - 2T\right]$$

$$+ \frac{(D_{c_2} + \sigma D'_{c_2})\beta_2}{2T}\left[(1 + p_2)(\eta_2 - T) + \frac{T^2}{2} - 2T\right] \qquad (7.12)$$

Therefore the total cost to the retailer is

$$\text{TC}_R = \frac{(h_{r_1} + \sigma h'_{r_1})\beta_1}{2T}\left[(1 + p_1)(\eta_1 - T) + \frac{T^2}{2}\right]$$

$$+ \frac{(h_{r_2} + \sigma h'_{r_2})\beta_2}{2T}\left[(1 + p_2)(\eta_2 - T) + \frac{T^2}{2}\right]$$

$$+ \frac{(d_{c_1} + D_{c_1} + \sigma D'_{c_1})\beta_1}{2T}\left[(1 + p_1)(\eta_1 - T) + \frac{T^2}{2} - 2T\right] + \frac{O_{r_1}}{T} + \frac{O_{r_2}}{T}$$

$$+ \frac{(d_{c_2} + D_{c_2} + \sigma D'_{c_2})\beta_2}{2T}\left[(1 + p_2)(\eta_2 - T) + \frac{T^2}{2} - 2T\right] \qquad (7.13)$$

7.4.1.5   Revenue

The selling price of the products is decided by the retailer. The revenue for every unit product is obtained by subtracting the wholesale price from the selling price. Since, the demand for the products A_1 and A_2 are respectively, β_1 and β_2, the total revenue of the retailer is calculated as follows:

$$\text{RR}_R = (S_{P_1} - \alpha_1)\beta_1 + (S_{P_2} - \alpha_2)\beta_2 \qquad (7.14)$$

The retailer's total profit after selling the products is obtained as follows:

$$\text{TP}_R = \text{RR}_R - \text{TC}_R$$

$$\text{TP}_R = (S_{P_1} - \alpha_1)\beta_1 + (S_{P_2} - \alpha_2)\beta_2 - \left[\frac{(d_{c_2} + D_{c_2} + \sigma D'_{c_2})\beta_2}{2T}\left[(1 + p_2)(\eta_2 - T)\right.\right.$$

$$+ \frac{T^2}{2} - 2T\right] + \frac{(h_{r_1} + \sigma h'_{r_1})\beta_1}{2T}\left[(1 + p_1)(\eta_1 - T) + \frac{T^2}{2}\right]$$

$$+ \frac{(h_{r_2} + \sigma h'_{r_2})\beta_2}{2T}\left[(1 + p_2)(\eta_2 - T) + \frac{T^2}{2}\right] + \frac{O_{r_1}}{T} + \frac{O_{r_2}}{T}$$

$$\left. + \frac{(d_{c_1} + D_{c_1} + \sigma D'_{c_1})\beta_1}{2T}\left[(1 + p_1)(\eta_1 - T) + \frac{T^2}{2} - 2T\right]\right] \qquad (7.15)$$

7.4.2 MATHEMATICAL MODEL FOR MANUFACTURER-I

The inventory volume of manufacturer-I is depicted in Figure 7.2. Adopting a flexible production system, he produces item A_1 of lot size q_1 in each setup to satisfy the demand of the retailer in 'x' shipments with a batch size Q_1 following the SSMD policy. Which means that $q_1 = xQ_1$. The demand of the retailer being β_1 per year for item A_1, the cycle length of the manufacturer-I is $\frac{xQ_1}{\beta_1}$. The cumulative inventory level of manufacturer-I is given in Figure 7.2. Now, the elementary description of their different costs per cycle of him are given as follows:

7.4.2.1 Setup Cost

At the beginning of the manufacturing cycle, manufacturer-I needs a basic setup for the production process. For this purpose, he has to invest some cost which is considered to be fixed in this chapter. Therefore, the setup cost along with the carbon tax per cycle of the manufacturer-I is described as follows:

$$\text{SC}_{\text{MI}} = \left(\frac{(K_{A_1} + \sigma K'_{A_1})}{xT} \right) \tag{7.16}$$

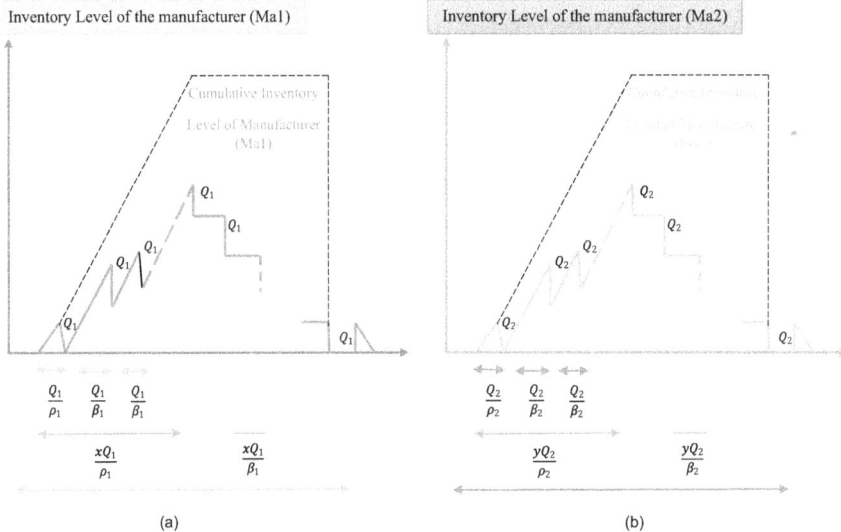

Figure 7.2 Manufacturer's Inventory level. (a) Inventory of level manufacturer-I and (b) inventory of level manufacturer-II.

7.4.2.2 Manufacturing Cost

The manufacturing cost per unit product A_1 of manufacturer-I is $P_{c_1} = c_1 + v_1\rho_1 + \frac{u_1}{\rho_1}$ (Sarkar et al., 2011). Where c_1 is the raw material cost for producing per unit product, u_1 represents fixed labor and energy cost which is constant, and v_1 is the tool or die cost. Therefore, the manufacturing cost per cycle for producing $q_1 = xQ_1$ units of A_1 the product is

$$
\begin{aligned}
MC_{MI} &= \frac{\beta_1}{xT} P_{c1}(\rho_1) \\
&= \frac{\beta_1}{xT}\left(c_1 + v_1\rho_1 + \frac{u_1}{\rho_1}\right)
\end{aligned}
\tag{7.17}
$$

7.4.2.3 Holding Cost

Rather than delivering all products in a single shipment, he executes it in 'x' shipment of equal batch size Q_1 to the retailer following the SSMD policy. For this process, he has to hold products for a certain time. The activities associated with holding the product is caused by emitting some carbon into the environment. Therefore, the holding cost of manufacturer-I per cycle including carbon tax is as follows:

$$
\begin{aligned}
HC_{MI} &= (H_{A_1} + \sigma H'_{A_1})\left[\left\{xQ_1\left(\frac{Q_1}{\rho_1} + (x-1)\frac{Q_1}{\beta_1}\right) - \frac{1}{2}(xQ_1)\left(\frac{xQ_1}{\rho_1}\right)\right\}\right. \\
&\left. - \left\{(1+2+..+(x-1))\frac{Q_1^2}{\beta_1}\right\}\right]\left(\frac{\beta_1}{xQ_1}\right) \\
&= \frac{(H_{A_1} + \sigma H'_{A_1})\beta_1 T}{2}\left[x\left(1-\frac{\beta_1}{\rho_1}\right) + \frac{2\beta_1}{\rho_1} - 1\right]
\end{aligned}
\tag{7.18}
$$

7.4.2.4 Transportation Cost

In this study, it is taken that all the transportation cost is accommodated by the manufacturer. The transportation consists of two components, namely (1) fixed cost to transport product and (2) variable cost to transport product. Whereas the variability in transportation is taken as the distance between retailer and manufacturer-I, He releases the inventory in equal 'x' shipments to the retailer, and therefore the transportation cost including carbon tax is derived as follows:

$$
TC_{MI} = \frac{1}{xT}\left(xF_T + x\beta_1 T\frac{f_1(T_c + \sigma C_e)}{\tau}\right)
\tag{7.19}
$$

Therefore the total cost for manufacturer-I is

$$
\begin{aligned}
TC_{MI} &= \left(\frac{(K_{A_1} + \sigma K'_{A_1})}{xT}\right) + \frac{(H_{A_1} + \sigma H'_{A_1})\beta_1 T}{2}\left[x\left(1-\frac{\beta_1}{\rho_1}\right) + \frac{2\beta_1}{\rho_1} - 1\right] \\
&+ \frac{\beta_1}{xT}\left(c_1 + v_1\rho_1 + \frac{u_1}{\rho_1}\right) + \frac{1}{T}\left(F_T + \beta_1 T\frac{f_1(T_c + \sigma C_e)}{\tau}\right)
\end{aligned}
\tag{7.20}
$$

7.4.2.5 Revenue

Manufacturer-I decides the wholesale price α_1 of the product A_1 per unit. The revenue comes when he releases the demand β_1 units to the retailer. Therefore, the revenue for the manufacturer-I is

$$R_{\text{MI}} = \beta_1 \alpha_1 \tag{7.21}$$

For Manufacturer-I, the profit gained is obtained by subtracting the total cost and the revenue, which is

$$\text{TP}_{\text{MI}} = R_{\text{MI}} - \text{TC}_{\text{MI}}$$

$$\text{TP}_{\text{MI}} = \beta_1 \alpha_1 - \left[\left(\frac{(K_{A_1} + \sigma K'_{A_1})}{xT} \right) + \frac{(H_{A_1} + \sigma H'_{A_1})\beta_1 T}{2} \left[x\left(1 - \frac{\beta_1}{\rho_1}\right) + \frac{2\beta_1}{\rho_1} - 1 \right] \right.$$
$$\left. + \frac{\beta_1}{xT}\left(c_1 + v_1\rho_1 + \frac{u_1}{\rho_1}\right) + \frac{1}{T}\left(F_T + \beta_1 T \frac{f_1(T_c + \sigma C_e)}{\tau}\right) \right] \tag{7.22}$$

7.4.3 MATHEMATICAL MODEL FOR MANUFACTURER-II

[] The inventory level of manufacturer-II is depicted in Figure 7.1b. Adopting a flexible production system, she produces item A_2 of lot size q_2 in each setup and satisfies the retailer's demand in 'y' shipments with a batch size Q_2 following the SSMD policy. Which means that $q_2 = yQ_2$. The demand of retailer being β_2 per year for item A_2, the cycle length of the manufacturer-II is $\frac{yQ_2}{\beta_2}$. The cumulative inventory level of manufacturer-II is given in Figure 7.2. Now, the elementary description of different costs per cycle of her is given as follows:

7.4.3.1 Setup Cost

At the beginning of the manufacturing cycle, manufacturer-II needs a basic setup for the production process. For this purpose, she has to invest some cost which is considered to be fixed in this chapter. Therefore, the setup cost along with the carbon tax per cycle of the manufacturer-II is described as follows:

$$\text{SC}_{\text{MII}} = \left(\frac{(K_{A_2} + \sigma K'_{A_2})}{yT} \right) \tag{7.23}$$

7.4.3.2 Manufacturing Cost

The manufacturing cost per unit product A_2 of manufacturer-II is $P_{c_2} = c_2 + v_2\rho_2 + \frac{u_2}{\rho_2}$ (Sarkar et al., 2011). Where c_2 is the raw material cost for producing per unit product, u_2 represents fixed labor and energy cost which is constant, and v_2 is the tool or die cost. Therefore, the manufacturing cost per cycle for producing $q_2 = yQ_2$ units of A_2 the product is

$$MC_{MII} = \frac{\beta_2}{yT}P_{c2}(\rho_2)$$

$$= \frac{\beta_2}{yT}\left(c_2 + v_2\rho_2 + \frac{u_2}{\rho_2}\right) \tag{7.24}$$

7.4.3.3　Holding Cost

Rather than delivering all products in a single shipment, she executes it in 'y' shipment of equal batch size Q_2 to the retailer following the SSMD policy. For this process, she has to hold products for a certain time. Activities associated with holding product is caused by emitting some carbon into the environment. Therefore, the holding cost of manufacturer-II per cycle including carbon tax is as follows:

$$HC_{MII} = (H_{A_2} + \sigma H'_{A_2})\left[\left\{yQ_2\left(\frac{Q_2}{\rho_2} + (y-1)\frac{Q_2}{\beta_1}\right) - \frac{1}{2}(yQ_2)\left(\frac{yQ_2}{\rho_2}\right)\right\}\right.$$
$$\left. - \left\{(1+2+..+(y-1))\frac{Q_2^2}{\beta_1}\right\}\right]\left(\frac{\beta_1}{yQ_2}\right)$$
$$= \frac{(H_{A_2} + \sigma H'_{A_2})\beta_2 T}{2}\left[y\left(1 - \frac{\beta_2}{\rho_2}\right) + \frac{2\beta_2}{\rho_2} - 1\right] \tag{7.25}$$

7.4.3.4　Transportation Cost

In this study, it is taken that all the transportation cost is accommodated by the manufacturer. The transportation consists of two components, namely (1) a fixed cost for transporting product and (2) a variable cost for transporting the product. Whereas the variability in transportation is taken over the distance between the retailer and manufacturer-II. She releases the inventory in equal 'y' shipments to the retailer, therefore the transportation cost including carbon tax is derived as follows:

$$TC_{MII} = \frac{1}{yT}\left(yF_T + y\beta_2 T\frac{f_2(T_c + \sigma C_e)}{\tau}\right) \tag{7.26}$$

Therefore the total cost of manufacturer-II is

$$TC_{MII} = \left(\frac{(K_{A_2} + \sigma K'_{A_2})}{yT}\right) + \frac{(H_{A_2} + \sigma H'_{A_2})\beta_2 T}{2}\left[y\left(1 - \frac{\beta_2}{\rho}\right) + \frac{2\beta_2}{\rho} - 1\right]$$
$$+ \frac{\beta_2}{yT}\left(c_2 + v_2\rho_2 + \frac{u_2}{\rho_2}\right) + \frac{1}{T}\left(F_T + \beta_2 T\frac{f_2(T_c + \sigma C_e)}{\tau}\right) \tag{7.27}$$

7.4.3.5 Revenue

Manufacturer-II decides the wholesale price α_2 of the product A_2 per unit. Revenue comes when she releases the demand β_2 units to the retailer. Therefore, the revenue of the manufacturer-II is

$$R_{\text{MII}} = \beta_2 \alpha_2 \tag{7.28}$$

The profit gained by the manufacturer-II is obtained subtracting the total cost and the revenue, which is

$$
\begin{aligned}
\text{TP}_{\text{MII}} &= R_{\text{MII}} - \text{TC}_{\text{MII}} \\
&= \beta_2 \alpha_2 - \left[\left(\frac{(K_{A_2} + \sigma K'_{A_2})}{yT} \right) \right. \\
&\quad + \frac{(H_{A_2} + \sigma H'_{A_2})\beta_2 T}{2} \left[y \left(1 - \frac{\beta_2}{\rho} \right) + \frac{2\beta_2}{\rho_2} - 1 \right] \\
&\quad + \left. \frac{\beta_2}{yT} \left(c_2 + v_2 \rho_2 + \frac{u_2}{\rho_2} \right) + \frac{1}{T} \left(F_T + \beta_2 T \frac{f_2(T_c + \sigma C_e)}{\tau} \right) \right]
\end{aligned} \tag{7.29}
$$

Therefore, the supply chain's profit

$$
\begin{aligned}
&\text{STP}(\rho_1, \rho_2, T, x, y) \\
&= S_{P_1}\beta_1 + S_{P_2}\beta_2 - \left(\frac{(d_{c_2} + D_{c_2} + \sigma D'_{c_2})\beta_2}{2T} \left[(1+p_2)(\eta_2 - T) + \frac{T^2}{2} - 2T \right] + \frac{O_{r_1}}{T} \right. \\
&\quad + \frac{(h_{r_1} + \sigma h'_{r_1})\beta_1}{2T} \left[(1+p_1)(\eta_1 - T) + \frac{T^2}{2} \right] + \frac{(h_{r_2} + \sigma h'_{r_2})\beta_2}{2T} \left[(1+p_2)(\eta_2 - T) + \frac{T^2}{2} \right] \\
&\quad + \frac{(d_{c_1} + D_{c_1} + \sigma D'_{c_1})\beta_1}{2T} \left[(1+p_1)(\eta_1 - T) + \frac{T^2}{2} - 2T \right] + \frac{\beta_1}{xT} \left(c_1 + v_1 \rho_1 + \frac{u_1}{\rho_1} \right) \\
&\quad + \frac{(H_{A_1} + \sigma H'_{A_1})\beta_1 T}{2} \left[(x-1) + \frac{\beta_1}{\rho_1}(2-x) \right] + \frac{1}{T} \left(F_T + \beta_1 T \frac{f_1(T_c + \sigma C_e)}{\tau} \right) \\
&\quad + \frac{\beta_2}{yT} \left(c_2 + v_2 \rho_2 + \frac{u_2}{\rho_2} \right) + \frac{(H_{A_2} + \sigma H'_{A_2})\beta_2 T}{2} \left[(y-1) + \frac{\beta_2}{\rho_2}(2-y) \right] + \frac{O_{r_2}}{T} \\
&\quad + \left. \frac{1}{T} \left(F_T + \beta_2 T \frac{f_2(T_c + \sigma C_e)}{\tau} \right) + \left(\frac{(K_{A_1} + \sigma K'_{A_1})}{xT} \right) + \left(\frac{(K_{A_2} + \sigma K'_{A_2})}{yT} \right) \right)
\end{aligned}
$$

$$= S_{p_1}\beta_1 + S_{p_2}\beta_2 - \left[\beta_1 X_{11} + \beta_2 X_{12} + \frac{1}{T}\left\{\beta_1 Y_{11}\ln\left(\frac{1+p_1}{1+p_1-T}\right)\right.\right.$$

$$+\beta_2 Y_{12}\ln\left(\frac{1+p_2}{1+p_2-T}\right) + Y_{13} + \frac{\beta_1}{x}\left(c_1 + v_1\rho_1 + \frac{u_1}{\rho_1}\right) + \frac{\beta_2}{y}\left(c_2 + v_2\rho_2 + \frac{u_2}{\rho_2}\right)\right\}$$

$$+T\left\{\beta_1 Z_{11} + \beta_2 Z_{12} + \frac{\beta_1^2}{\rho_1}Z_{13} + \frac{\beta_2^2}{\rho_2}Z_{14}\right\}\right] \tag{7.30}$$

The values of X_{11}, X_{12}, Y_{11}, Y_{12}, Y_{13}, Z_{11}, Z_{12}, Z_{13} nd Z_{14} are given in Appendix A.

7.5 SOLUTION METHODOLOGY

In this chapter, we have formulated a profit maximization problem for a two-player sustainable supply network management participating two manufacturers and a re-tailer. The profit function of the supply chain presented by equation (7.26) is non-linear. The whole supply network profit is maximized by optimizing the number of lots sent (integer decision variable) to the retailer by the manufacturer, cycle length, and the production rate of the manufacturers. Thus, this is a mixed-integer non-linear (MINLOP) optimization problem and is solved by using the classical optimization method. First of all, the total profit function (STP) is differentiated for non-integer decision variable (ρ_1, ρ_2, and T) and equate to zero for getting optimal results of those decision variables. Both first-order and second-order derivatives are provided in Appendix B. The optimum values of the decision variables are given as follows:

$$\rho_1^* = \sqrt{\frac{u_1 + \beta_1 x Z_3 T^2}{v_1}}$$

$$\rho_2^* = \sqrt{\frac{u_2 + \beta_2 y Z_4 T^2}{v_2}}$$

$$T^* = \sqrt{\frac{\beta_1 Y_1\ln\left(\frac{1+p_1}{1+p_1-T}\right) + \beta_2 Y_2\ln\left(\frac{1+p_2}{1+p_2-T}\right) + Y_3 + \frac{\beta_1}{x}\left(c_1 + v_1\rho_1 + \frac{u_1}{\rho_1}\right)}{\frac{\beta_1 Y_1}{T(1+p_1-T)} + \frac{\beta_2 Y_2}{T(1+p_2-T)} + \beta_1 Z_1 + \beta_2 Z_2 + \frac{\beta_1^2 Z_3}{\rho_1} + \frac{\beta_2^2}{\rho_2}Z_4}}$$

Further the concavity of the function STP at ρ_1^*, ρ_2^*, and T^* is examined using Hessian matrix which is given in Appendix C. Also, concavity of STP is presented graphically in Figure 7.3 for non-integer variable.

7.6 NUMERICAL EXAMPLE

For validation of the presented model, two separate examples are deliberated. One for the substitutable product i.e., for positive cross-price elasticity of demand and another for complementary product i.e., for negative cross-price elasticity of demand.

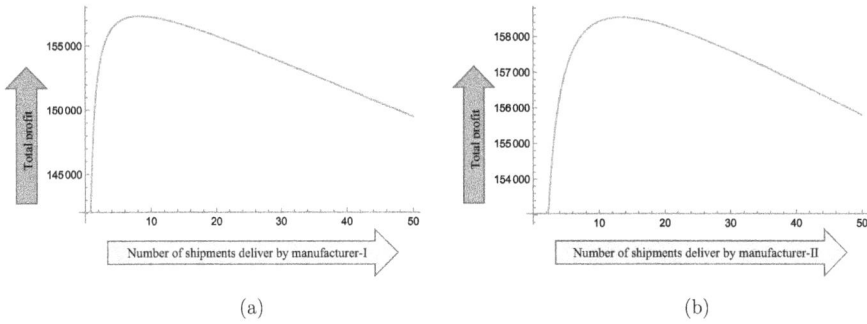

(a)   (b)

Figure 7.3   Depict concavity of profit function for integer variable x, y. (a) Optimality with respect to number of shipment of manufacturer-I and (b) optimality with respect to number of shipment of manufacturer- II.

The parametric values for the examples are obtained from Yadav et al. (2021). The optimal solution of the mentioned decision variables is gained by conducting Mathematica 12.3 software.

Example 7.1: Positive Cross-Price Elasticity

For the rationality of the present model, it is examined numerically. The values of the used parameter for the example are enlisted in Table 7.2. Solving the problem

Table 7.2
Parameters for Example 7.1

Parameter	Value	Parameter	Value	Parameter	Value
H_{A_1}	10 unit/year	H'_{A_1}	0.001 ton/unit	H_{A_2}	9.000 unit/year
H'_{A_2}	0.002 ton/unit	K_{A_1}	850.0 \$/setup	K'_{A_1}	0.002 ton/setup
K_{A_2}	750.0 \$/setup	K'_{A_2}	0.002 ton/setup	σ	2 \$/ton
F_T	5 \$	f_1	400 km	f_2	410 km
T_c	0.05 \$/truck	C_e	0.0015 ton/km	α_1	110 \$/unit
α_2	120 \$/unit	τ	200	u_1	6000
u_2	5900	v_1	0.020	v_2	0.020
c_1	0.300	c_2	0.400	s_{p_1}	240 \$/unit
s_{p_2}	250 \$/unit	O_{r_1}	35 \$/order	O_{r_2}	32 \$/order
h_{r_1}	8 \$/unit/year	h'_{r_1}	0.001 ton/unit	h_{r_2}	7 \$/unit/year
h'_{r_2}	0.001 ton/year	d_{c_1}	29 \$/unit	d_{c_2}	28 \$/unit
D_{c_1}	10.00 \$/unit	D'_{c_1}	0.008 ton/unit	D_{c_2}	9.000 \$/unit
D'_{c_2}	0.008 ton/unit	p_1	6 year	p_2	5 year
E_1	250 units	E_2	310 units	γ_1	100
γ'_1	150	γ_2	0.21	γ'_2	0.31
a_1	0.49	a_2	0.41	m_1	0.20
m_2	0.20				

Table 7.3

Optimal Result for Example 7.1

Variable	ρ_1	ρ_2	T	x	y	STP
Optimal Value	429	407	0.45	4	4	157,255

Table 7.4

Effect of SSMD Policy

	x	y	T	Carbon Emission	STP
SSSD	1	1	.94	2.22	141,779
SSMD	4	4	.45	2.40	157,255

optimum solution of the mentioned variables is given in Table 7.3. The concavity of the profit function is also given in Figure 7.4.

In the existing literature only Yadav et al. (2021) consider a sustainable supply network taking deterioration and cross-price elasticity (both positive and negative). But they formed a continuous review model with a constant deterioration rate and the demand depending on selling price linearly.

Table 7.4. provides how the adaptation of the SSMD policy effect on the cycle time of retailer, carbon emission for holding product and to transport, and system total profit for example 1.

Example 7.2: Negative Cross-Price Elasticity
To assure the validity of the proposed model is examined numerically. The values of the parameters related to this example are enlisted in Table 7.5. The optimum values of the decision variables are given in Table 7.6. The system profit function is concave

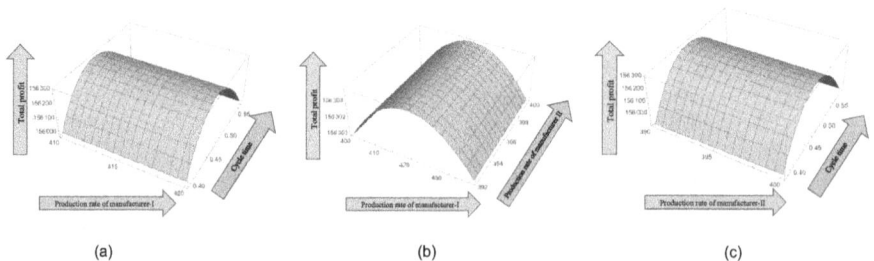

(a)   (b)   (c)

Figure 7.4 Concavity of profit function for Example 7.1. (a) Optimality with respect to cycle time and production rate of Ma_1, (b) optimality with respect to production rate of Ma_1 and production rate of Ma_2, and (c) optimality with respect to production rate of Ma_2 and cycle time.

Table 7.5
Parameters for Example 7.2

Parameter	Value	Parameter	Value	Parameter	Value
H_{A_1}	10 unit/year	H'_{A_1}	0.001 ton/unit	H_{A_2}	9.000 unit/year
H'_{A_2}	0.002 ton/unit	K_{A_1}	800.0 \$/setup	K'_{A_1}	0.002 ton/setup
K_{A_2}	650.0 \$/setup	K'_{A_2}	0.002 ton/setup	σ	2 \$/ton
F_T	5 \$	f_1	400 km	f_2	410 km
T_c	0.05 \$/truck	C_e	0.0015 ton/km	α_1	110 \$/unit
α_2	120 \$/unit	τ	200	u_1	5000
u_2	5100	v_1	0.020	v_2	0.020
c_1	0.300	c_2	0.400	s_{p_1}	350 \$/unit
s_{p_2}	280 \$/unit	O_{r_1}	65 \$/order	O_{r_2}	62 \$/order
h_{r_1}	7 \$/unit/year	h'_{r_1}	0.001 ton/unit	h_{r_2}	6 \$/unit/year
h'_{r_2}	0.001 ton/year	d_{c_1}	39 \$/unit	d_{c_2}	37 \$/unit
D_{c_1}	15 \$/unit	D'_{c_1}	0.008 ton/unit	D_{c_2}	12 \$/unit
D'_{c_2}	0.008 ton/unit	p_1	4 year	p_2	3 year
E_1	250 units	E_2	320 units	γ_1	100
γ'_1	150	γ_2	0.21	γ'_2	0.31
a_1	0.49	a_2	0.41	m_1	0.20
m_2	0.20				

Table 7.6
Optimal Results for Example 7.2

Variable	ρ_1	ρ_2	T	x	y	STP
Optimal Value	378	368	0.35	5	5	165,888

and is depicted in Figure 7.5. From the result, we can conclude that the increment of cycle time causes for increasing deterioration rate of the product. The effect of cycle time on deterioration is presented in Figure 7.6.

Table 7.7. explains how the adaptation of the SSMD policy effect on cycle time of the retailer, carbon emission for holding product and to transport, and system total profit for example 2.

7.6.1　DISCUSSION OF THE NUMERICAL EXAMPLE

The numerical result provides the strategy for both substitutable and complementary products such that the system can gain more profit. The result shows that the profit will be maximum in the case of substitutable products for high positive cross-price elasticity i.e. if the retailer store similar types of products with the original one. The profit will be maximum for complimentary products for low negative cross-price elasticity i.e. if the retailer stores two less dependent products. The result also provides that manufacturing cost, as well as the self-price elasticity factor, affect

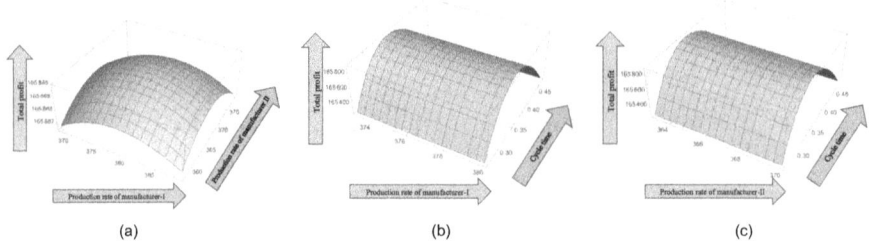

(a)   (b)   (c)

Figure 7.5 Concavity of profit function for Example 7.2. (a) Optimality with respect to manufacturing rate of Ma_2 and the production rate of Ma_1, (b) Optimality with respect to the manufacturing rate of Ma_1 and cycle time, and (c) optimality with respect to manufacturing rate of Ma_2 and cycle time.

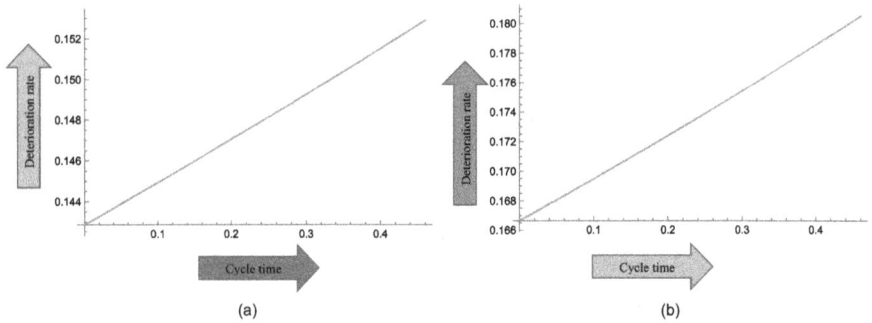

(a)   (b)

Figure 7.6 Cycle time vs deterioration. (a) For first product (A_1) and (b) For second product (A_2).

Table 7.7
Effect of SSMD Policy

	x	y	T	Carbon Emission	STP
SSSD	1	1	.85	1.86	151,116
SSMD	5	5	.35	1.77	165,888

---

the system profit significantly. From Figure 7.6 we see that the deterioration rate increases with the cycle time. Also, the SSMD policy reduces cycle time and system emission.

7.7 SENSITIVITY ANALYSIS

In this section, we discuss how the total system profit is influenced by the change in some crucial cost and scaling parameters. The parameters are changed by −50%,

−25%, +25%, and +50%. The changes in system total profit due to changes in parameters are enlisted in the Tables 7.8 and 7.9. We conclude the following from the Tables 7.8 and 7.9. The sensitivity for crucial parameters is given in Figure 7.7.

1. The selling price for both cases is the most sensorial parameter. It has a direct impact on the system's profit. An increment in selling price increases the total profit and decrement in it decrease the total profit.
2. Another most sensorial parameter is the basic demand of the market. A small change in it causes extensive variation in the total profit. From the table, we can see that the total profit increases when the basic demand increases significantly and vice versa.
3. The price coefficient parameter has a prior effect on the profit function of the system. For substitutable products, the profit increases for decrement in price coefficient and vice versa. While for complementary products, the system profit increases for increment in self-price coefficient and vice versa.
4. The cross-price elasticity coefficient also affects system profit sufficiently. For negative cross-price elasticity, system profit increases for the decrement in the coefficient and vice versa. The reverse result happens for positive elasticity coefficient.
5. In a manufacturing system, the manufacturing cost's scaling parameter plays a vital role. For an increment in the development cost, manufacturing cost as well as system profit decreases and vice versa.
6. In the supply chain, the manufacturer's holding cost has a crucial role in system profit. When holding cost decreases system profit decreases and for increment in holding cost, system profit decreases significantly as provided in the sensitivity table.
7. Besides these above-mentioned parameters, some parameters like manufacturer's setup cost, retailer's ordering cost, holding cost, maximum lifetime of inventory have less impact on the total profit function as provided in the sensitivity table.

The effect of number of shipment on system profit, cycle time, and production rate are given in Figure 7.8.

7.8   MANAGERIAL INSIGHT

This section contains some recommendations for the managers of industries under different aspects.

From the result of the study, we have noticed that the negative cross-price elasticity factor is inversely related to the system's total profit, i.e., the increase in negative elasticity significantly decreases the profit. Thus, if one product is more dependent on another product i.e., complementing to each other and then the selling of one will be fully inclined to the other. In that case, the unavailability of one product will decrease the selling of another product. For this reason, managers of companies need to follow proper strategies about storing products at the retailer's house.

Table 7.8

Sensitivity Analysis of Input Parameters for Substitutable Product

Parameter	Changes (%)	STP (%)	Parameter	Changes (%)	STP (%)
H_{A_1}	−50	0.36	H_{A_2}	−50	0.32
	−25	0.17		−25	0.16
	+25	−0.16		+25	−0.14
	+50	−0.32		+50	−0.24
K_{A_1}	−50	0.15	K_{A_2}	−50	0.13
	−25	0.07		−25	0.06
	+25	−0.07		+25	−0.06
	+50	−0.14		+50	−0.13
S_{p_1}	−50	−23.47	S_{p_2}	−50	−29.05
	−25	−11.67		−25	−14.44
	+25	11.93		+25	14.70
	+50	23.73		+50	29.28
γ_2	−50	−2.08	γ_2'	−50	−2.56
	−25	−1.04		−25	−1.28
	+25	1.04		+25	1.28
	+50	2.08		+50	2.57
u_1	−50	0.64	u_2	−50	0.70
	−25	0.38		−25	0.49
	+25	−0.30		+25	−0.39
	+50	−0.59		+50	−0.71
B_1	−50	0.02	B_2	−50	0.02
	−25	0.01		−25	0.01
	+25	−0.02		+25	−0.02
	+50	−0.01		+25	−0.01
h_{r_1}	−50	0.16	h_{r_2}	−50	0.18
	−25	0.08		−25	0.09
	+25	−0.08		+25	−0.09
	+50	−0.16		+50	−0.18
P_1	−50	−0.03	P_2	−50	−0.05
	−25	−0.01		−25	−0.02
	+25	0.01		+25	0.01
	+50	0.02		+50	0.02
E_1	−50	−20.11	E_2	−50	−25.92
	−25	−10.08		−25	−13.00
	+25	10.14		+25	13.10
	+50	20.37		+50	26.37
a_1	−50	−2.08	a_2	−50	−2.56
	−25	−1.04		−25	−1.28
	+25	1.04		+25	1.28
	+50	2.08		+50	2.57

Table 7.9
Sensitivity Analysis of Input Parameters for Complementary Product

Parameter	Changes (%)	STP (%)	Parameter	Changes (%)	STP (%)
H_{A_1}	−50	0.29	H_{A_2}	-50	0.26
	−25	0.14		−25	0.12
	+25	−0.13		+25	−0.12
	+50	−0.24		+50	−0.18
K_{A_1}	−50	0.14	K_{A_2}	−50	0.11
	−25	0.07		−25	0.05
	+25	−0.07		+25	−0.05
	+50	−0.13		+50	−0.12
S_{p_1}	−50	−19.21	S_{p_2}	−50	−19.95
	−25	−9.60		−25	−9.97
	+25	9.61		+25	9.98
	+50	19.22		+50	19.95
γ_2	−50	3.30	γ_2'	−50	4.12
	−25	1.65		−25	2.06
	+25	−1.64		+25	−2.06
	+50	−3.30		+50	−4.12
u_1	−50	0.70	u_2	−50	0.89
	−25	0.28		−25	0.37
	+25	−0.22		+25	−0.30
	+50	−0.69		+50	−0.83
B_1	−50	0.05	B_2	−50	0.05
	−25	0.03		−25	0.03
	+25	−0.03		+25	−0.02
	+50	−0.11		+25	−0.05
h_{r_1}	−50	0.06	h_{r_2}	−50	0.09
	−25	0.04		−25	0.05
	+25	−0.04		+25	−0.05
	+50	−0.08		+50	−0.09
d_{c_1}	−50	−2.10	d_{c_2}	−50	−2.47
	−25	−1.06		−25	−1.24
	+25	1.06		+25	1.23
	+50	2.14		+50	2.53
D_{c_1}	−50	−0.81	D_{c_2}	−50	−0.81
	−25	−0.41		−25	−0.40
	+25	0.41		+25	0.41
	+50	0.82		+50	0.81
P_1	−50	−0.03	P_2	−50	−0.03
	−25	−0.01		−25	−0.01
	+25	0.01		+25	0.01
	+50	0.01		+50	0.01

(*Continued*)

Table 7.9 (*Continued*)

Sensitivity Analysis of Input Parameters for Complementary Product

Parameter	Changes (%)	STP (%)	Parameter	Changes (%)	STP (%)
E_1	−50	−28.49	E_2	−50	−29.44
	−25	−14.28		−25	−14.77
	+25	14.36		+25	16.12
	+50	28.86		+50	29.96
b_1	−50	3.30	b_2	−50	4.13
	−25	1.65		−25	2.06
	+25	−1.65		+25	−2.06
	+50	−3.30		+50	−4.12

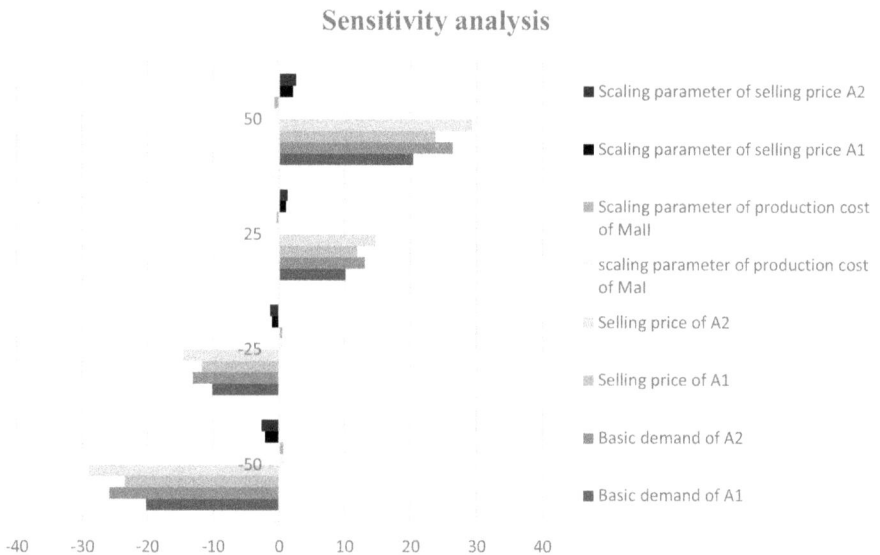

Figure 7.7   Sensitivity analysis for some crucial parameters.

It can be noticed from the result that the substitutability factor has a direct impact on the system's total profit, i.e., the increase in the positive elasticity factor will be an increase in total. That is, in the absence of the customer's choice product, the customer got satisfied with a similar type of the product. Thus, this study suggests the managers of the industry follow the strategy of storing similar types of products at their retail store to overcome the situation of unavailability of the original product. This helps the industry to retain its brand image and helps to gain more profit.

In the light of inventory management, the supply chain managers should be careful about product deterioration as this study ensures that product deterioration increases as time flies away. Thus, the more time inventory is held at the retailer store

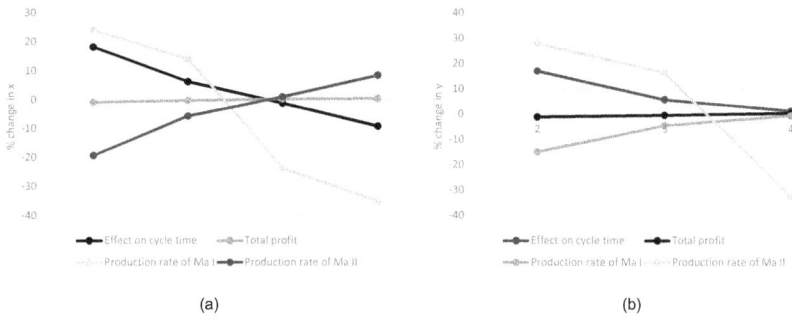

Figure 7.8 Effect of the number of shipments. (a) For manufacture-I and (b) for manufacturer-II.

more deteriorated inventory will generate in the system which is harmful to system profit. Hence the managers should adopt an SSMD policy rather than an SSSD policy for delivering products to the retailer.

In a traditional manufacturing system, the products are generated at a constant rate but the demand for that product may not be the same though out of time. In that case, the supply chain confronted a high loss. For example, the demand for winter clothes are for the pick at the winter time and down the other time. A flexible manufacturing system provides the opportunity for producing less or more products depending on the market demand by governing the manufacturing rate. It also provides the facility to reduce deteriorated products which happens when storing products for a long time.

Different activities associated with supply chain members like transporting items, setting up for manufacturing, holding products, and disposing of the defective and deteriorated items are caused because of environmental pollution. For this, governments of different nations are imposing environmental rules and regulations. Supply chain managers are highly recommended to follow those rules and regulations for a less pollutant environment.

7.9 CONCLUSIONS

This research deals with a two-player sustainable supply network management with the participation of two manufacturers and a single retailer. From the result, we found that if the two manufacturers produced complementary products, then they should produce with low negative cross-price elasticity of products on demand while for the substitutable products, it should be a highly positive cross-price elasticity on demand. The major finding of the research is the product's deterioration. The result suggests that an increment in the retailer cycle length will increase the product deterioration significantly which harms the system profit. For the reduction of retailer's cycle length, supply chain managers are suggested to adapt the SSMD policy for delivering products to the retailer rather than the SSSD policy. Adaptation of the SSMD policy will increase the number of transportation and so increase the emission to the environment. But a reduced number transportation will increase holding period and so the emission due to the same. As a net result, the adaptation of SSMD reduced the whole system emission by approx 8% to the environment and increased the profit by 10%. The deteriorated items increase the hazard to the society, but by following Government rules those items are disposed of allowing some extra cost, which helps the society.

This research can be extended considering multi-retailer, a setup cost reduction for manufacturers under a flexible production system. In this research deterioration cost is taken only for the retailer so, in the future, this can be applied for the manufacturer also. As the production system is variable, this can be extended considering random defective production generation and inspection policy.

APPENDIX A.

$$X_{11} = \frac{f_1(T_c + \sigma C_e)}{\tau} - \frac{(3+p_1)(d_{c_1} + D_{c_1} + \sigma D'_{c_1}) + (h_{r_1} + \sigma h'_{r_1})(1+p_1)}{2}$$

$$X_{12} = \frac{f_2(T_c + \sigma C_e)}{\tau} - \frac{(3+p_2)(d_{c_2} + D_{c_2} + \sigma D'_{c_2}) + (h_{r_2} + \sigma h'_{r_2})(1+p_2)}{2}$$

$$Y_{11} = \frac{(d_{c_1} + D_{c_1} + \sigma D'_{c_1} + h_{r_1} + \sigma h'_{r_1})(1+p_1)^2}{2}$$

$$Y_{12} = \frac{(d_{c_2} + D_{c_2} + \sigma D'_{c_2} + h_{r_2} + \sigma h'_{r_2})(1+p_2)^2}{2}$$

$$Y_{13} = \frac{(K_{A_1} + \sigma K'_{A_1})}{x} + \frac{(K_{A_2} + \sigma K'_{A_2})}{y} + O_{r_1} + O_{r_2} + 2F_T$$

$$Z_{11} = \frac{(d_{c_1} + D_{c_1} + \sigma D'_{c_1} + h_{r_1} + \sigma h'_{r_1})}{4} + \frac{(H_{A_1} + \sigma H'_{A_1})(x-1)}{2}$$

$$Z_{12} = \frac{(d_{c_2} + D_{c_2} + \sigma D'_{c_2} + h_{r_2} + \sigma h'_{r_2})}{4} + \frac{(H_{A_2} + \sigma H'_{A_2})(y-1)}{2}$$

$$Z_{13} = \frac{(H_{A_1} + \sigma H'_{A_1})(2-x)}{2}$$

$$Z_{14} = \frac{(H_{A_2} + \sigma H'_{A_2})(2-y)}{2}$$

APPENDIX B: FIRST AND SECOND-ORDER DERIVATIVE

The first-order derivatives are:

$$\frac{\partial STP}{\partial \rho_1} = -\left[\frac{\beta_1}{xT}\left(v_1 - \frac{u_1}{\rho_1^2}\right) - \frac{\beta_1^2 Z_{13} T}{\rho_1^2}\right]$$

$$\frac{\partial STP}{\partial \rho_2} = -\left[\frac{\beta_2}{yT}\left(v_2 - \frac{u_2}{\rho_2^2}\right) - \frac{\beta_2^2 Z_{14} T}{\rho_2^2}\right]$$

$$\frac{\partial STP}{\partial T} = -\left[-\frac{\beta_1 Y_{11}}{T^2} \ln\left(\frac{1+p_1}{1+p_1-T}\right) + \frac{\beta_1 Y_{11}}{T} \frac{1}{(1+p_1-T)} + \frac{\beta_2 Y_{12}}{T} \frac{1}{(1+p_2-T)} \right.$$

$$-\frac{Y_{13}}{T^2} - \frac{\beta_1}{xT^2}\left(c_1 + v_1\rho_1 + \frac{u_1}{\rho_1}\right) + \left(\beta_1 Z_{11} + \beta_2 Z_{12} + \frac{\beta_1^2}{\rho_1}Z_{13} + \frac{\beta_2^2}{\rho_2}Z_{14}\right)$$

$$\left. -\frac{\beta_2 Y_{12}}{T^2} \ln\left(\frac{1+p_2}{1+p_2-T}\right) - \frac{\beta_2}{yT^2}\left(c_2 + v_2\rho_2 + \frac{u_2}{\rho_2}\right) \right]$$

The second-order derivatives are:

$$\frac{\partial^2 STP}{\partial \rho_1^2} = -\frac{1}{\rho_1^3}\left[\frac{2u_1\beta_1}{xT} + 2\beta_1^2 Z_{13}T\right] = -\Theta_1, \text{say}$$

$$\frac{\partial^2 STP}{\partial \rho_2^2} = -\frac{1}{\rho_2^3}\left[\frac{2u_2\beta_2}{yT} + 2\beta_2^2 Z_{14}T\right] = -\Theta_2, \text{say}$$

$$\frac{\partial^2 STP}{\partial T^2} = -\left[\frac{2\beta_1 Y_{11}}{T^3} \ln\left(\frac{1+p_1}{1+p_1-T}\right) + \frac{2\beta_2 Y_{12}}{T^3} \ln\left(\frac{1+p_2}{1+p_2-T}\right) + \frac{2Y_{13}}{T^3} \right.$$

$$+\frac{2\beta_1}{xT^3}\left(c_1 + v_1\rho_1 + \frac{u_1}{\rho_1}\right) + \frac{2\beta_2}{yT^3}\left(c_2 + v_2\rho_2 + \frac{u_2}{\rho_2}\right) - \frac{\beta_1 Y_{11}(1+p_1)}{T(1+p_1-T)}$$

$$\left. -\frac{\beta_2 Y_{12}(1+p_2)}{T(1+p_2-T)} = -\Theta_3, \text{say} \right.$$

$$\frac{\partial^2 STP}{\partial \rho_1 \partial \rho_2} = 0$$

$$\frac{\partial^2 STP}{\partial \rho_2 \partial \rho_1} = 0$$

$$\frac{\partial^2 STP}{\partial T \partial \rho_1} = \frac{\partial^2 STP}{\partial \rho_1 \partial T} = \frac{\beta_1}{xT^2}\left(v_1 - \frac{u_1}{\rho_1^2}\right) + \frac{\beta_1^2 Z_{13}}{\rho_1^2} = \Theta_4, \text{say}$$

$$\frac{\partial^2 STP}{\partial T \partial \rho_2} = \frac{\partial^2 STP}{\partial \rho_2 \partial T} = \frac{\beta_2}{yT^2}\left(v_2 - \frac{u_2}{\rho_2^2}\right) + \frac{\beta_2^2 Z_{14}}{\rho_2^2} = \Theta_5, \text{say}$$

APPENDIX C: HESSIAN MATRIX

The optimality of the profit function is checked using Hessian matrix in the following:

$$|H_{11}| = \left|\frac{\partial^2 STP}{\partial \rho_1^2}\right| = -\frac{1}{\rho_1^3}\left[\frac{2u_1\beta_1}{xT} + 2\beta_1^2 Z_{13}T\right] < 0$$

$$|H_{22}| = \begin{vmatrix} \frac{\partial^2 \text{STP}}{\partial \rho_1^2} & \frac{\partial^2 \text{STP}}{\partial \rho_1 \partial \rho_2} \\ \frac{\partial^2 \text{STP}}{\partial \rho_2 \partial \rho_1} & \frac{\partial^2 \text{STP}}{\partial \rho_2^2} \end{vmatrix}, = \begin{vmatrix} -\Theta_1 & 0 \\ 0 & -\Theta_2 \end{vmatrix} = \Theta_1 \Theta_2 > 0$$

$$|H_{33}| = \begin{vmatrix} \frac{\partial^2 \text{STP}}{\partial \rho_1^2} & \frac{\partial^2 \text{STP}}{\partial \rho_1 \partial \rho_2} & \frac{\partial^2 \text{STP}}{\partial \rho_1 \partial T} \\ \frac{\partial^2 \text{STP}}{\partial \rho_2 \partial \rho_1} & \frac{\partial^2 \text{STP}}{\partial \rho_2^2} & \frac{\partial^2 \text{STP}}{\partial \rho_2 \partial T} \\ \frac{\partial^2 \text{STP}}{\partial T \partial \rho_1} & \frac{\partial^2 \text{STP}}{\partial T \partial \rho_2} & \frac{\partial^2 \text{STP}}{\partial T^2} \end{vmatrix} . = \begin{vmatrix} -\Theta_1 & 0 & \Theta_4 \\ 0 & -\Theta_2 & \Theta_5 \\ \Theta_4 & \Theta_5 & -\Theta_3 \end{vmatrix}.$$

$$= \Theta_1 (\Theta_5^2 - \Theta_2 \Theta_3) + \Theta_2 \Theta_4^2$$

REFERENCES

Bai, Q., Gong, Y. Y., Jin, M., and Xu, X. (2019). Effects of carbon emission reduction on supply chain coordination with vendor-managed deteriorating product inventory. *International Journal of Production Economics*, 208: 83–99.

Basiri, Z., and Heydari, J. (2017). A mathematical model for green supply chain coordination with substitutable products. *Journal of Cleaner Production*, 145(1): 232–249.

Bhuniya, S., Pareek, S., Sarkar, B., and Set, B. K. (2021). A smart production process for the optimum energy consumption with maintenance policy under a supply chain management. *Processes*, 9(1): 19.

Bhuniya, S., Sarkar, B., and Pareek, S. (2019). Multi-product production system with the reduced failure rate and the optimum energy consumption under variable demand. *Mathematics*, 7(5): 465.

Chang, H. (2014). An analysis of production inventory models with deteriorating items in a two-echelon supply chain. *Applied Mathematical Modelling*, 38(3): 1187–1191.

Daryanto, Y., MingWee, H., and Astanti, R. D. (2019). Three-echelon supply chain model considering carbon emission and item deterioration. *Transportation Research Part E: Logistics and Transportation Review*, 122: 368–383.

Dey, B. K., Sarkar, B., Sarkar, M., and Pareek, S. (2019). An integrated inventory model involving discrete setup cost reduction, variable safety factor, selling-price dependent demand, and investment. *RAIRO-Operations Research*, 53(1): 39–57.

Fang, F., Nguyen, T. D., and Currie, C. S. M. (2021). Joint pricing and inventory decisions for substitutable and perishable products under demand uncertainty. *European Journal of Operational Research*, 293(2): 594–602.

Feng, L. Chan, Y. L., and Cárdenas-Barrón, L. E. (2017). Pricing and lot-sizing polices for perishable goods when the demanddepends on selling price, displayed stocks, and expiration date. *International Production Production Economics*, 185: 11–20.

Gan, S. S., Suparno, I. N. P., and Widodo, B. (2018). Pricing decisions for short life-cycle product in a closed-loop supply chainwith random yield and random demands. *Operations Research Perspectives*, 5: 174–190.

Geda, A., Ghosh, V., Karamemis, G., and Vakharia, A. (2020). Coordination strategies and analysis of waste management supply chain. *Journal of Cleaner Production*, 256: 120298.

Ghiami, Y., and Williams, T. (2015). A two-echelon production-inventory model for deteriorating items with multiple buyers. *International Journal of Production Economics*, 159: 233–240.

Giri, B. C., Pal, H., and Maiti, T. (2017). A vendor-buyer supply chain model for time-dependent deteriorating item with preservation technology investment. *International Journal of Mathematics in Operational Research*, 10(4): 431–449.

Giri, R. N., Mondal, S. K., and Maiti, M. (2020). Bundle pricing strategies for two complementary products with different channel powers. *Annals of Operations Research*, 287: 701–725.

Halat, K., Hafezalkotob, A., and Sayadi, M. K. (2021). Cooperative inventory games in multi-echelon supply chains under carbon tax policy: Vertical or horizontal? *Applied Mathematical Modelling*, 99: 166–203.

Haung, H., He, Y., and Li, D. (2018). Pricing and inventory decisions in the food supply chain with production disruption and controllable deterioration. *Journal of Cleaner Production*, 180: 280–296.

Hemmati, M., Ghomi, S. M. T. F., and Sajadieh, M. S. (2018). Inventory of complementary products with stock-dependent demand under vendor-managed inventory with consignment policy. *International Journal of Science and Technology*, 25(4): 2347–2360.

Huang, H., and Ke, H. (2014). Pricing decision problem for substitutable products based on uncertainty theory. *Journal of Intelligent Manufacturing*, 28: 503–514.

Jalali, H., Ansaipoor, A. H., and Giovanni, P. D. (2020). Closed-loop supply chains with complementary products. *International Journal of Production Economics*, 229: 107757.

Kaur, M., Pareek, S., and Tripathi, R. P. (2016). Optimal ordering policy with non- increasing demand for time dependent deterioration under fixed life time production and permissible delay in payments. *International Journal of Operations Research*, 13(2): 035–046.

Li, G., He, X., Zhou, J., and Wu, H. (2019). Pricing, replenishment and preservation technology investment decisions for non-instantaneous deterioration item. *Omega*, 84: 114–126.

Li, R., Chan, Y. L., Chang, C. T., and Cárdenas-Barrón, L. E. (2017). Pricing and lot-sizing policies for perishable products with advance-cash-credit payments by a discounted cash-flow analysis. *International Journal of Production Economics*, 193: 578–589.

Liu, L., Zhao, Q., and Goh, M. (2020). Perishable material sourcing and final product pricing decisions for two-echelon supply chain under price-sensitive demand. *Computers & Industrial Engineering*, 156: 107260.

Liu, Z., Lang, L., Hu, B., Shi, L., Huang, B., and Zhao, Y. (2021). Emission reduction decision of agricultural supply chain considering carbon tax and investment cooperation. *Journal of Cleaner Production*, 294: 126305.

Luo, R., Zhou, L., Song, Y., and Fan, T. (2022). Evaluating the impact of carbon tax policy on manufacturing and remanufacturing decisions in a closed-loop supply chain. *International Journal of Production Economics*, 245: 108408.

Maihami, R., Govindan, K., and Fattahi, M. (2019). The inventory and pricing decisions in a three-echelon supply chain of deteriorating items under probabilistic environment. *Transportation Research Part E: Logistics and Transportation Review*, 131: 118–138.

Mishra, U., Wu, J. Z., Tsao, Y. C., and Tseng, M. L. (2020). Sustainable inventory system with controllable non-instantaneous deterioration and environmental emission rates. *Journal of Cleaner Production*, 244: 118807.

Moubed, M., Boroumandzad, Y., and Nadizadeh, A. (2021). A dynamic model for deteriorating products in a closed-loop supply chain. *Simulation Modelling Practice and Theory*, 108: 102269.

Ngendakuriyo, F., and Taboubo, S. (2016). Pricing strategies of complementary products in distribution channels: A dynamic approach. *Dynamic Games and Applications*, 7: 48–66.

Palanivel, M., and Uthayakumar, R. (2017). A production-inventory model with promotional effort, variable production cost and probabilistic deterioration. *International Journal of System Assurance Engineering and Management*, 8: 290–300.

Pan, Q., He, X., Skouri, K., Chen, S. C., and Teng, J. T. (2018). An inventory replenishment system with two inventory-based substitutable products. *International Journal of Production Economics*, 204: 135–147.

Pattnaik, M., and Gahan, P. (2020). Preservation effort effects on retailers and manufacturers in integrated multi-deteriorating item discrete supply chain model. *OPSEARCH*, 58: 276–329.

Poormoaied, S., and Atan, Z. (2020). A continuous review policy for two complementary products with interrelated demand. *Computers & Industrial Engineering*, 150: 106980.

Rout, C., Paul, A., Kumar, R. S., Chakraborty, D., and Goswami, A. (2020). Supply chain for deteriorating item and imperfect production under different carbon emission regulations. *Journal of Clenear Production*, 272: 122170.

Roy, A., Sana, S. S., and Chaudhuri, K. (2018). Optimal pricing of competing retailers under uncertain demand-a two layer supply chain model. *Annals of Operations Research*, 260: 481–500.

Sarkar, B., Gangulay, B., Sarkar, M., and Pareek, S. (2016). Effect of variable transportation and carbon emission in a three-echelon supply chain model. *Transportation Research Part E: Logistics and Transportation Review*, 91: 112–128.

Sarkar, B., Mridha, B., Pareek, S., Sarkar, M., and Thangavelu, L. (2021b). A flexible biofuel and bioenergy production system with transportation disruption under a sustainable supply chain network. *Journal of Clenear Production*, 317: 128079.

Sarkar, B., Omair, M., and Choi, S. B. (2018). A multi-objective optimization of energy, economic, and carbon emission in a production model under sustainable supply chain management. *Applied Sciences*, 8(10): 1744.

Sarkar, B., Sarkar, M., Ganguly, B., and Cárdenas-Barón, L. E. (2021a). Combined effects of carbon emission and production quality improvement for fixed lifetime products in a sustainable supply chain management. *International Journal of Production Economics*, 231: 107867.

Sarkar, B., and Saren, S. (2017). Ordering and transfer policy and variable deterioration for a warehouse model. *Hacettepe Journal of Mathematics and Statistics*, 46(5): 985–1014.

Sarkar, M., and Lee, Y. H. (2017). Optimum pricing strategy for complementary products with reservation price in a supply chain model. *American Institute of Mathematical Sciences*, 13(3): 1553–1586.

Sebatjanea, M., and Adetunji, O. (2020). A three-echelon supply chain for economic growing quantity model with price- and freshness-dependent demand: Pricing, ordering and shipment decisions. *Operations Research Perspectives*, 7: 100153.

Shah, N. H., Chaudhari, U., and Jani, M. Y. (2016). An integrated production-inventory model with preservation technology investment for time-varying deteriorating item under time and price sensitive demand. *International Journal of Inventory Research*, 3(1): 81–98.

Saha, S., Nielsen, I. E., and Moon, I. (2021). Strategic inventory and pricing decision for substitutable products. *Computers & Industrial Engineering*, 160: 107570.

Shaikh, A. A., and Cárdenas-Barrón, L. E. (2020). An EOQ inventory model for non-instantaneous deteriorating products with advertisement and price sensitive demand under oredr quantity dependent trade credit. *Revista Investigacion Operacional*, 41(2): 168–187.

Sing, S. R., Khurana, D., and Tayal, S. (2017). An economic order quantity model for deteriorating products having stock dependent demand with trade credit period and preservation technology. *Uncertain Supply Chain Management*, 4: 29–42.

Taleizadeh, A. A., Babaei, M. S., Sana, S. S., and Sarkar, B. (2019). Pricing decision within an inventory model for complementary and substitutable products. *Mathematics*, 7: 568.

Tayyab, M., and Sarkar, B. (2016). Optimal batch quantity in a cleaner multi-stage lean production system with random defective rate. *Journal of Cleaner Production*, 139: 922–934.

Tiwari, S., Cárdenas-Barrón, L. E., Malik, A. I., and Jaggi, C. K. (2021). Retailer's credit and inventory decisions for imperfect quality and deteriorating items under two-level trade credit. *Computers & Operations Research*, 138: 105617.

Wang, L., Song, H., and Wang, Y. (2017). Pricing and service decisions of complementary products in a dual-channel supply chain. *Computers & Industrial Engineering*, 105: 223–233.

Wang, F., Diabat, A., and Wu, L. (2021). Supply chain coordination with competing suppliers under price-sensitive stochastic demand. *International Journal of Production Economics*, 234: 108020.

Wei, J., Zhao, J., and Li, Y. (2015). Price and warranty period decisions for complementary products with horizontal firms' cooperation/noncooperation strategies. *Journal of Cleaner Production*, 105: 86–102.

Xu, J., Chen, Y., and Bai, Q. (2016). A two-echelon sustainable supply chain coordination under cap-and-trade regulation. *Journal of Cleaner Production*, 135: 42–56.

Yadav, D., Kumari, R., Kumar, N., and Sarkar, B. (2021). Reduction of waste and carbon emission through the selection of items with cross-price elasticity of demand to form a sustainable supply chain with preservation technology. *Journal of Clenar Production*, 297: 126298.

Yan, C., Banerjee, A., and Yang, L. (2014). An integrated production–distribution model for a deteriorating inventory item. *International Journal of Production Economics*, 133(1): 228–232.

Yu, W., Shang, H., and Han, R. (2020). The impact of carbon emissions tax on vertical centralized supply chain channel structure. *Computers & Industrial Engineering*, 141(3–4): 106303.

Zhao, J., Wei, J., and Li, Y. (2014). Pricing decisions for substitutable products in a two-echelon supply chain with firms' different channel powers. *International Journal of Production Economics*, 153: 243–252.

Notes

[1] https://comparecamp.com/food-waste-statistics/#::text=The%20Big%20Picture %201%20Around%20one-third%20of%20the,food%20to%20feed%20the%20 hungry.%20More%20items...%20 (accessed 20th January 2022 at 01:10 AM).

[2] https://www.google.com/search?q=amount+of+e-+waste+in+india+yearly+ generated&oq=amount+of+e-+waste+in+india+yearly+generated&aqs=chrome.. 69i57j33i22i29i30l2.15736j0j7&sourceid=chrome&ie=UTF-8.

[3] https://www.downtoearth.org.in/news/waste/solid-waste-management-rules-2016- 53443.

[4] https://earth.org/what-countries-have-a-carbon tax/ (accessed 26th October 2021 at 09:20 AM).

8 Framing the Slip Flow of TiO$_2$ Nanofluid Past an Inclined Porous Plate Coexistence of Solar Radiation: An Application of Differential Equation

Tanmoy Chakraborty
SRM Institute of Science and Technology

Arunava Majumder
Lovely Professional University

CONTENTS

DOI: 10.1201/9781003227847-8

8.1 INTRODUCTION

Solar energy is one of the immense sources of renewable energy on our planet which plays the most important role in our daily life. Since it is not consumable energy. Nowadays researchers are showing their interest in using this energy in different technologies such as solar vehicles, solar electricity, solar heating, solar photovoltaic cells, etc. The concept of the use of solar energy-absorbent particles in liquid was pioneered by Hunt (1978). Suspensions of nanoparticles (diameters of nano-materials lie within 50–100 nm) (Ag, Cu, Au, etc., or their oxides or nanotubes) in liquid, recognized as nanofluid, have made extensive interest for their potential to make a dramatic enhancement in the heat transport mechanism. Hence the research of heat transport coexisting solar radiation has acknowledged considerable attention among scientists and engineers in the recent era. Many researchers (Choi and Eastman 1995; Gangadhar et al. 2020; Waini et al. 2020; Roy and Pop 2020) have acknowledged the efficiency of nanofluids in heat transport mechanisms owing to their effectively high thermal conductivity. The upshots for thermal radiation on the heat transport problems have been represented by Hossain and Takhar (1996) and Kandasamy et al. (2013) studied Cu-nanofluid flow past a permeable wedge coexisting of thermal stratification owing to solar energy radiation (Rashidi et al. 2013) analyzed the entropy generation of nanofluid due to a rotating porous disk.

In engineering systems, the transfer of heat in free convection drift is reasonably vital for its wide range of applications. Numerous researchers (Tiwari and Das 2007; Wang and Majumder 2007; Shehzad et al. 2015; Das et al. 2017; Chakraborty et al. 2018; Ullah et al. 2021; Molli and Naikoti 2020) have reported the consequences of natural convection heat transport considering different flow situations in diverse geometries. Free convective heat transports of nanofluids in the occurrence of solar energy radiation have ample applications in automotive technology, solar-based power systems, building heating, solar technology, etc. An informative review of work has been done by Karak and Pramuanjaroenkij (2009) on heat transfer owing to free convection in nanofluids (Gorla and Chamka, 2011) investigated the free convection flow of nanofluid through a flat plate immersed in an absorbent medium. The double-diffusive natural convective heat transport phenomenon of nanofluid over an inclined plate was analyzed by Das et al. (2019).

In most of the earlier researches, no-slip boundary conditions are entertained but this is no longer valid from the study of interfacial thermodynamic equilibrium. According to the experimental values of Knudsen number (Kn) it is confirmed that when K_n lies between [0.01, 0.1] the flow is considered as a slip flow. When Kn deviates from the limit ($Kn < 0.01$) of continuum flow then no-slip condition failed to justify the flow behavior at the boundary owing to the small collision frequency. In many practical situations like flow through micro or nanochannels, specifically at high-speed flow through the micro-electro-mechanical-system (MEMS) the particles at the wall do not receive the velocity of the wall. Hence the particle receives a small tangential velocity, so slip occurs. Also if the boundary wall is continuously heated from surrounding then inside the boundary layer thermal jump occurs (Navier 1823) introduced a momentum boundary condition to describe this slip phenomenon.

According to his hypothesis, the fluid velocity at the wall is in proportional relation with applied tangential stress. The suggestion of (Navier 1823) was incorporated by several researchers (Uddin et al. 2016; Das et al. 2016; Zheng et al. 2013; Yazdi et al. 2011). Many exciting articles on heat transfer with slip flow can be found in the studies (Das et al. 2015, 2016; Khan et al. 2021; Wahid et al. 2021).

Heat transport owing to convection and fluid passing through a permeable medium is of immense attention because of its several applications in engineering and geophysical fields. Many researches (Cheng and Minkowycz 1977; Lai and Kulacki 1991; Ahmad et al. 2021; Raizah et al. 2021) deal with the convective flow within the porous medium. But many of the former researches on absorbent media have used Darcy's law. According to that the volume-averaged velocity is proportional to the pressure gradient. Owing to significant applications of Darcy-Forchheimer effects on convective fluid flow through different geometrical surfaces are found in the researches (Vafai and Tien 1982; Hong et al. 1987; Kaviany 1987; Chakraborty et al. 2017; Rami et al. 2001; Ishak 2010; Mukhopadhyay et al. 2012; Fathalah and Elsayed 1980; Chamka 1997; Chamka et al. 2002).

Inspired by the prior researches, here we have investigated the slip flow of TiO_2-H_2O nanofluid over a slanted absorbent plate entrenched in a non-Darcy permeable medium in the occurrence of solar radiation. The numerical upshots reveal some significant effects of pertinent parameters on the flow which are presented vividly through graphs, bar diagrams, and tables. Also, the results are in excellent harmony with the existing research papers under identical conditions. To the best of our knowledge, no such work has been reported yet, hence we can claim the uniqueness of the problem. This chapter is an effort to bring some applications of differential equations in the emerging fields of science and engineering.

8.2 FORMULATION OF MATHEMATICAL MODEL

8.2.1 FLOW ANALYSIS

Here, we suppose a two dimensional, viscous, steady, mixed convective laminar drift of a TiO_2-H_2O nanofluid through a heated inclined plate drenched in a fluid soaked absorbent medium. The permeable flat plate is heated by solar radiation and of electrically non-conducting, non-absorbing and non-reflecting. The physical abstract is offered through Figure 8.1 where along the plate x-axis is considered and y-axis is perpendicular to it and in outward direction toward the flow. The drift being confined to y ¿ 0. The solar radiation is acting perpendicularly to the plate. Thermal conductivity of TiO_2 nanoparticles is significantly higher than water. Also among all metallic/metallic oxide nanoparticles, TiO_2 nanoparticles are easily available nanoparticles and their heat transport mechanism is excellent. Moreover, TiO_2 nanoparticles are heat-absorbent oxide nanoparticles therefore it absorbs the incident solar heat flux Q_r and passes it to the adjacent fluid due to convection. Under the prior supposition, the fundamental partial differential boundary layer equations of our prescribed model are (Chakraborty et al. 2017; Mukhopadhyay et al. 2012):

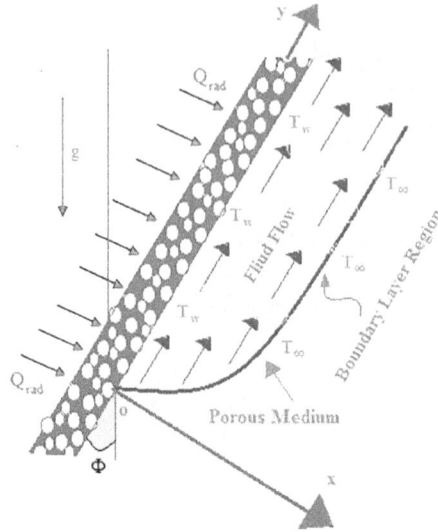

Figure 8.1 Boundary layer flow of TiO_2 nanofluid over an inclined stretching sheet embedded in a porous medium in the presence of solar radiation.

$$\frac{\partial u}{\partial x} + \frac{\partial w}{\partial y} = 0 \tag{8.1}$$

$$u\frac{\partial u}{\partial x} + w\frac{\partial u}{\partial y} = \nu_{nf}\frac{\partial^2 u}{\partial y^2} + \frac{(\rho\beta)_{nf}}{\rho_{nf}}g(T - T_\infty)\cos\Phi - \frac{\nu_{nf}}{D}(u - u_\infty) - \frac{F'}{\sqrt{D}}(u^2 - u_\infty^2) \tag{8.2}$$

$$u\frac{\partial T}{\partial x} + w\frac{\partial T}{\partial y} = \alpha_{nf}\frac{\partial^2 T}{\partial y^2} - \frac{1}{(\rho C_p)_{nf}}\frac{\partial Q_r}{\partial y} \tag{8.3}$$

Here the velocity components are u, w along x and y direction respectively; the suffixes 'f' and 'nf' represents for primary fluid and nanofluid respectively; $\nu_f = \frac{\mu_f}{\rho_f}$ represents the kinematic viscosity; $D = D_0 x$ symbolizes the Darcy permeability of the absorbent medium with D_0 is the initial permeability; $F' = \frac{F'_0}{\sqrt{x}}$ exemplifies the Forchheimer resistance factor with F'_0 is the Forchheimer constant; u_∞ stands for the stream velocity away from wall and C_p signifies the specific heat at constant pressure. With the aid of Rosseland approximation for nonlinear radiation (Sparrow and Cess 1978) we can convey $Q_r = -\frac{4\sigma}{3k^*}\frac{\partial T^4}{\partial y}$ where σ denotes the Stefan-Boltzman constant and k^* symbolizes the mean absorption coefficient. Now if we impose the dimensionless form of temperature T i.e. $\theta = \frac{T - T_\infty}{T_w - T_\infty}$ as in equation (8.9) then we receive $T = (T_w - T_\infty)(\theta + r_T)$, where r_T represents temperature ratio factor. Here the plate makes an angle of inclination with the vertical axis as Φ whose value is considered within $[0°, 60°]$. Also g denotes the acceleration due to gravity.

8.2.2 THERMOPHYSICAL PROPERTIES OF THE NANOFLUIDS

In this chapter, we have taken TiO_2-H_2O nanofluid which has different thermophysical properties than H_2O. The mathematical models of density (ρ_{nf}), dynamic viscosity (μ_{nf}), thermal expansion coefficient $(\rho\beta)_{nf}$, effective heat capacity $(\rho C_p)_{nf}$, and thermal diffusivity (α_{nf}) (Das et al. 2017; Kaviany 1987; Chakraborty et al. 2017) of the nanofluids are described as below:

$$\left.\begin{array}{l} \rho_{nf} = (1-\varphi)\rho_f + \varphi\rho_s, \quad \mu_{nf} = \frac{\mu_f}{(1-\varphi)^{2.5}}, \quad (\rho\beta)_{nf} = (1-\varphi)(\rho\beta)_f + \varphi(\rho\beta)_s, \\ (\rho C_p)_{nf} = (1-\varphi)(\rho C_p)_f + \varphi(\rho C_p)_s, \quad \alpha_{nf} = \frac{\kappa_{nf}}{(\rho C_p)_{nf}} \end{array}\right\}$$

Maxwell (1891) pioneered the mathematical model of measuring the effective thermal conductivity (κ_{nf}) of nanofluid. In the experimental studies it is found that the model described by Maxwell (1891) is well fitted for homogeneous and low-volume concentrated nanofluid with arbitrarily spread and uniformly sized non-contracting spherical particles and it is formulated as:

$$\frac{\kappa_{nf}}{\kappa_f} = \left\{ \frac{(\kappa_s + 2\kappa_f) - 2\varphi(\kappa_f - \kappa_s)}{(\kappa_s + 2\kappa_f) + 2\varphi(\kappa_f - \kappa_s)} \right\}$$

Here φ symbolizes for the volume fraction of nanoparticles; when $\varphi = 0$ then it becomes regular fluid. Here κ_f and κ_s exemplify the thermal conductivities of the primary fluid and nanoparticles; β_f and β_s signify the thermal expansion coefficients of the primary fluid and nanoparticles. The experimental values of the thermophysical properties of the TiO_2-H_2O which are used in this chapter are produced in Table 8.1.

8.2.3 BOUNDARY CONDITIONS

The apposite boundary conditions from the physical standpoint of the present problem can be described as:

$$\left.\begin{array}{l} u = u_s(x), \; w = \pm w_s(x), \; T = T_w + T_s \, at \, y = 0 \\ u \rightarrow u_\infty(x), \; T \rightarrow T_\infty \; as \; y \rightarrow \infty \end{array}\right\} \qquad (8.4)$$

Table 8.1

All the Thermophysical Properties of H_2O and TiO_2 Nanoparticles [8,12] Are Presented Below

Physical Properties	H_2O	TiO_2
$\rho(kg/m^3)$	997.1	4250
$C_p(J/kg\,K)$	4179	686.2
$\kappa(W/m\,K)$	0.613	8.9538
$\beta \times 10^{-5}(K^{-1})$	21	0.9

These values are collected from experimental studies and used in this chapter for numerical computation.

The wall temperature considered as the constant value T_w. The far away temperature is assumed to be constant T_∞. Here, u_∞ is assumed as uniform free stream velocity. It is noteworthy to mention that $T_w > T_\infty$ aids the flow due to buoyancy effects have a positive component along u_∞ on the contrary $T_w < T_\infty$ oppose the flow since buoyancy effects are in reverse direction of u_∞. Since the plate is permeable in nature and soaked in absorbent medium so, fluid passes through the holes of the plate owing to suction/injection. Near the wall $w_s(x) = w_0/\sqrt{x}$ denotes the velocity of injection ($w_0 < 0$) or suction ($w_0 > 0$) of the fluid. Here $u_s(x)$ exemplifies the slip velocity which is in proportion to local wall shear stress and defined as $u_s(x) = l_v \frac{\partial u}{\partial y}$ where $l_v = l'_v \sqrt{x}$ signifies hydrodynamic slip factor; T_s denotes the thermal slip and defined as $T_s = l_T \frac{\partial T}{\partial y}$ where $l_T = l'_T \sqrt{x}$ represents thermal slip factor. In case of $l'_T = l'_v = 0$ exemplifies no-slip flow.

8.2.4   TRANSFORMED EQUATIONS

Now setting stream function ψ to satisfy the continuity equation (8.1) and defined as

$$u = \frac{\partial \psi}{\partial y}, \quad w = -\frac{\partial \psi}{\partial x} \Bigg\} \tag{8.5}$$

Then equation (8.2) and (8.3) along with boundary conditions (8.4) takes the following form

$$\frac{\partial \psi}{\partial y}\frac{\partial^2 \psi}{\partial x \partial y} - \frac{\partial \psi}{\partial x}\frac{\partial^2 \psi}{\partial y^2} = v_{nf}\frac{\partial^3 \psi}{\partial y^3} + \frac{(\rho \beta)_{nf}}{\rho_{nf}}g(T - T_\infty)\cos\Phi$$
$$- \frac{v_{nf}}{D}\left(\frac{\partial \psi}{\partial y} - u_\infty\right) - \frac{F'}{\sqrt{D}}\left(\left(\frac{\partial \psi}{\partial y}\right)^2 - u_\infty^2\right) \tag{8.6}$$

$$\frac{\partial \psi}{\partial y}\frac{\partial T}{\partial x} - \frac{\partial \psi}{\partial x}\frac{\partial T}{\partial y} = \alpha_{nf}\frac{\partial^2 T}{\partial y^2} - \frac{1}{(\rho C_p)_{nf}}\frac{\partial Q_r}{\partial y} \tag{8.7}$$

Associated boundary condition is

$$\frac{\partial \psi}{\partial y} = l'_v \sqrt{x}\frac{\partial^2 \psi}{\partial y^2}, \frac{\partial \psi}{\partial x} = \mp w_s(x), \ T = T_w + l'_T\sqrt{x}\frac{\partial T}{\partial y} \text{ at } y = 0 \ \Bigg\}$$
$$\frac{\partial \psi}{\partial y} \rightarrow u_\infty(x), \ T \rightarrow T_\infty \text{ as } y \rightarrow \infty \tag{8.8}$$

Now to transform the leading PDEs (equations 8.6 and 8.7) into ODEs we introduce similarity transformations as below (Das et al. 2015; Mukhopadhyay et al. 2012):

$$\theta = \frac{T - T_\infty}{T_w - T_\infty}, \quad \eta = y\sqrt{\frac{u_\infty}{v_f x}}, \quad \psi = \sqrt{u_\infty v_f x}f(\eta) \Bigg\} \tag{8.9}$$

Using the relations (8.9) the governing equations (8.6) and (8.7) and the boundary conditions are converted into the following form:

$$f''' + (1-\varphi)^{\frac{5}{2}}\left[\begin{array}{l} 0.5ff''\left(1-\varphi+\varphi\frac{\rho_s}{\rho_f}\right) + \left(1-\varphi+\varphi\frac{(\rho\beta)_s}{(\rho\beta)_f}\right)\delta\theta\cos\Phi \\ -Da\left(1-\varphi+\varphi\frac{\rho_s}{\rho_f}\right)(f'-1) - F_s\left(1-\varphi+\varphi\frac{\rho_s}{\rho_f}\right)(f'^2-1) \end{array}\right] = 0 \tag{8.10}$$

$$\frac{\kappa_{nf}}{\kappa_f}\theta'' + \frac{4}{3}R\left\{(r_T+\theta)^3\,\theta'\right\}' + \frac{1}{2}\Pr f\theta'\left(1-\varphi+\varphi\frac{(\rho C_p)_s}{(\rho C_p)_f}\right) = 0 \qquad (8.11)$$

and the boundary conditions convert to

$$\left.\begin{array}{l} f(\eta) = f_w,\ f'(\eta) = \xi f''(\eta),\ \theta(\eta) = 1 + \zeta\theta'(\eta)\ \text{ at } \eta = 0 \\ f'(\eta) = 1, \theta(\eta) = 0 \text{ at } \eta \to \infty \end{array}\right\} \qquad (8.12)$$

Here f represents a non-dimensional stream function; $f'(\eta)$ signifies the non-dimensional velocity near the wall; $\theta(\eta)$ exemplifies the non-dimensional temperature distribution across the boundary region. The following formulas represent the physical parameters involved in the problem which are the buoyancy parameter (δ), porosity parameter (Da), inertial parameter (F_s), conduction radiation parameter (R), temperature ratio factor (r_T), Prandtl number (Pr), suction/injection parameter (f_w), velocity slip parameter (ξ), thermal slip parameter (ζ) and they are symbolized as:

$$\left.\begin{array}{l} \delta = \frac{gx(\rho\beta)_f(T_w-T_\infty)}{u_\infty^2\rho_f},\ \mathrm{Da} = \frac{v_f}{D_0u_\infty},\ F_s = \frac{F_0'}{\sqrt{D}},\ R = \frac{4\sigma(T_w-T_\infty)^3}{k^*(\rho C_p)_f}, \\[2mm] r_T = \frac{T_\infty}{(T_w-T_\infty)},\ \Pr = \frac{\mu_f(\rho C_p)_f}{\rho_f\kappa_f},\ f_w = \frac{\pm 2w_0}{\sqrt{u_\infty v_f}},\ \xi = l'_v\sqrt{\frac{u_\infty}{v_f}},\ \zeta = l'_T\sqrt{\frac{u_\infty}{v_f}} \end{array}\right\} \qquad (8.13)$$

It is noteworthy to mention that $\delta > 0$ boosts the flow whereas $\delta < 0$ oppose the flow and $\delta = 0$ represents no prevalence of free convection. Here throughout this chapter, the prime denotes differentiation with respect to η.

8.2.5 PARAMETERS OF ENGINEERING SIGNIFICANCE

The parameters of engineering and the physical importance of the current problem are the local Nusselt number (Nu) and the local skin friction coefficient (C_f).

Now the C_f (rate of shear stress) is defined as

$$C_f = \frac{\mu_{nf}}{\rho_f u_\infty^2}\left(\frac{\partial u}{\partial y}\right)_{y=0} = \frac{1}{(1-\varphi)^{2.5}}(\mathrm{Re}_x)^{-\frac{1}{2}} f''(0) \qquad (8.14)$$

where $\mathrm{Re}_x = \frac{u_\infty x}{v_f}$ signifies the local Reynolds number. Hence the reduced skin friction coefficient C_{fr} can be expressed as

$$C_{fr} = \mathrm{Re}_x^{\frac{1}{2}} C_f = \frac{1}{(1-\varphi)^{2.5}} f''(0) \qquad (8.15)$$

The Nu (rate of heat transfer) is defined as

$$\begin{aligned} Nu &= \frac{x\kappa_{nf}}{\kappa_f(T_w-T_\infty)}\left(\frac{\partial T}{\partial y}\right)_{y=0} - \frac{4\sigma}{3k^*}\left(\frac{\partial T^4}{\partial y}\right)_{y=0} \\ &= -(\mathrm{Re}_x)^{\frac{1}{2}}\frac{\kappa_{nf}}{\kappa_f}\theta'(0)\left[1+\frac{4}{3}R(r_T+\theta(0))^3\right] \end{aligned} \qquad (8.16)$$

Thus the reduced Nusselt number Nu_r can be expressed as

$$Nu_r = Nu\,(\mathrm{Re}_x)^{-\frac{1}{2}} = -\frac{\kappa_{nf}}{\kappa_f}\theta'(0)\left[1+\frac{4}{3}R(r_T+\theta(0))^3\right] \qquad (8.17)$$

8.3 NUMERICAL PROCEDURE

8.3.1 METHOD OF SOLUTION

To avoid the complexity in the derivation of the solution we have used the following notations throughout the calculation as:

$$A_1 = (1-\varphi)^{\frac{5}{2}}\left(1-\varphi+\varphi\frac{\rho_s}{\rho_f}\right), \quad A_2 = (1-\varphi)^{\frac{5}{2}}\left(1-\varphi+\varphi\frac{(\rho\beta)_s}{(\rho\beta)_f}\right), \\ A_3 = \left(1-\varphi+\varphi\frac{(\rho C_p)_s}{(\rho C_p)_f}\right), \quad A_4 = \frac{\kappa_{nf}}{\kappa_f} \qquad (8.18)$$

Using equation (8.18) the governing equations (8.10) and (8.11) becomes

$$f''' + \frac{A_1}{2}ff'' + A_2\delta\theta\cos\Phi - \mathrm{Da}A_1\left(f'-1\right) - F_sA_1\left(f'^2-1\right) = 0 \qquad (8.19)$$

$$A_5\theta'' + \frac{4}{3}R\left\{(r_T+\theta)^3\,\theta'\right\}' + \frac{A_4}{2}\Pr f\theta' = 0 \qquad (8.20)$$

Now the equations (8.19) and (8.20) converted to a system of simultaneous first order equations as below

$$\begin{aligned} f' &= T_1, \\ T_1' &= T_2, \\ T_2' &= -\frac{A_1}{2}fT_2 - A_2\delta\theta\cos\Phi + \mathrm{Da}A_1\left(T_1-1\right) + F_sA_1\left(T_1^2-1\right), \\ \theta' &= T_3, \\ T_3' &= -\frac{4R\left(r_T+\theta\right)^2 T_3^2 + \frac{A_4}{2}\Pr fT_3}{A_5 + \frac{4}{3}R\left(r_T+\theta\right)^3} \end{aligned} \qquad (8.21)$$

the given initial conditions are $f(0) = f_w$, $T_1(0) = \xi T_2(0)$, $\theta(0) = 1 + \zeta T_3(0)$. To carry out the integration by Runge-Kutta 4th order scheme, let us consider $T_2(0) = p$, $T_3(0) = q$ where p,q are the constant which are to be determined using the boundary conditions $T_1(\infty) = 1, \theta(\infty) = 0$. The approximation of p,q can be done by the Gill-based shooting practice. The mesh size $\Delta\eta = 0.0001$ is implemented to gain the numerical fallouts by replacing the ambient boundary condition by $\eta = 12$. The procedure is performed until the desired fallouts are correct up to the preferred degree of exactness, say 10^{-9}. The calculation has performed by computer algebra software Maple 17.

8.3.2 CODE VERIFICATION

To examine the validity of the present numerical code, the acquired outcomes have been compared with Mukhopadhyay et al. (2012) under equal conditions. Graphical comparison is drawn for dimensionless temperature and velocity profiles of the present model $\varphi = \delta = \xi = \zeta = 0$ and $\Phi = \frac{\pi}{2}$ as demonstrated in Figure 8.2a and b. One can notice from the figures that the present numerical fallouts and the existing ones are in good accord, hence justifying the application of the current code for this model.

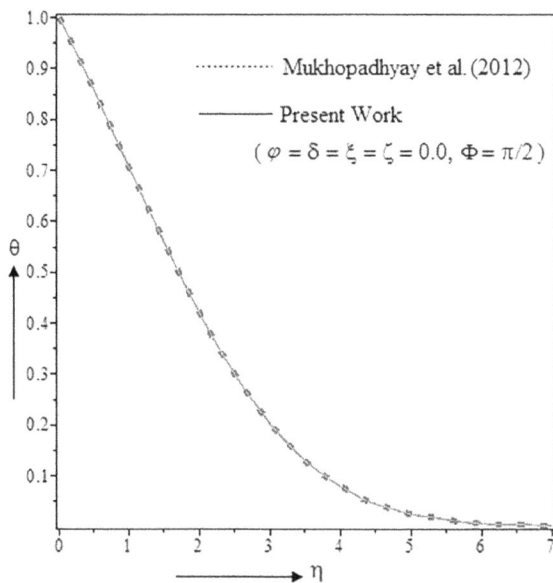

(a)

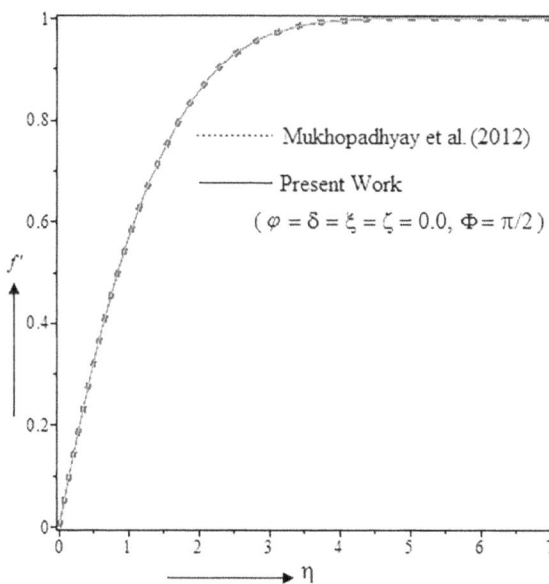

(b)

Figure 8.2 Comparisons of dimensionless temperature and velocity profiles with Mukhopadhyay et al (2012) and the present study are drawn in panels (a) and (b) respectively under identical conditions. Both the diagrams show that the results are in excellent agreement and hence justify the use of the present code.

Table 8.2

Variations of Pertinent Parameters on Nu_r and C_{fr} Has Been Observed in This Table

ξ	ζ	Da	F_s	R	Nu_r		C_{fr}	
					$H_2O(\varphi = 0.0)$	$TiO_2\text{-}H_2O(\varphi = 0.1)$	$H_2O(\varphi = 0.0)$	$TiO_2\text{-}H_2O(\varphi = 0.1)$
0.1	0.2	0.1	0.2	0.2	0.476740	0.509564	0.900261	1.450000
0.3	—	—	—	—	0.494904	0.525160	0.792553	1.285855
0.5	—	—	—	—	0.508280	0.536814	0.703012	1.147337
0.7	—	—	—	—	0.518403	0.545737	0.629186	1.031711
0.3	0.2	—	—	—	0.494904	0.525160	0.792553	1.285855
—	0.5	—	—	—	0.435998	0.470374	0.781896	1.277908
—	0.8	—	—	—	0.387200	0.424949	0.772948	1.270972
—	1.2	—	—	—	0.335303	0.375868	0.763322	1.263081
—	0.2	0.1	—	—	0.494904	0.525160	0.792553	1.285855
—	—	0.4	—	—	0.503147	0.533699	0.882831	1.451209
—	—	0.7	—	—	0.509193	0.539836	0.955571	1.583338
—	—	1.0	—	—	0.513913	0.544557	1.016856	1.694099
—	—	0.1	0.2	—	0.494904	0.525160	0.792553	1.285855
—	—	—	0.4	—	0.503731	0.534281	0.884567	1.453075
—	—	—	0.6	—	0.510159	0.540795	0.959472	1.588473
—	—	—	0.8	—	0.515146	0.545774	1.023022	1.702942
—	—	—	0.2	0.2	0.494904	0.525160	0.792553	1.285855
—	—	—	—	0.6	0.361015	0.425440	0.803979	1.292171
—	—	—	—	0.9	0.306166	0.375819	0.810615	1.296156
—	—	—	—	1.5	0.242427	0.310072	0.820799	1.302726

To obtain the variation, some default values of some parameters are considered as $\delta = 0.2, r_T = 0.2,$ $Pr = 6.7, f_w = 0.5, \Phi = 45°$.

8.4 RESULTS AND DISCUSSIONS

This portion deals with the understanding of the effects of pertinent parameters in the governing flow field. Numerical fallouts are thus obtained and presented here through graphs and charts. The computation has been done on the basis of the RK-4 method coupled with the shooting technique as discussed earlier in the previous section and the parametric values are considered as $\delta = 0.2, \Phi = \frac{\pi}{4}, f_w = 0.5, Pr = 6.7, \varphi = 0.1, \xi = 0.3, \zeta = 0.2, Da = 0.3, F_s = 0.2, r_T = 0.2, R = 0.2$ unless otherwise specified.

8.4.1 INFLUENCE OF VELOCITY SLIP PARAMETER (ξ)

The consequence of ξ on $f'(\eta)$ and $\theta(\eta)$ for both H_2O and $TiO_2\text{-}H_2O$ in presence of other parameters is observed in Figure 8.3a and b respectively. It is vividly clear from Figure 8.3a that with the rising values of ξ the fluid velocity $f'(\eta)$ increases for both fluids but the rate of increment is faster for H_2O in comparison to $TiO_2\text{-}H_2O$

nanofluid. It is noteworthy to mention from Table 8.2 and Figure 8.4a that with the mounting values of ξ, C_{fr} declines for both the fluids. Moreover for $0.1 \leq \xi \leq 0.7$, the rate of shear stress for H_2O and TiO_2-H_2O are reduced by 30.11% and 28.85% respectively. But a reverse result is obtained from Figure 8.3b with the same increment of ξ. The dimensionless temperature $\theta(\eta)$ lessens with the escalating values of ξ consequently the thermal boundary layer width declines for both the fluids. It is noteworthy that the rate of declination in thermal boundary layer width is quite faster for H_2O in rather than TiO_2-H_2O nanofluid.

Table 8.2 and Figure 8.4b reveal that with the hike in ξ i.e. $0.1 \leq \xi \leq 0.7$, Nu_r boosts for H_2O by 8.74% and for TiO_2-H_2O by 7.1%. The physics behind these phenomena can be understood as ξ increases, the slip at the wall enhances. Hence the relative velocity of the fluid particles to the wall increase and particles got a tendency to slip away from the wall, and consequently, the momentum boundary layer width enhances. Since fluid particles received higher velocity to the wall so in the presence of buoyancy force they pass the temperature due to convection from the wall rapidly and cause the decrement in the thermal boundary layer.

8.4.2   INFLUENCE OF THERMAL SLIP PARAMETER (ζ)

Figure 8.5a and b describe the influence of ζ on velocity and temperature profiles of both H_2O and TiO_2-H_2O in existence of other physical parameters. It is clear from Figure 8.5a that with the mounting values of ζ the fluid velocity $f'(\eta)$ declines rapidly for both the fluids also the rate of decrement is faster for TiO_2-H_2O nanofluid. One can note that the variation is significant within the region when $1 \leq \eta \leq 3$ (not accurately determined) and for $\eta > 3$, fluid velocity tends asymptotically to satisfy far field boundary conditions.

Now Table 8.2 and Figure 8.6a show that C_{fr} reduces with the escalating values of ζ. For H_2O the rate of reduction in C_{fr} is 3.69% whereas for TiO_2-H_2O nanofluid the reduction rate is 1.77% when ζ ranges from 0.2 to 1.2. Figure 8.5b illustrates that with the increasing values of ζ the fluid temperature within the boundary region declines; hence causing the reduction in thermal boundary layer width. The rate of reduction is slower for TiO_2-H_2O nanofluid in comparison to H_2O. Also, Table 8.2 and Figure 8.6b reveal that as ζ ranges from 0.2 to 1.2 the rate of heat transfer for H_2O is lessened by 32.25% whereas for TiO_2-H_2O nanofluid the reduction rate is 28.43%. These fallouts can be understood as the enhancement of thermal slip causes less amount of heat transfer from the wall to the liquid medium, consequently the thermal boundary layer width decays.

8.4.3   INFLUENCE OF POROSITY PARAMETER (DA)

The influence of Da on $f'(\eta)$ and $\theta(\eta)$ for both the fluids is illustrated through Figure 8.7a and b respectively. It is clear from Figure 8.7a that with the rising values of Da the fluid velocity boosts rapidly for both the fluids; also the rate of increment is slower for TiO_2-H_2O nanofluid. Table 8.2 and Figure 8.8a indicate that with the rising values of Da the C_{fr} for both the fluids enhances. When Da augments from 0.1

(a)

(b)

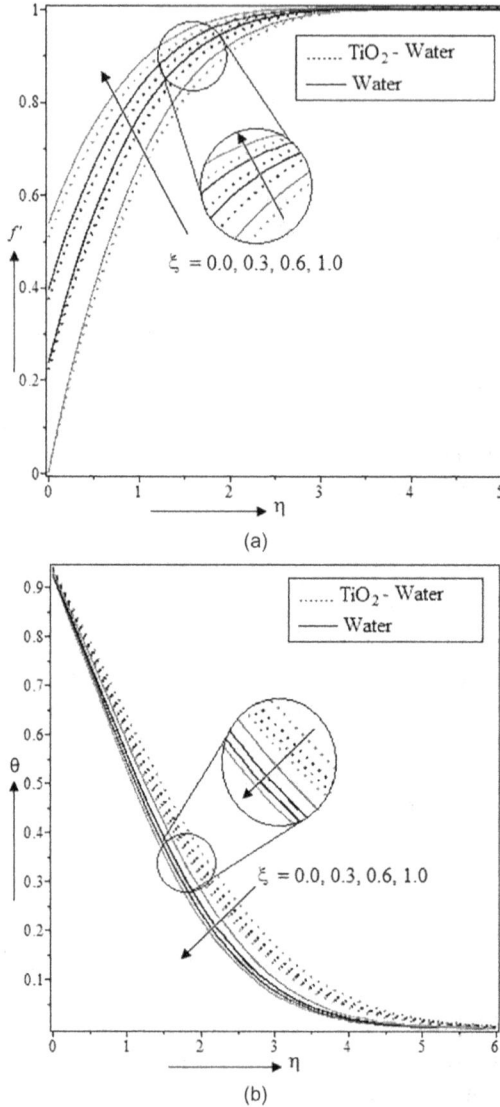

Figure 8.3   Variation of ξ on $f'(\eta)$ and $\theta(\eta)$ are presented through (a) and (b) respectively for both the fluids where the default values of other parameters are $\delta = 0.2, \Phi = \frac{\pi}{4}, f_w = 0.5, \mathrm{Pr} = 6.7, \varphi = 0.1, \zeta = 0.2, Da = 0.3, F_s = 0.2, r_T = 0.2, R = 0.2$.

to 1.0, C_{fr} increase for H_2O by 28.30% and for TiO_2-H_2O by 31.75%. A reverse ef-fect is noticed in Figure 8.7b, i.e. with the same rise in values of Da the temperature profile for both the fluids declines. The rate of declination of thermal boundary layer width is slower for TiO_2-H_2O nanofluid. The effect is significant when $2 \leq \eta \leq 3$ (not accurately determined) and temperature reaches its highest peak when $Da = 0$.

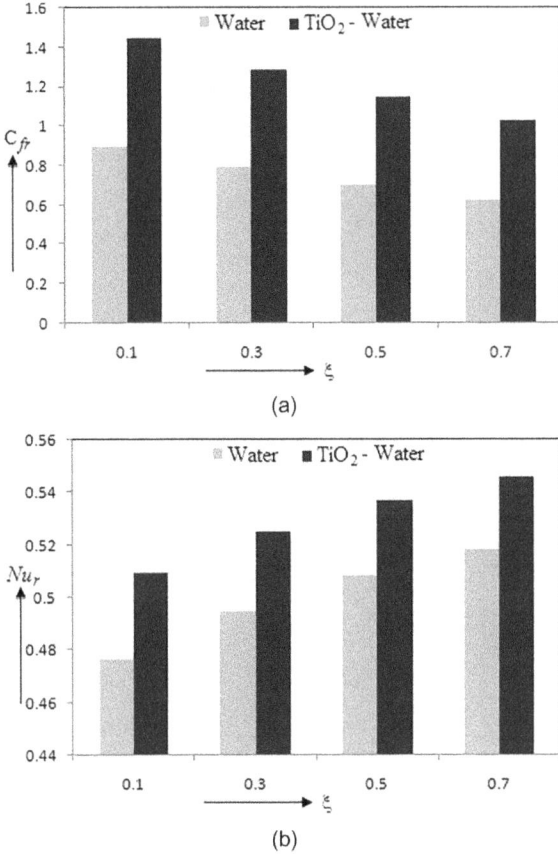

Figure 8.4 Variation of ξ on C_{fr} and Nu_r represented through bar diagrams (a) and (b) respectively for both types of fluids. For TiO$_2$ nanofluid the variation is significantly high.

Now the result from Table 8.2 and Figure 8.8b depict that with the rise of Da from 0.1 to 1.0 the rate of heat transfer for H$_2$O is increased by 3.84% whereas for TiO$_2$-H$_2$O it is 3.69%. The physics behind these outcomes is with the augmentation of Da the medium becomes more absorbent. Since the Darcian body force is inversely proportional to the permeability of the medium, so as a consequence increase of Da results the decrement in Darcian body force. This body force has a propensity to reduce the speed of the fluid particles by increasing the friction between them. Hence increment in Da produce less drag in fluid motion and increases the momentum boundary layer width. Since the fluid particles gain higher velocity near the wall therefore they took the heat from the regime and move away quickly which cause the reduction in thermal boundary layer width.

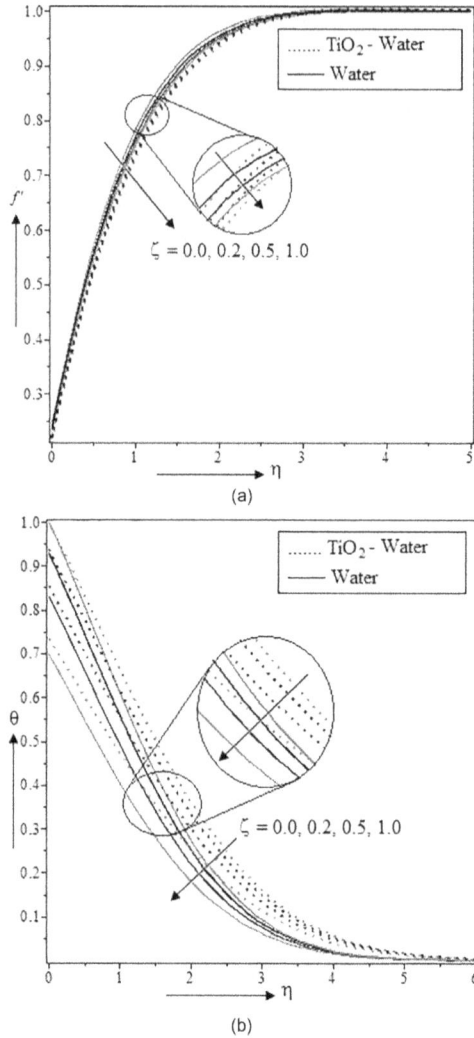

Figure 8.5   Variation of ζ on $f'(\eta)$ and $\theta(\eta)$ are presented through (a) and (b) respectively for both the fluids where the default values of other parameters are $\delta = 0.2, \Phi = \frac{\pi}{4}, f_w = 0.5, \Pr = 6.7, \varphi = 0.1, \xi = 0.3, \text{Da} = 0.3, F_s = 0.2, r_T = 0.2, R = 0.2.$

8.4.4   INFLUENCE OF INERTIAL PARAMETER (F_s)

The effect of F_s on $f'(\eta)$ and $\theta(\eta)$ for both the fluids is illustrated through Figure 8.9a and b respectively. Figure 8.9a deliberates that with the increment of F_s the fluid velocity significantly increases for both the fluids whereas a reverse effect is observed in Figure 8.9b i.e. the temperature of the fluid decays with growing values of F_s. Moreover it is clear from Figure 8.9a that the growth rate in velocity profile is slower for TiO_2-H_2O nanofluid whereas in temperature distribution the rate of

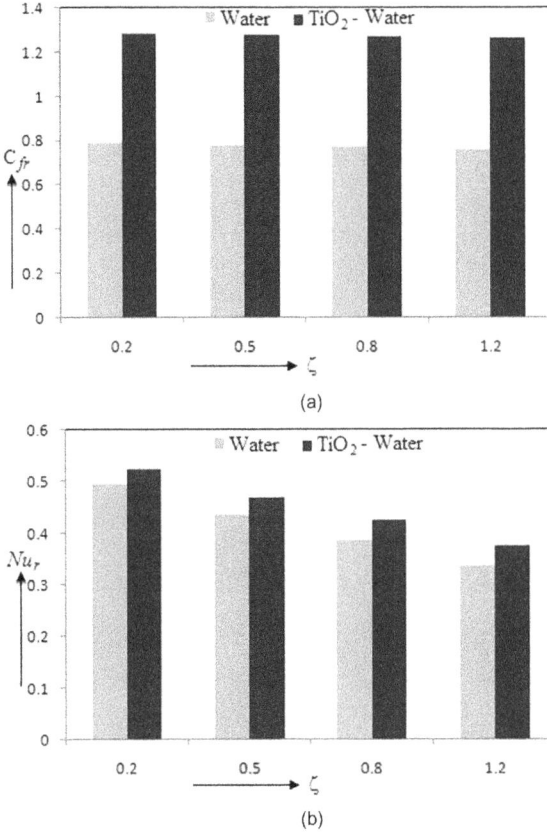

Figure 8.6 Variation of ζ on C_{fr} and Nu_r represented through bar diagrams (a) and (b) respectively for both types of fluids. For TiO_2 nanofluid the variation is significantly high.

decrement for H_2O is faster. It is noteworthy to mention that $f'(\eta)$ attains its highest peak when $F_s = 1.2$. Table 8.2 and Figure 8.10a illustrate that as F_s increases from 0.2 to 0.8 the skin friction coefficient for H_2O is enhanced by 29.08% and for TiO_2-H_2O nanofluid it is increased by 32.44%. Also from Table 8.2 and Figure 8.10b it is clear that with the same range of increment in F_s, Nu_r for TiO_2-H_2O is increased by 3.93% whereas for H_2O the increment is 4.09%. This happens due to the Forchheimer factor brings the velocity square term as an inertial effect in momentum equation. It helps to reduce the internal drag produce by the porous regime. Consequently, significant augmentation in momentum boundary layer width is noticed. Whereas the thermal boundary layer width becomes thinner.

(a)

(b)

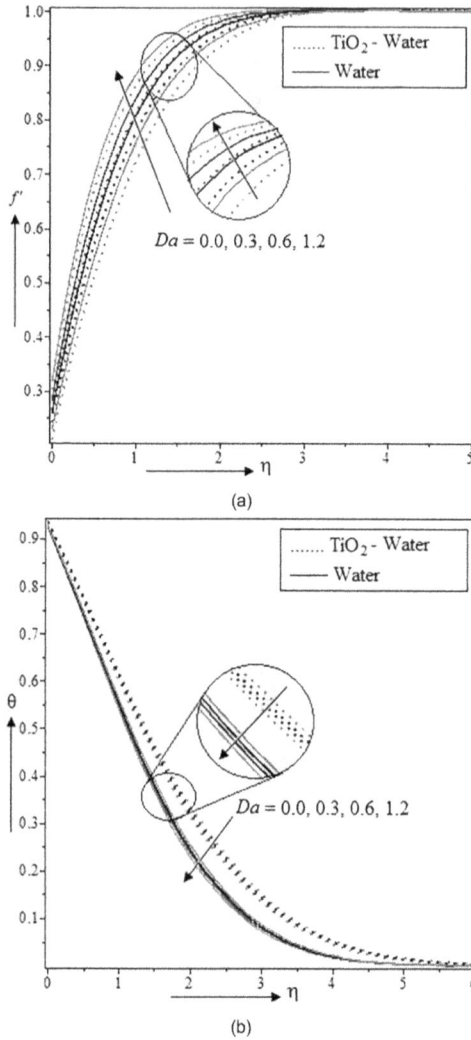

Figure 8.7 Variation of Da on $f'(\eta)$ and $\theta(\eta)$ are presented through (a) and (b) respectively for both the fluids where the default values of other parameters are $\delta = 0.2, \Phi = \frac{\pi}{4}, f_w = 0.5, \Pr = 6.7, \varphi = 0.1, \xi = 0.3, \zeta = 0.2, F_s = 0.2, r_T = 0.2, R = 0.2$.

8.4.5   INFLUENCE OF RADIATION PARAMETER (R)

To visualize the significance of solar radiation on the flow field here we have presented the effects of R on $f'(\eta)$ and $\theta(\eta)$ graphically through Figure 8.11a and b respectively.

(a)

(b)

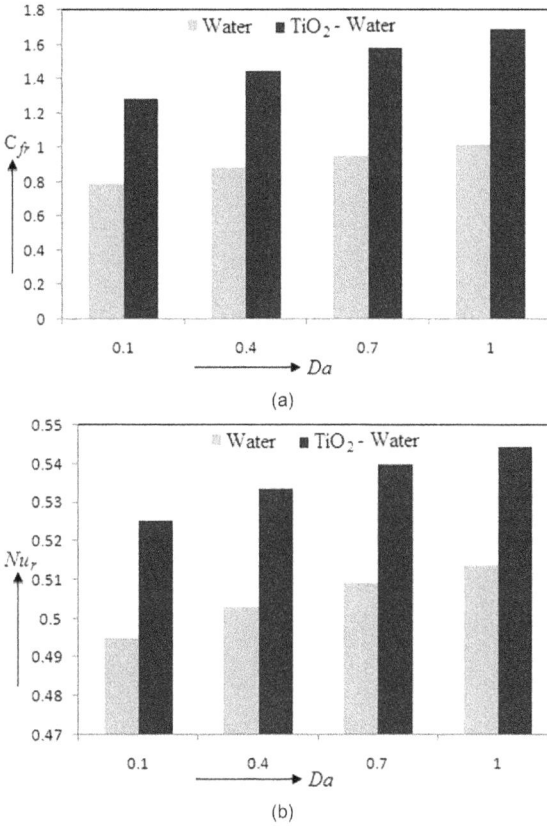

Figure 8.8 Variation of Da on C_{fr} and Nu_r represented through bar diagrams (a) and (b) respectively for both types of fluids. For TiO_2 nanofluid the variation is significantly high.

8.5   CONCLUSIONS

This chapter is an effort to show an application of differential equations in the field of advanced fluid dynamics. Here we have made a mathematical model which exhibits the influence of solar radiation on TiO_2 nanofluid past a slanted absorbent plate entrenched in a Darcy-Forchheimer permeable medium in the occurrence of slip conditions. The noteworthy fallouts gained in this chapter are enlisted below:

- A significant change in C_{fr} and Nu_r for TiO_2 nanofluid is noticed in comparison to ordinary water. So, TiO_2 nanofluid can act as a better coolant than normal water.
- The hydrodynamic boundary layer boosts with the escalating values of R, F_s, Da and ξ whereas the opposite effect is noticed for ζ.
- The widths of the thermal boundary layer boost with R but reverse results are observed for F_s, Da, ξ, ζ.

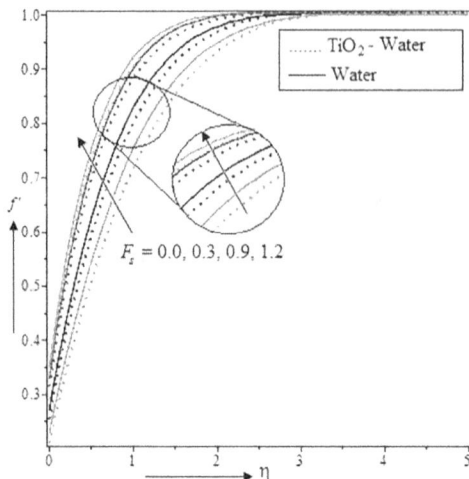

(a)

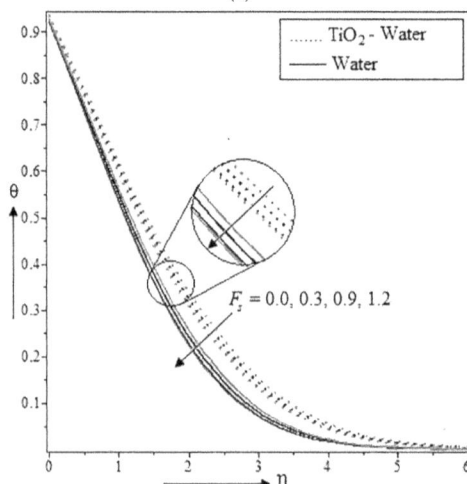

(b)

Figure 8.9   Variations of F_s on $f'(\eta)$ and $\theta(\eta)$ are presented through (a) and (b) respectively for both the fluids where the default values of other parameters are $\delta = 0.2, \Phi = \frac{\pi}{4}, f_w = 0.5, \text{Pr} = 6.7, \varphi = 0.1, \xi = 0.3, \zeta = 0.2, Da = 0.3, r_T = 0.2, R = 0.2$.

- This chapter shows how the heat transfers can be controlled within the boundary layer by regulating the physical parameters of the flow.

It is hoped that the results to be gained in this chapter, although theoretical in nature, must have several applications in different regions of engineering, especially in the technologies based on solar energy. Moreover, this work can be extended by considering magnetohydrodynamics (MHD) flow features and the hybrid nanofluid model. Also, the effect of viscous dissipation can be incorporated into future work in

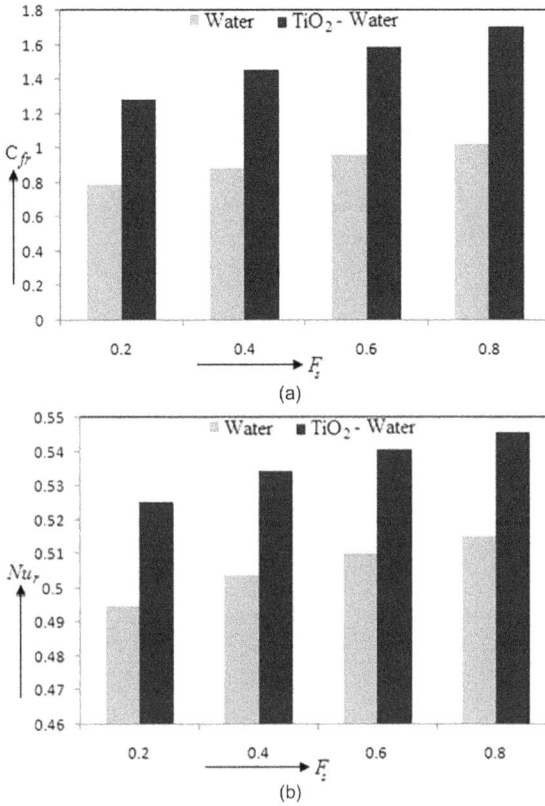

Figure 8.10 Variation of F_s on C_{fr} and Nu_r represented through bar diagrams (a) and (b) respectively for both types of fluids. For TiO_2 nanofluid the variation is significantly high.

this chapter. Figure 8.11a shows that with the enhancement of R the velocity profile boosts and causes the enlargement in momentum boundary layer width. One can notice that lesser augmentation is observed in fluid velocity for TiO_2-H_2O nanofluid and fluid velocity attains its highest peak for $R = 1.2$. Similar effect is observe for temperature profile i.e. thermal boundary layer width significantly boosts with mounting values of R but for this case augmentation in temperature is faster for TiO_2-H_2O nanofluid. Now Table 8.2 and Figure 8.12a deliberate that when $0.2 \leq R \leq 1.2$, C_{fr} increases for H_2O by 3.56% whereas for the same range of R, C_{fr} augments a little for TiO_2-H_2O nanofluid by 1.31%. But significant decrement is observed in Nu_r for both types of fluids for the same increment in R observed I Figure 8.12b. Nu_r for H_2O and TiO_2-H_2O nanofluid are decreased by 51.02% and 40.98%, respectively. These phenomena can be justified as the radiation causes release of heat energy from the wall to the fluid. As radiation increases the wall temperature becomes high and r_T becomes small and causes nonlinear radiation.

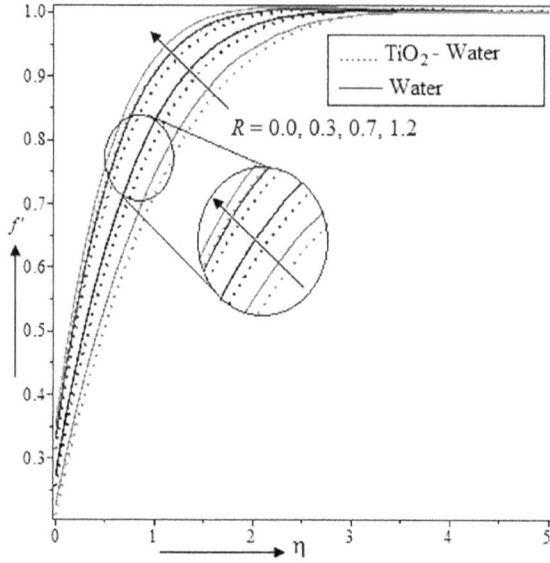

(a)

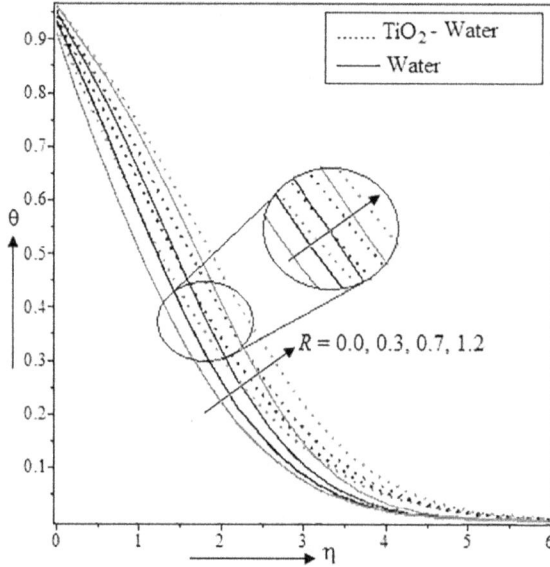

(b)

Figure 8.11 Variations of R on $f'(\eta)$ and $\theta(\eta)$ are presented through (a) and (b) respectively for both the fluids where the default values of other parameters are $\delta = 0.2, \Phi = \frac{\pi}{4}, f_w = 0.5, \mathrm{Pr} = 6.7, \varphi = 0.1, \xi = 0.3, \zeta = 0.2, Da = 0.3, F_s = 0.2, r_T = 0.2$.

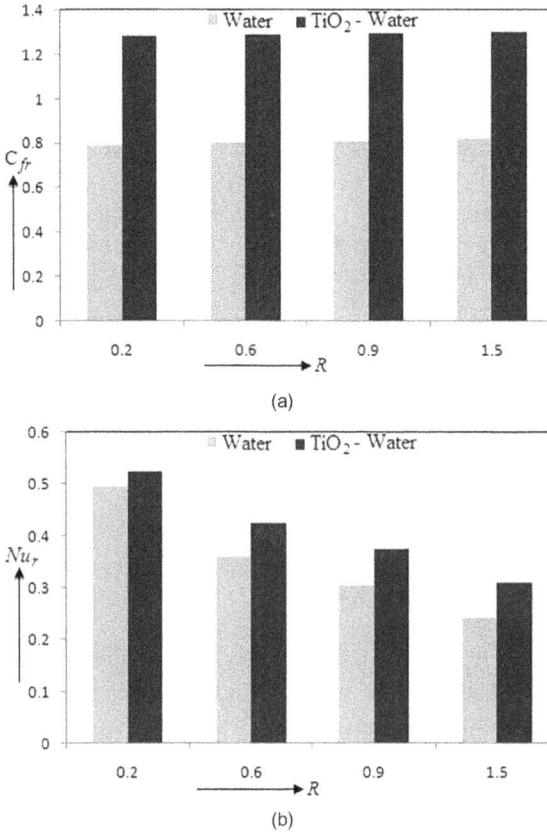

Figure 8.12 Variations of R on C_{fr} and Nu_r are represented through bar diagrams (a) and (b), respectively, for both types of fluids. For TiO$_2$ nanofluid the variation is significantly high.

REFERENCES

Ahmad, I., Muhammad, F., and Tariq, J. (2021). Unsteady flow of walters-b magneto-nanofluid over a bidirectional stretching surface in a porous medium with heat generation. *Special Topics & Reviews in Porous Media: An International Journal* 12: 3.

Chakraborty, T., Das, K., and Kundu, P.K. (2017). Ag-water nanofluid flow over an inclined porous plate embedded in a no-Darcy porous medium due to solar radiation. *Journal of Mechanical Science and Technology* 31: 2443–2449.

Chakraborty, T., Das, K., and Kundu, P.K. (2018). Framing the impact of external magnetic field on bioconvection of a nanofluid flow containing gyrotactic microorganisms with convective boundary conditions. *Alexandria Engineering Journal* 57: 61–71.

Chamka, A.J. (1997). Solar radiation assisted natural convection in uniform porous medium supported by a vertical flat plate. *ASME Journal of Heat Transfer* 119: 35–43.

Chamka, A.J., Issa C., and Khanafer, K. (2002). Natural convection from an inclined plate embedded in a variable porosity porous medium due to solar radiation. *International Journal of Thermal Science* 41: 73–81.

Cheng, P., and Minkowycz, W. (1977). Free convection about a vertical flat plate embedded in a porous medium with application to heat transfer from a dike. *Journal of Geophysical Research* 82: 2040–2044.

Choi, S.U.S, and Eastman, J.A. (1995). Enhancing thermal conductivity of fluids with nanoparticles. No. ANL/MSD/CP-84938; CONF-951135-29. Argonne National Laboratory, Illinois.

Das, K., Chakraborty T., and Kundu P.K. (2019). Lie Group transformation for double diffusive free convection nanofluid flow over an inclined plate. *Proceeding of National Academy of Science India Section A: Physical Science* 89: 387–396.

Das, K., Chakraborty, T., and Kundu, P.K. (2016). Slip effects of nanofluid flow over a non linear permeable stretching surface with chemical reaction. *Proceedings of the Institution of Mechanical Engineers Part C: Journal of Mechanical Engineering Science* 230: 2473–2482.

Das, K., Chakraborty, T., and Kundu, P.K. (2017). Analytical exploration of a TiO_2 nanofluid along a rotating disk with homogeneous-heterogeneous chemical reactions and non-uniform heat source/sink. *European Physical Journal Plus* 132: 555–577.

Das, K., Sk, M.T., and Kundu, P.K. (2016). Multiple slip effects on bioconvection of nanofluid flow containing gyrotactic microorganisms and nanoparticles. *Journal of Molecular Liquids* 220: 518–526.

Das, S., Jana, R.N., and Makinde, O.D. (2015). Magnetohydrodynamic mixed convective slip flow over an inclined porous plate with viscous dissipation and Joule heating. *Alexandria Engineering Journal* 54: 251–261.

Fathalah, K.A., and Elsayed, M.M. (1980). Natural convection due to solar radiation over a non absorbing plate with and without heat losses. *International Journal of Heat and Fluid Flow* 2: 41–45.

Gangadhar, K. Bhargavi, D.N., Kannan, T., Venkata Subba Rao, M., and Chamkha, A.J. (2020). Transverse MHD flow of Al_2O_3–Cu/H_2O hybrid nanofluid with active radiation: A novel hybrid model. *Mathematical Methods in Applied Sciences*. DOI: 10.1002/mma.6671.

Gorla, R.S.R., and Chamka, A. (2011). Natural convective boundary layer flow over a horizontal plate embedded in a porous medium saturated with nanofluid. *Journal of Modern Physics* 2: 62–71.

Hong, J.T., Yamada, Y., and Tien, C.L. (1987). Effects of non-Darcian and non-uniform porosity on vertical plate natural convection in porous media, *ASME Journal of Heat Transfer* 109: 356–362.

Hossain, M.A., and Takhar, H.S. (1996). Radiation effect on mixed convection along a vertical plate with uniform surface temperature. *International Journal of Heat and Mass Transfer* 31: 243–248.

Hunt, A.J. (1978). Small particle heat exchangers. *Journal of Renew Sustainable Energy*, Lawrence Berkeley Laboratory report no. LBL 7841.

Ishak, A. (2010). Similarity solutions for flow and heat transfer over a permeable surface with convective boundary conditions. *Applied Mathematics and Computation* 217: 837–842.

Kandasamy, R., Muhaimin, I., Khamis, A.B., and Roslan, R. (2013). Unsteady Heimenz flow of Cu-nanofluid over a porous wedge in the presence of thermal stratification due to solar energy radiation: Lie group transformation. *International Journal of Thermal Science* 65: 196–205.

Karak, S., and Pramuanjaroenkij, A. (2009). Review of convective heat transfer enhancement with nanofluids. *International Journal of Heat and Mass Transfer* 52: 3187–3196.

Kaviany, M. (1987). Boundary layer treatment of forced convection heat transfer from a semi infinite flat plate embedded in porous media. *ASME Journal of Heat Transfer* 109: 345–349.

Khan, M.I., Waqas, H., Khan, S.U., Imran, M., Chu, Y.M., Abbasi, A., and Kadry, S. (2021). Slip flow of micropolar nanofluid over a porous rotating disk with motile microorganisms, nonlinear thermal radiation and activation energy. *International Communications in Heat and Mass Transfer* 122: 105161.

Lai, F.C., and Kulacki, F.A. (1991). Non-Darcy mixed convection along a vertical wall in a saturated porous medium. *International Journal of Heat and Mass Transfer* 113: 252–255.

Maxwell, J.C.A. (1891). *A Treatise on Electricity and Magnetism*: Vol. 2 Unabridged, third ed., Clarendon Press, Oxford, UK.

Molli, S., and Naikoti, K. (2020). MHD natural convective flow of Cu-water nanofluid over a past infinite vertical plate with the presence of time dependent boundary condition. *International Journal of Thermofluid Science and Technology* 7: 1–18.

Mukhopadhyay, S., De, P.R., Bhattacharyya, K., and Layek, G.C. (2012). Forced convective flow and heat transfer over a porous plate in a Darcy-Forchhimer medium in presence of radiation. *Meccanica* 47: 153–161.

Navier, C.L.M.H. (1823). Mmoire sur les lois du mouvement des fluids. *Memories Academie Royale des Science Institute France* 6: 389–440.

Raizah, Z.A.S., Abdelraheem M.A., and Sameh E.A. (2021). Natural convection flow of a nanofluid-filled V-shaped cavity saturated with a heterogeneous porous medium: Incompressible smoothed particle hydrodynamics analysis. *Ain Shams Engineering Journal* 12: 2033–2046.

Rami, Y.J., Fawzi, A., and Rub, F.A. (2001). Darcy-Forchhimer mixed convection heat and mass transfer in fluid saturated porous media. *International Journal of Numerical Methods for Heat and Fluid Flow* 11: 600–618.

Rashidi, M.M., Abelman, S., and Mehr N.F. (2013). Entropy generation in steady MHD flow due to a rotating porous disk in a nanofluid. *International Journal of Heat and Mass Transfer* 62: 515–525.

Roy, N.C., and Pop, I. (2020). Flow and heat transfer of a second-grade hybrid nanofluid over a permeable stretching/shrinking sheet. *European Physical Journal Plus* 135: 768.

Shehzad, S.A., Hayat, T., Alsaedi, A., and Ahmad, B. (2015). Effects of thermophoresis and thermal radiation in mixed convection three-dimensional flow of Jeffrey fluid. *Applied Mathematics and Mechanics* 36: 655–668.

Sparrow, E.M., and Cess, R.D. (1978). *Radiation Heat Transfer.* Taylor & Francis, New York.

Tiwari, R.K., and Das, M.K. (2007). Heat transfer augmentation in a two sided lid-driven differentially heated square cavity utilizing nanofluids. *International Journal of Heat and Mass Transfer* 50: 2002–2018.

Uddin, M.J., Kabir, M.N., and Anwar, B.O. (2016). Computational investigation of Stefan blowing and multiple-slip effects on buoyancy driven bioconvection nanofluid flow with microorganisms. *International Journal of Heat and Mass Transfer* 95: 116–130.

Ullah, N., Nadeem, S., and Khan, A.U. (2021). Finite element simulations for natural convective flow of nanofluid in a rectangular cavity having corrugated heated rods. *Journal of Thermal Analysis and Calorimetry* 143: 4169–4181.

Vafai, K., and Tien, C.L. (1982). Boundary and inertia effect on convection mass transfer in porous media. *International Journal of Heat and Mass Transfer* 25: 1183–1190.

Wahid, N.S., Arifin, N.M., Khashi'ie, N.S., and Pop, I. (2021). Hybrid nanofluid slip flow over an exponentially stretching/shrinking permeable sheet with heat generation. *Mathematics* 9: 30.

Waini, I., Ishak, A., and Pop, I. (2020). Hybrid nanofluid flow induced by an exponentially shrinking sheet. *Chinese Journal of Physics* 68: 468–482.

Wang, X.Q., and Mujumder, A.S. (2007). Heat transfer characteristics of nanofluids: A review. *International Journal of Thermal Science* 46: 1–19.

Yazdi, M.H., Abdullah, S., Hashim, I., and Sopian, K. (2011). Slip MHD liquid flow and heat transfer over non-linear permeable stretching surface with chemical reaction. *International Journal of Heat and Mass Transfer* 54: 3214–3225.

Zheng, L., Zhang, C., Zhang, X., and Zhang, J. (2013). Flow and radiation heat transfer of a nanofluid over a stretching sheet with velocity slip and temperature jump in porous medium. *Journal of Franklin Institute* 350: 990–1007.

9 Advection-Diffusion Equations and Its Applications in Sciences and Engineering

Varun Joshi and Mamta Kapoor
Lovely Professional University

CONTENTS

DOI: 10.1201/9781003227847-9

9.1 INTRODUCTION

Partial differential equations are considered basic tools of several mathematical models regarding chemical, physical and biological phenomena as well as associated applications that have also been provided in the field of economics, financial forecasting, and other numerous fields. It is a need of time to approximate solutions of such Partial Differential Equations (PDEs) from a numerical aspect to investigate predictions of diversified mathematical models, as finding exact solutions is cumbersome to obtain in most complex cases.

While dealing with ADE, the initial question triggered into the mind is regarding the necessity of dealing with the ADE or it can be said like why so much effort is to be implemented regarding the ADE. The notion about it is that the ADE is the base of many important phenomena in sciences and engineering, and at the same time, it helps in the study of the more advanced and complex Navier-Stokes equation (Brooks and Hughes 1982; Johnson and Saranen1986; Niemi et al. 2013). Naturally, the process of transportation occurs in the fluid because of the mixture of the advection process as well as the diffusion process.

ADE elaborates on quantities like velocity, vorticity, mass, heat, energy, etc. Solutions of mentioned equation model the variety of phenomena such as draining film, the spread of pollutants in rivers and streams, thermal pollution in river systems, dispersion of dissolved salts in groundwater, etc. Delayed progress has been considered toward analytical solutions of ADE along with complicated Initial Conditions (ICs) and Boundary Conditions (B.C.s). Besides, many analytical solutions are not easy to fetch. So, a great deal of effort has been provided for the development of efficient and stable numerical regimes. Numerous numerical regimes are proposed to discuss physical phenomena via ADE in a variety of disciplines. Difficulties are arising in the numerical approximation of ADE results due to dominant advection that is for moderately high Peclect number.

9.1.1 LITERATURE REVIEW

Several researchers have developed a variety of analytical and numerical regimes to tackle the solution of ADE. Some of this outstanding work regarding the solution of ADE is given ahead.

- Korkmaz and Dağ (2012) implemented a numerical scheme named cubic B-spline-based differential quadrature method (DQM) to solve ADE.
- Korkmaz and Dağ (2012) implemented a numerical scheme using quartic and quintic B-spline in DQM to deal with the numerical approximation of ADE.
- Dhawan et al. (2012) gave a numerical approximation of ADE by FEM and B-spline-based method.
- Dehghan (2004) implemented the notion of fully explicit and fully implicit finite approximation schemes (two-level) to solve 3D ADE.

- Bhrawy and Baleanu (2013) implemented the notion of a collocation method related to a spectral Legendre-Gaus-Lobatto concept to numerically approximate space fractional ADE based on variable coefficients.
- Bhatt et al. (2018) implemented the notion of the exponential differencing method to solve 3D ADE.
- Company R et al. (2009) applied a scheme considered a space-time conversation element and solution element regime to solve ADE.
- Dağ et al. (2006) implemented the notion of the least square FE method to deal with the numerical approximation of ADE.
- Tajadodi (2020) implemented an Atangana-Baleanu derivative-based numerical scheme for the numerical approximation of the fractional ADE.
- Lin et al. (2011) employed the Taylor-Galerkin scheme to get a numerical approximation of the ADE.
- Du Toit et al. (2018) implemented the concept of a positivity-preserving scheme for the numerical solution of 2D ADE.
- Gurarslan et al. (2013) implemented the notion of the compact FD method of sixth order for the numerical approximation of ADE.
- Irk et al. (2015) implemented an extended cubic-B-spline-based method for getting a numerical approximation of ADE, where a comparative study of B-spline and extended B-spline-based collocation regimes is provided.
- Liu et al. (2018) employed a meshless method using a radial basis function for the numerical approximation of time-fractional ADE.
- Liu et al. (2019) implemented the notion of DQM using RBF for the solution of 2D variable-order time-fractional ADE.
- Sakai (1999) presented the notion of a non-oscillatory numerical regime for the solution of 2D ADE (unsteady in nature).
- Liu et al. (2007) presented stability and convergence aspects of different difference methods regarding space-time fractional ADE.
- Al-Jawary et al. (2018) presented the notion of a semi-analytical iterative regime to tackle with analytical and numerical results of ADE.
- Du and Liu (2020) implemented the Lattice-Boltzmann model for solving fractional ADE.
- Gharehbaghi (2016) implemented the notion of an explicit and implicit form of DQM for solving 1D time-dependent ADE with variable coefficients.
- Jannelli et al. (2019) gave the notion of Lie transformations to get analytical and numerical results of ADE (time and space fractional).
- Lian et al. (2016) employed a Petrov-Galerkin FE regime for the solution of fractional ADE.
- Sakai and Kimura (2005) implemented the notion of Cole-Hopf transformation to solve non-linear ADE.
- Rubbab et al. (2021) provided the numerical solution regarding the ADE with the Fabrizi time-fractional derivative.
- Aghdam et al. (2021) provided an approach for the solution of the space-time fractional ADE.

- Mirzaee et al. (2021) implemented FD and spline approximations for the solution of the fractional stochastic ADE.
- Abbaszadeh and Dehghan (2021) employed the Crank Nicolson/interpolating stabilized element-free Galerkin approach for the solution of the fractional Galilei invariant ADE.
- Brahim (2021) provided the numerical technique for the solution of the Reaction-ADE for time-space conformable fractional derivatives.

9.2   CLASSIFICATION

9.2.1   1D ADE

ADE in one dimension has the following form (Korkmaz and Dağ 2012):

$$\frac{\partial U}{\partial t} + \alpha \frac{\partial U}{\partial x} - \lambda \frac{\partial^2 U}{\partial x^2} = 0 \tag{9.1}$$

Where, $x \in [0, 1]$, t and x are considered as the time and space variables, U is known as the concentration of a pollutant at the given point x, α is known as the velocity of present water flow, λ is considered as a coefficient of diffusion in the direction of space derivative.

9.2.1.1   Initial Condition

$$U(x,\ 0) = f_0(x),\ x \in [0,1]$$

9.2.1.2   Boundary Conditions

$$U(0,\ t)\ =\ g_1(t) \text{ and } U(l,t) = g_2(t)$$

l is considered as the length of the taken channel. $\frac{\partial U}{\partial x}$ is the advection term, $\frac{\partial^2 U}{\partial x^2}$ is the given diffusion term. In most of the problems related to the environment, $U(x,t)$ is considered as the contaminant or pollutant material at the point x and at the time t.

9.2.2   2D ADE

The 2DAdvection-Diffusion equation is defined as follows (Du Toit et al. 2018):

$$\frac{\partial U}{\partial t} = A \frac{\partial^2 U}{\partial x^2} + B \frac{\partial^2 U}{\partial x\, \partial y} + C \frac{\partial^2 U}{\partial y^2} + D \frac{\partial U}{\partial x} + E \frac{\partial U}{\partial y} + F\,U \tag{9.2}$$

Where $U = U(x,y,t)$

9.2.3   2D UNSTEADY ADE

2D unsteady ADE is given as follows (Sakai 1999):

$$\frac{\partial V}{\partial t} + u \frac{\partial V}{\partial x} + v \frac{\partial V}{\partial y} = \nu \left[\frac{\partial^2 V}{\partial x^2} + \frac{\partial^2 V}{\partial y^2} \right] \tag{9.3}$$

t is the time, x is the spatial coordinate, and ν is known as the coefficient of diffusion.

9.2.4   3D ADE

$$U_t + \beta_x U_x + \beta_y U_y + \beta_z U_z = \alpha_x U_{xx} + \alpha_y U_{yy} \alpha_z U_{zz} \qquad (9.4)$$

Where, $x, y, z \in [0, 1]$, $t \in [0, T]$

9.2.4.1   Initial Condition

$u(x, y, z, 0) = F(x, y, z)$,
$x \in [0, 1]$, $y \in [0, 1]$, $z \in [0, 1]$

9.2.4.2   Boundary Conditions

$U(0, y, z, t) = G_0(y; z; t)$, $t \in (0, T]$
$U(1, y, z, t) = G_1(y; z; t)$, $t \in (0, T]$
$U(x, 0, z, t) = H_0(x; z; t)$, $t \in (0, T]$
$U(x, 1, z, t) = H_1(x; z; t)$, $t \in (0, T]$
$U(x, y, 0, t) = I_0(x; y; t)$, $t \in (0, T]$
$U(x, y, 1, t) = I_1(x; y; t)$, $t \in (0, T]$

Where, E, G_0, G_1, H_1, H_2, I_1 and I_2 are the known functions, the function U is the unknown function. Where the point to be noticed is that $U(x, y, z, t)$ is a scalar variable, β_x, β_y and β_z all are non-negative and non-zero, considered as constant speeds of advection, α_x, α_y and α_z all are also non-negative and non-zero and known as constant of diffusivities in x-direction, y-direction, and z-direction.

9.2.5   FRACTIONAL ADE

Space-time fractional diffusion equation (FDE) is given as follows (Liu et al. 2007):

$$\frac{\partial^\alpha u(x, t)}{\partial t^\alpha} = -v_1(x, t) D_x^\beta u(x, t) + v_2(x, t) D_x^\gamma u(x, t) + f(x, t) \qquad (9.5)$$

Where, $0 < t < T, 0 < x < L$
$u(x, 0) = \psi(x)$
$u(0, t) = \psi_1$
$u(L, t) = \psi_2(t)$
u is considered as solute concentration, $v_1(x, t)$ and $v_2(x, t)$ are the fluid velocity and coefficient of diffusion, respectively. $\frac{\partial^\alpha u(x, t)}{\partial t^\alpha}$ is the time-fractional derivative known as fractional derivative (Caputo sense) with order α, $0 ¡ \alpha \leq 1$.

9.3   APPLICATIONS

9.3.1   APPLICATIONS OF ADE: IN FIELD OF SCIENCES

There are an enormous number of applications of ADE in different fields of sciences and engineering. Basically, ADE is used to elaborate heat, mass, vorticity, velocity, etc. Noye (1990). Heat transfer regarding the draining film was presented by Isenberg and Gutfinger (1973). ADE is also modeled for the transportation of water in soils Parlange (1980). Many more applications of ADE are explored. ADE has been used as the governing equation to solve a range of engineering as well as chemistry problems like tracers dispersion in porous media Fattah and Hoopes (1985); it has also been used for the intrusion of salty water into freshwater aquifers as well as in discussions of pollutant spreads in streams and rivers by Chatwin and Allen (1985). Applications, like contaminated dispersion is then shallow lakes Salmon et al. (1980), it has also been used in the concept of chemicals absorption into beds Lapidus and Amundston (1952), it has been related to the notion of solute spread in a liquid flowing via a tube and transport of pollutants having long-range in atmosphere Zlatev et al. (1984). Naturally, the occurrence of transportation in the case of fluids is due to the amalgamation of advection and diffusion. ADE represents different physical processes as a domain of physical oceanography.

9.3.2 SOLUTIONS OF ADE: AN ANALYTICAL APPROACH

9.3.2.1 Development of Recurrence Relation for the Solution of 1D ADE by HPM

The 1DAdvection-Diffusion equation is provided as follows:

$$U_t + \alpha\, U_x - \lambda U_{xx} = 0 \tag{9.6}$$

9.3.2.1.1 Initial Condition

$U(x,\, 0) = U_0(x)$

9.3.2.1.2 Boundary Conditions

$U(0,t) = f(t)$
$U(L,t) = g(t)$

Constructed Homotopy as follows:

$$(1-p)[L(v) - L(v_0)] + p\,[A(v) - f(r)] = 0$$
$$\Rightarrow (1-p)\left[\frac{\partial v}{\partial t} + \alpha\frac{\partial v}{\partial x} - \left(\frac{\partial u_0}{\partial t} + \alpha\frac{\partial u_0}{\partial x}\right)\right] + p\left[\frac{\partial v}{\partial t} + \alpha\frac{\partial v}{\partial x} - \lambda\frac{\partial^2 v}{\partial x^{\Rightarrow 2}}\right] = 0$$
$$\Rightarrow (1-p)\frac{\partial v}{\partial t} + \alpha(1-p)\frac{\partial v}{\partial x} - (1-p)\left[\frac{\partial u_0}{\partial t} + \alpha\frac{\partial u_0}{\partial x}\right] + p\frac{\partial v}{\partial t} + \alpha p\frac{\partial v}{\partial x} - \lambda p\frac{\partial^2 v}{\partial x^2} = 0$$
$$\Rightarrow \frac{\partial v}{\partial t} - p\frac{\partial v}{\partial t} + \alpha\frac{\partial v}{\partial x} - \alpha p\frac{\partial v}{\partial x} - (1-p)\left[\frac{\partial u_0}{\partial t} + \alpha\frac{\partial u_0}{\partial x}\right] + p\frac{\partial v}{\partial t} + \alpha p\frac{\partial v}{\partial x} - \lambda p\frac{\partial^2 v}{\partial x^2} = 0$$
$$\Rightarrow \frac{\partial v}{\partial t} + \alpha\frac{\partial v}{\partial x} = (1-p)\left[\frac{\partial u_0}{\partial t} + \alpha\frac{\partial u_0}{\partial x}\right] + \lambda\, p\frac{\partial^2 v}{\partial x^2}$$

L.H.S. of the above equation is as follows:

$\frac{\partial v}{\partial t} + \alpha\frac{\partial v}{\partial x}$

$$\Rightarrow \frac{\partial}{\partial t}\left[v_0 + p\, v_1 + p^2 v_2 + p^3 v_3 + \cdots\right] + \alpha\,\frac{\partial}{\partial x}\left[v_0 + p\, v_1 + p^2 v_2 + p^3 v_3 + \cdots\right]$$

R.H.S. is provided as follows:

$$(1-p)\left[\frac{\partial u_0}{\partial t}+\alpha\frac{\partial u_0}{\partial x}\right]+\lambda\ p\frac{\partial^2 v}{\partial x^2}$$

$$\Rightarrow \frac{\partial u_0}{\partial t}+\alpha\frac{\partial u_0}{\partial x}-p\left(\frac{\partial u_0}{\partial t}+\alpha\frac{\partial u_0}{\partial x}\right)+\lambda p\left[\frac{\partial^2 v_0}{\partial x^2}+p\frac{\partial^2 v_1}{\partial x^2}+p^2\frac{\partial^2 v_2}{\partial x^2}+p^3\frac{\partial^2 v_3}{\partial x^2}+\cdots\right]$$

On comparing different powers of p:

$$p^0: \frac{\partial v_0}{\partial t}+\alpha\frac{\partial v_0}{\partial x}=\frac{\partial u_0}{\partial t}+\alpha\frac{\partial u_0}{\partial x}$$

$$\Rightarrow \frac{\partial v_0}{\partial t}+\alpha\frac{\partial v_0}{\partial x}-\frac{\partial u_0}{\partial t}-\alpha\frac{\partial u_0}{\partial x}=0$$

$$p^1: \frac{\partial v_1}{\partial t}+\alpha\frac{\partial v_1}{\partial x}=-\left[\frac{\partial u_0}{\partial t}+\alpha\frac{\partial u_0}{\partial x}\right]+\lambda\frac{\partial^2 v_0}{\partial x^2}$$

$$\Rightarrow \frac{\partial v_1}{\partial t}+\alpha\frac{\partial v_1}{\partial x}+\left[\frac{\partial u_0}{\partial t}+\alpha\frac{\partial u_0}{\partial x}\right]-\lambda\frac{\partial^2 v_0}{\partial x^2}=0$$

$$p^2: \frac{\partial v_2}{\partial t}+\alpha\frac{\partial v_2}{\partial x}=\lambda\frac{\partial^2 v_1}{\partial x^2}$$

$$\Rightarrow \frac{\partial v_2}{\partial t}+\alpha\frac{\partial v_2}{\partial x}-\lambda\frac{\partial^2 v_1}{\partial x^2}=0$$

$$p^3: \frac{\partial v_3}{\partial t}+\alpha\frac{\partial v_3}{\partial x}=\lambda\frac{\partial^2 v_2}{\partial x^2}$$

$$\Rightarrow \frac{\partial v_3}{\partial t}+\alpha\frac{\partial v_3}{\partial x}-\lambda\frac{\partial^2 v_2}{\partial x^2}=0$$

In general, the developed recurrence relation is as follows:

$$\frac{\partial v_j}{\partial t}+\alpha\frac{\partial v_j}{\partial x}-\lambda\frac{\partial^2 v_{j-1}}{\partial x^2}=0$$

$$\Rightarrow \frac{\partial v_j}{\partial t}=\lambda\frac{\partial^2 v_{j-1}}{\partial x^2}-\alpha\frac{\partial v_j}{\partial x}$$

$$\Rightarrow v_j=\int_0^t\left[\lambda\frac{\partial^2 v_{j-1}}{\partial x^2}-\alpha\frac{\partial v_j}{\partial x}\right]dt$$

9.3.2.2 Development of Recurrence Relation for Solution of 2DADE by HPM

2D unsteady ADE is provided as:

$$\frac{\partial U}{\partial t}+u\frac{\partial U}{\partial x}+v\frac{\partial U}{\partial y}=\mu\left[\frac{\partial^2 U}{\partial x^2}+\frac{\partial^2 U}{\partial y^2}\right] \tag{9.7}$$

Constructing Homotopy:

$$(1-p)\left[L(v)-L(u_0)\right]+p\left[A(v)-f(r)\right]=0$$

$$(1-p)\left[\frac{\partial v}{\partial t}-\frac{\partial u_0}{\partial t}\right]+p\left[\frac{\partial v}{\partial t}+u\frac{\partial v}{\partial x}+v\frac{\partial v}{\partial y}-\mu\left(\frac{\partial^2 U}{\partial x^2}+\frac{\partial^2 U}{\partial y^2}\right)\right]=0\ \frac{\partial v}{\partial t}-p\frac{\partial v}{\partial t}-\frac{\partial u_0}{\partial t}+$$

$$p\frac{\partial u_0}{\partial t} + p\frac{\partial v}{\partial t} + p\left[\left(u\frac{\partial v}{\partial x} + v\frac{\partial v}{\partial y}\right) - \mu\left(\frac{\partial^2 U}{\partial x^2} + \frac{\partial^2 U}{\partial y^2}\right)\right] = 0$$

$$\frac{\partial v}{\partial t} - \frac{\partial u_0}{\partial t} = p\left[-\frac{\partial u_0}{\partial t} - \left(u\frac{\partial v}{\partial x} + v\frac{\partial v}{\partial y}\right) + \mu\left(\frac{\partial^2 U}{\partial x^2} + \frac{\partial^2 U}{\partial y^2}\right)\right]$$

Considered,

$$v = v_0 + p\, v_1 + p^2 v_2 + p^3 v_3 + \cdots$$

L.H.S. is considered as follows:

$$\frac{\partial v}{\partial t} - \frac{\partial u_0}{\partial t}$$

$$\Rightarrow \frac{\partial}{\partial t}\left[v_0 + p\, v_1 + p^2 v_2 + p^3 v_3 + \cdots\right] - \frac{\partial u_0}{\partial t}$$

$$\Rightarrow \frac{\partial v_0}{\partial t} + p\frac{\partial v_1}{\partial t} + p^2\frac{\partial v_2}{\partial t} + p^3\frac{\partial v_3}{\partial t} + \cdots \frac{\partial u_0}{\partial t}$$

R.H.S. is considered as follows:

$$-p\left[\frac{\partial u_0}{\partial t} + \left(u\frac{\partial v}{\partial x} + v\frac{\partial v}{\partial y}\right) - \mu\left(\frac{\partial^2 U}{\partial x^2} + \frac{\partial^2 U}{\partial y^2}\right)\right]$$

Comparing different powers of p:

$$p^0: \frac{\partial v_0}{\partial t} - \frac{\partial u_0}{\partial t} = 0$$

$$p^1: \frac{\partial v_1}{\partial t} + \frac{\partial u_0}{\partial t} + \left[u\frac{\partial v_0}{\partial x} + v\frac{\partial v_0}{\partial y}\right] - \mu\left[\frac{\partial^2 v_0}{\partial x^2} + \frac{\partial^2 v_0}{\partial y^2}\right] = 0$$

$$p^2: \frac{\partial v_2}{\partial t} + \left[u\frac{\partial v_1}{\partial x} + v\frac{\partial v_1}{\partial y}\right] - \mu\left[\frac{\partial^2 v_1}{\partial x^2} + \frac{\partial^2 v_1}{\partial y^2}\right] = 0$$

$$p^3: \frac{\partial v_3}{\partial t} + \left[u\frac{\partial v_2}{\partial x} + v\frac{\partial v_2}{\partial y}\right] - \mu\left[\frac{\partial^2 v_2}{\partial x^2} + \frac{\partial^2 v_2}{\partial y^2}\right] = 0$$

In general,

$$\frac{\partial v_j}{\partial t} + \left[u\frac{\partial v_{j-1}}{\partial x} + v\frac{\partial v_{j-1}}{\partial y}\right] - \mu\left[\frac{\partial^2 v_{j-1}}{\partial x^2} + \frac{\partial^2 v_{j-1}}{\partial y^2}\right] = 0$$

EXAMPLES:

Example 9.1

In the present example, the series and exact solutions of equation (9.8) are matched in Figures 9.1 and 9.2 at the time levels $t = 0.1, 0.5,$ and 0.8 for different numbers of grid points by using the Sumudu HPM concept. Considered homogeneous ADE as follows (Singh et al. 2011; Wazwaz 2017; Khan and Austin 2010):

$$u_t + u\, u_x = 0 \tag{9.8}$$

I.C.: $u(x, 0) = -x$

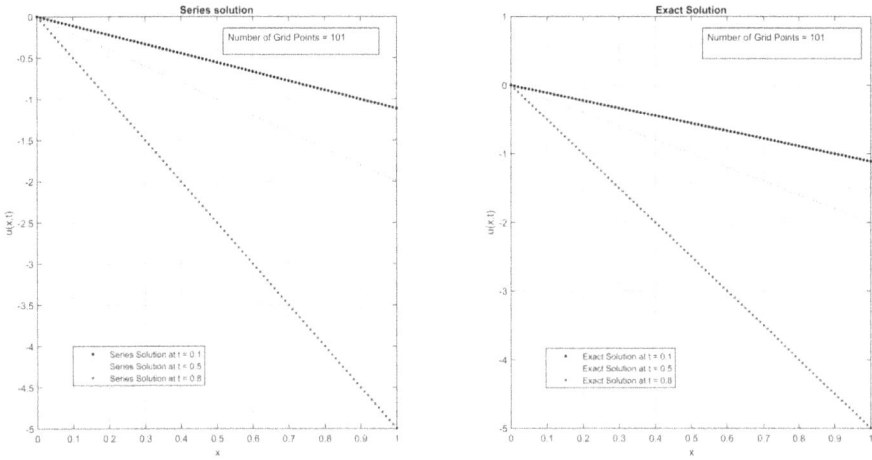

Figure 9.1 Comparison of series solution and exact solution by implementing Sumudu HPM at different time levels for $N = 101$.

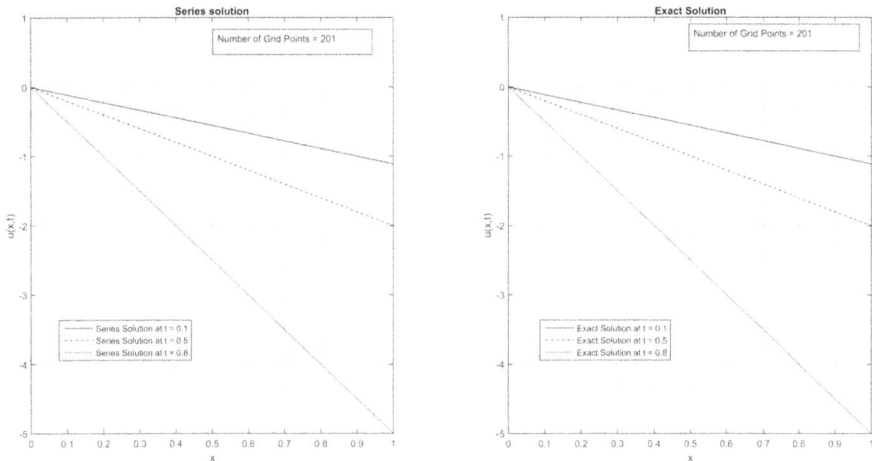

Figure 9.2 Comparison of series solution and exact solution by implementing Sumudu HPM at different time levels for $N = 201$.

Applying Sumudu transform in equation (9.8)

$$S[u_t] + S[u \, u_x] = 0$$

$$\frac{1}{u} S[u(x, t) - u(x, 0)] + S[u \, u_x] = 0$$

$$S[u(x,t)] = u(x,0) - uS[u \, u_x]$$

$$S[u(x,t)] = -x - uS[u \, u_x]$$

$$u(x,t) = -x - S^{-1}[uS[u \, u_x]] \tag{9.9}$$

Applying HPM in equation (9.9)

$$\sum_{n=0}^{\infty} p^n u_n(x,t) = -x - p \, S^{-1}\left[u S \left\{ \sum_{n=0}^{\infty} p^n A_n(u) \right\} \right] \qquad (9.10)$$

Where, $A_n(u)$ is considered as the non-linear term.

$A_n(u) = u \, u_x$
$A_0(u) = u_0 \, (u_0)_x = x$
$A_1(u) = u_0 \, (u_1)_x + u_1 \, (u_0)_x = 2xt$
$A_2(u) = u_0 \, (u_2)_x + u_1 \, (u_1)_x + u_2 \, (u_0)_x = 3 \, x \, t^2$

On comparing powers of p in equation (9.10)

$p^0 : u_0(x,t) = -x$

$p^1 : u_1(x, t) = -S^{-1}[u \, S\{A_0(u)\}]$
$u_1(x, t) = -S^{-1}[u \, S\{x\}]$
$u_1(x, t) = -x S^{-1}[u \, S\{1\}]$
$u_1(x, t) = -x S^{-1}[u]$
$u_1(x, t) = -xt$

$p^2 : u_2(x, t) = -S^{-1}[u \, S\{A_1(u)\}]$
$u_2(x, t) = -S^{-1}[u \, S\{2xt\}]$
$u_2(x, t) = -2x \, S^{-1}[u \, S\{t\}]$
$u_2(x, t) = -2x \, S^{-1}[u^2]$
$u_2(x, t) = -x \, t^2$

$p^3 : u_3(x, t) = -S^{-1}[u \, S\{A_2(u)\}]$
$u_3(x, t) = -S^{-1}[u \, S\{3xt^2\}]$
$u_3(x, t) = -3x S^{-1}[u \, S\{t^2\}]$
$u_3(x, t) = -3x S^{-1}[u \, (2u^2)]$
$u_3(x, t) = -6x[S^{-1}[u^3]]$
$u_3(x, t) = -6xt^3$

The solution is given as follows:

$$u(x, t) = u_0(x, t) + u_1(x, t) + u_2(x, t) + u_3(x, t) + \cdots$$
$$= -x - xt - xt^2 - xt^3 - \cdots$$
$$= -x\left[1 + t + t^2 + t^3 - \cdots\right]$$
$$= -\frac{x}{(1-t)}$$

Example 9.2
In the present example, the series and exact solutions of equation (9.11) are matched in Figures 9.3 and 9.4 at the time levels $t = 0.1, 0.5,$ and 0.8 for a different number of grid points Sumudu HPM concept.

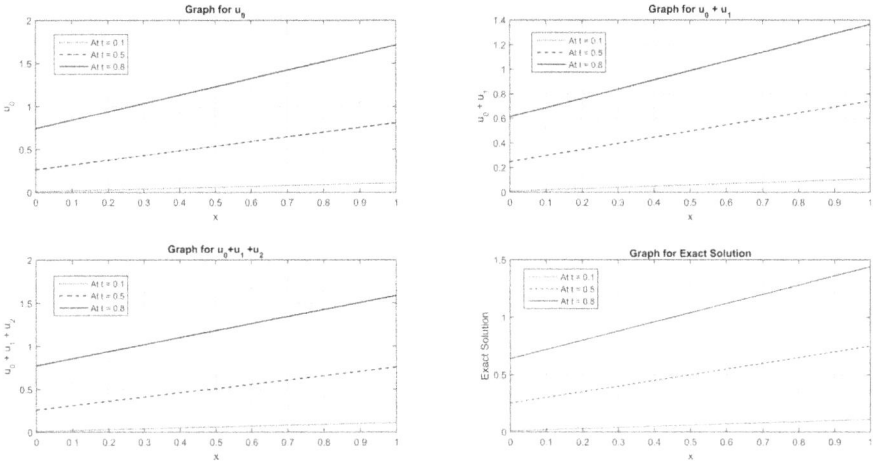

Figure 9.3 Comparison of sum series solution and exact solution by implementing Sumudu HPM at different time levels for $N = 101$.

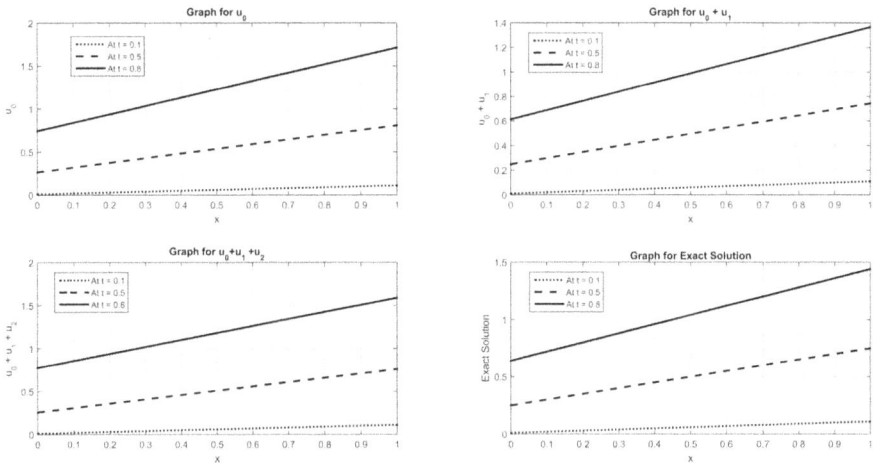

Figure 9.4 Comparison of series solution and exact solution by implementing Sumudu HPM at different time levels for $N = 201$.

Considered non-homogeneous ADE as follows (Singh et al. 2011; Wazwaz 2007; Khan and Austin 2010):

$$u_t + u\,u_x = 2t + x + t^3 + x\,t^2 \tag{9.11}$$

I.C.: $u(x,0) = 0$

Applying Sumudu Transform in equation (9.11):

$$S\left[u_t + u\, u_x\right] = S\left[2t + x + t^3 + x\, t^2\right]$$

$$\frac{1}{u}\left[S\left[u\left(x,t\right)\right] - u\left(x,0\right)\right] = S\left[2t + x + t^3 + x\, t^2\right] - S\left[u\, u_x\right]$$

$$S\left[u\left(x,t\right)\right] - u\left(x,0\right) = u\left[S\left[2t + x + t^3 + x\, t^2\right] - S\left[u\, u_x\right]\right]$$

$$S\left[u\left(x,t\right)\right] = u\left(x,0\right) + u\left[S\left[2t + x + t^3 + x\, t^2\right] - S\left[u\, u_x\right]\right]$$

$$u\left(x,t\right) = u\left(x,0\right) + S^{-1}\left[u\left[S\left[2t + x + t^3 + x\, t^2\right] - S\left[u\, u_x\right]\right]\right]$$

$$u\left(x,t\right) = u\left(x,0\right) + S^{-1}\left[uS\left\{2t + x + t^3 + x\, t^2\right\}\right] - S^{-1}\left[uS\left\{u\, u_x\right\}\right]$$

$$u\left(x,t\right) = S^{-1}\left[u\left\{2u + x + 6\, u^3 + 2\, x\, u^2\right\}\right] - S^{-1}\left[uS\left\{u\, u_x\right\}\right]$$

$$u\left(x,t\right) = S^{-1}\left[2u^2 + xu + 6u^4 + 2xu^3\right] - S^{-1}\left[uS\left\{u\, u_x\right\}\right]$$

$$u\left(x,t\right) = \left[t^2 + xt + \left(\frac{t^4}{4}\right) + \left(\frac{xt^3}{3}\right)\right] - S^{-1}\left[uS\left\{u\, u_x\right\}\right] \qquad (9.12)$$

Applying HPM in equation (9.12):

$$\sum_{n=0}^{\infty} p^n u_n\left(x,t\right) = \left[t^2 + xt + \left(\frac{t^4}{4}\right) + \left(\frac{xt^3}{3}\right)\right] - p\, S^{-1}\left[uS\left\{\sum_{n=0}^{\infty} p^n A_n\left(u\right)\right\}\right] \qquad (9.13)$$

Where, $A_n\left(u\right)$ is considered as the non-linear term.

$A_n\left(u\right) = u\, u_x$
$A_0\left(u\right) = u_0\left(u_0\right)_x$
$A_1\left(u\right) = u_0\left(u_1\right)_x + u_1\left(u_0\right)_x$
$A_2\left(u\right) = u_0\left(u_2\right)_x + u_1\left(u_1\right)_x + u_2\left(u_0\right)_x$

Comparing powers of p in equation (9.13)

$p^0 : u_0\left(x,\, t\right) = t^2 + xt + \left(\frac{t^4}{4}\right) + \left(\frac{xt^3}{3}\right)$

$p^1 : u_1\left(x,t\right) = -\left(\frac{t^4}{4}\right) - \left(\frac{xt^3}{3}\right) - \frac{2}{15}xt^5 - \frac{7}{72}t^6 - \frac{1}{63}xt^7 - \frac{1}{98}t^8$

$p^2 : u_2\left(x,t\right) = \frac{5}{8064}t^{12} + \frac{2}{2079}xt^{11} + \frac{2783}{302400}t^{10} + \frac{38}{2835}xt^9 + \frac{143}{2880}t^8 + \frac{22}{315}xt^7 + \frac{7}{12}t^6 + \frac{2}{15}xt^5$

$u\left(x,\, t\right) = u_0\left(x,\, t\right) + u_1\left(x,\, t\right) + u_2\left(x,\, t\right) + \cdots$

$u\left(x,t\right) = \left[t^2 + xt + \left(\frac{t^4}{4}\right) + \left(\frac{xt^3}{3}\right)\right] + \left[-\left(\frac{t^4}{4}\right) - \left(\frac{xt}{3}\right) - \frac{2}{15}xt^5 - \frac{7}{72}t^6 - \frac{1}{63}xt^7 - \frac{1}{98}t^8\right]$
$+ \left[\frac{5}{8064}t^{12} + \frac{2}{2079}xt^{11} + \frac{2783}{302400}t^{10} + \frac{38}{2835}xt^9 + \frac{143}{2880}t^8 + \frac{22}{315}xt^7 + \frac{7}{12}t^6 + \frac{2}{15}xt^5\right] + \cdots$

The exact solution will be as follows:

$u\left(x,t\right) = t^2 + xt$

9.4 APPLICATION OF ADE: A NUMERICAL ASPECT

It is a cumbersome task to handle the solution of ADE from an analytical point of view. In order to deal with this problem, various numerical techniques have been developed. It can be said that PDEs are the foundation of different mathematical models of chemical, physical and biological phenomena and their applications in financial forecasting, economics, and other areas. It is a need for time to approximate the solutions of such PDEs numerically to fetch the predictions of the mathematical modeling. This concept of numerical approximation gets more critical when exact solutions are unavailable. Since ADE models are some important models, it is essential to have knowledge of numerical approximation whenever an exact solution is not available. A range of numerical techniques is developed to solve ADE equations numerically in order to get a better numerical approximation.

9.5 NUMERICAL METHODS DEVELOPED FOR SOLUTION OF ADE

- Korkmaz and Dağ (2012) provided the numerical solution of ADE using cubic B-spline-DQM. In their paper, cubic B-spline is used as a basis function to find weighting coefficients required in DQM. The formula for cubic B-spline is given as follows:

$$B_k(\theta) = \frac{1}{h^3} \begin{cases} (\theta - \theta_{k-2})^3, & [\theta_{k-2}, \theta_{k-1}) \\ h^3 + 3 h^2 (\theta - \theta_{k-1}) + 3h (\theta - \theta_{k-1})^2 - 3 (\theta - \theta_{k-2})^3, & [\theta_{k-1}, \theta_k) \\ h^3 + 3 h^2 (\theta_{k+1} - \theta) + 3h (\theta_{k+1} - \theta)^2 - 3 (\theta_{k+1} - \theta)^3, & [\theta_k, \theta_{k+1}) \\ (\theta_{k+2} - \theta)^3, & [\theta_{k+1}, \theta_{k+2}) \\ 0, & \text{elsewhere} \end{cases}$$

On applying the cubic B-spline in the DQM approximation, the following system of equations was obtained.
$A W[i] = V[i]$

$$\text{Where, } A = \begin{bmatrix} \begin{bmatrix} 1 & 4 & 1 & & & \\ & & \cdots & & & \\ 0 & 1 & 4 & 1 & & \\ \vdots & & \ddots & & \vdots & \\ & & & 1 & 4 & 1 & 0 \\ & & & & \cdots & & \\ & & & 0 & 1 & 4 & 1 \end{bmatrix} \end{bmatrix}, W[i] = \begin{bmatrix} w_{i,\,-1}^{(1)} \\ w_{i,\,0}^{(1)} \\ w_{i,\,1}^{(1)} \\ \\ w_{i,\,n+2}^{(1)} \end{bmatrix}$$

- Korkmaz and Dağ (2016) provided the solution for ADE using quartic, and quintic B-spline-based DQM. In their paper, quartic, and quintic B-splines were used as basis functions in DQM to find required weighting coefficients.

9.5.1   QUARTIC B-SPLINE

$$B_k(\theta) = \frac{1}{h^4} \begin{cases} (\theta - \theta_{k-2})^4, \ [\theta_{k-2}, \ \theta_{k-1}) \\ (\theta - \theta_{k-2})^4 - 5\,(\theta - \theta_{k-1})^4, \ [\theta_{k-1}, \ \theta_k) \\ (\theta - \theta_{k-2})^4 - 5\,(\theta - \theta_{k-1})^4 + 10\,(\theta - \theta_k)^4, \ [\theta_k, \ \theta_{k+1}) \\ (\theta_{k+3} - \theta)^4 - 5\,(\theta_{k-2} - \theta)^4, \ [\theta_{k+1}, \ \theta_{k+2}) \\ (\theta_{k+3} - \theta)^4, \ [\theta_{k+2}, \ \theta_{k+3}) \end{cases}$$

9.5.2   QUINTIC B-SPLINE

$$B_k(\theta) = \frac{1}{h^5} \begin{cases} \mu_1, \ [\theta_{k-3}, \theta_{k-2}) \\ \mu_1 - 6\,\mu_2, \ [\theta_{k-2}, \ \theta_{k-1}) \\ \mu_1 - 6\,\mu_2 + 15\mu_3, \ [\theta_{k-1}, \ \theta_k) \\ \mu_1 - 6\,\mu_2 + 15\mu_3 - 20\,(\theta - \theta_k)^5, \ [\theta_k, \ \theta_{k+1}) \\ \mu_1 - 6\,\mu_2 + 15\mu_3 - 20\,(\theta - \theta_k)^5 + 15\,(\theta - \theta_{k+1})^5, \ [\theta_{k+1}, \ \theta_{k+2}) \\ \mu_1 - 6\,\mu_2 + 15\mu_3 - 20\,(\theta - \theta_k)^5 + 15\,(\theta - \theta_{k+1})^5 - 6\,(\theta - \theta_{k+2})^5, \\ \quad [\theta_{k+2}, \ \theta_{k+3}) \end{cases}$$

Where $\mu_1 = (\theta - \theta_{i-3})^5$, $\mu_2 = (\theta - \theta_{i-2})^5$ and $\mu_3 = (\theta - \theta_{i-1})^5$

9.6   CONCLUSION

Advection-Diffusion plays a very important role in different aspects of science and engineering. Different ranges of ADEs have been discussed in this chapter, like one-dimensional ADE, two-dimensional ADE, two-dimensional unsteady ADE, three-dimensional ADE, and fractional ADE. In Section 9.2, classification regarding the ADE is provided. In Section 9.2.1, applications of ADE in the field of sciences are provided. In Section 9.2.2, solutions of ADE are provided using an analytical approach named HPM. In Section 9.2.3, applications of ADE from a numerical aspect are provided. In Section 9.3.4, numerical methods developed in the literature for the solution of ADE are mentioned. In Section 9.4, the crux of this discussion is provided as a conclusion.

REFERENCES

Abbaszadeh, M., and Dehghan, M. (2021). The Crank-Nicolson/interpolating stabilized element-free Galerkin method to investigate the fractional Galilei invariant advection-diffusion equation. *Mathematical Methods in the Applied Sciences*, 44(4): 2752–2768.

Aghdam, Y. E., Mesgrani, H., Javidi, M., & Nikan, O. (2021). A computational approach for the space-time fractional advection–diffusion equation arising in contaminant transport through porous media. *Engineering with Computers*, 37(4): 3615–3627.

Al-Jawary, M. A., Azeez, M. M., & Radhi, G. H. (2018). Analytical and numerical solutions for the nonlinear Burgers and advection–diffusion equations by using a semi-analytical iterative method. *Computers & Mathematics with Applications*, *76*(1): 155–171.

Bhatt, H. P., Khaliq, A. Q. M., & Wade, B. A. (2018). Efficient Krylov-based exponential time differencing method in application to 3D advection-diffusion-reaction systems. *Applied Mathematics and Computation*, *338*: 260–273.

Bhrawy, A. H., & Baleanu, D. (2013). A spectral Legendre–Gauss–Lobatto collocation method for a space-fractional advection diffusion equations with variable coefficients. *Reports on Mathematical Physics*, *72*(2): 219–233.

Brahim, N. (2021). Numerical approach of the nonlinear reaction-advection-diffusion equation with time-space conformable fractional derivatives. *AIP Conference Proceedings*, *2334*(1): 060012.

Brooks, A. N., & Hughes, T. J. (1982). Streamline upwind/Petrov-Galerkin formulations for convection dominated flows with particular emphasis on the incompressible Navier-Stokes equations. *Computer Methods in Applied Mechanics and Engineering*, *32*(1–3): 199–259.

Chatwin, P. C., & Allen, C. M. (1985). Mathematical models of dispersion in rivers and estuaries. *Annual Review of Fluid Mechanics*, *17*(1): 119–149.

Company R, Ponsoda, E., Romero, J. V., & Roselló, M. D. (2009). A second order numerical method for solving advection-diffusion models. *Mathematical and Computer Modelling*, *50*(5–6): 806–811.

Dağ, İ., Irk, D., & Tombul, M. (2006). Least-squares finite element method for the advection–diffusion equation. *Applied Mathematics and Computation*, *173*(1): 554–565.

Dehghan, M. (2004). Numerical solution of the three-dimensional advection–diffusion equation. *Applied Mathematics and Computation*, *150*(1): 5–19.

Dhawan, S., Kapoor, S., & Kumar, S. (2012). Numerical method for advection diffusion equation using F.E.M. and B-splines. *Journal of Computational Science*, *3*(5): 429–437.

Du, R., & Liu, Z. (2020). A lattice Boltzmann model for the fractional advection–diffusion equation coupled with incompressible Navier–Stokes equation. *Applied Mathematics Letters*, *101*: 106074.

Du Toit, E. J., O'Brien, M. R., & Vann, R. G. (2018). Positivity-preserving scheme for two-dimensional advection–diffusion equations including mixed derivatives. *Computer Physics Communications*, *228*: 61–68.

Fattah, Q. N., & Hoopes, J. A. (1985). Dispersion in anisotropic, homogeneous, porous media. *Journal of Hydraulic Engineering*, *111*(5): 810–827.

Gharehbaghi, A. (2016). Explicit and implicit forms of differential quadrature method for advection–diffusion equation with variable coefficients in semi-infinite domain. *Journal of Hydrology*, *541*: 935–940.

Gurarslan, G., Karahan, H., Alkaya, D., Sari, M., & Yasar, M. (2013). Numerical solution of advection-diffusion equation using a sixth-order compact finite difference method. *Mathematical Problems in Engineering*, *2013*: 7.

Irk, D., Dağ, İ., & Tombul, M. (2015). Extended cubic B-spline solution of the advection-diffusion equation. *KSCE Journal of Civil Engineering, 19*(4): 929–934.

Isenberg, J., & Gutfinger, C. (1973). Heat transfer to a draining film. *International Journal of Heat and Mass Transfer, 16*(2): 505–512.

Jannelli, A., Ruggieri, M., & Speciale, M. P. (2019). Analytical and numerical solutions of time and space fractional advection–diffusion–reaction equation. *Communications in Nonlinear Science and Numerical Simulation, 70*: 89–101.

Johnson, C., & Saranen, J. (1986). Streamline diffusion methods for the incompressible Euler and Navier-Stokes equations. *Mathematics of Computation, 47*(175): 1–18.

Khan, Y., & Austin, F. (2010). Application of the Laplace decomposition method to nonlinear homogeneous and non-homogenous advection equations. *ZeitschriftfuerNaturforschung A, 65*(10): 849–853.

Korkmaz, A., & Dağ, İ. (2012). Cubic B-spline differential quadrature methods for the advection-diffusion equation. *International Journal of Numerical Methods for Heat & Fluid Flow, 22*(8): 1021–1036.

Korkmaz, A., & Dağ, I. (2016). Quartic and quintic B-spline methods for advection–diffusion equation. *Applied Mathematics and Computation, 274:* 208–219.

Lapidus, L., & Amundson, N. R. (1952). Mathematics of adsorption in beds. VI. The effect of longitudinal diffusion in ion exchange and chromatographic columns. *The Journal of Physical Chemistry, 56*(8): 984–988.

Lian, Y., Ying, Y., Tang, S., Lin, S., Wagner, G. J., & Liu, W. K. (2016). A Petrov–Galerkin finite element method for the fractional advection–diffusion equation. *Computer Methods in Applied Mechanics and Engineering, 309*: 388–410.

Lin, Y., Xiong, H., Chen, M., Dağ, İ., Canivar, A., & Sahin, A. (2011). Taylor-Galerkin method for advection-diffusion equation. *Kybernetes, 40*(5–6): 762–777.

Liu, F., Zhuang, P., Anh, V., Turner, I., & Burrage, K. (2007). Stability and convergence of the difference methods for the space–time fractional advection–diffusion equation. *Applied Mathematics and Computation, 191*(1): 12–20.

Liu, J., Li, X., & Hu, X. (2019). A RBF-based differential quadrature method for solving two-dimensional variable-order time fractional advection-diffusion equation. *Journal of Computational Physics, 384*: 222–238.

Liu, Q., Mu, S., Liu, Q., Liu, B., Bi, X., Zhuang, P., ... & Gao, J. (2018). An RBF based meshless method for the distributed order time fractional advection–diffusion equation. *Engineering Analysis with Boundary Elements, 96*: 55–63.

Mirzaee, F., Sayevand, K., Rezaei, S., & Samadyar, N. (2021). Finite difference and spline approximation for solving fractional stochastic advection-diffusion equation. *Iranian Journal of Science and Technology, Transactions A: Science, 45*(2): 607–617.

Niemi, A. H., Collier, N. O., & Calo, V. M. (2013). Discontinuous Petrov–Galerkin method based on the optimal test space norm for steady transport problems in one space dimension. *Journal of Computational Science, 4*(3): 157–163.

Noye, B. J. (1990). Numerical solution of partial differential equations. Lecture Notes.

Parlange, J. Y. (1980). Water transport in soils. *Annual Review of Fluid Mechanics*, *12*(1): 77–102.

Rubbab, Q., Nazeer, M., Ahmad, F., Chu, Y. M., Khan, M. I., & Kadry, S. (2021). Numerical simulation of advection–diffusion equation with Caputo-Fabrizio time fractional derivative in cylindrical domains: Applications of pseudo-spectral collocation method. *Alexandria Engineering Journal*, *60*(1): 1731–1738.

Sakai, K. (1999). A nonoscillatory numerical scheme based on a general solution of 2-D unsteady advection–diffusion equations. *Journal of Computational and Applied Mathematics*, *108*(1–2): 145–156.

Sakai, K., & Kimura, I. (2005). A numerical scheme based on a solution of nonlinear advection–diffusion equations. *Journal of Computational and Applied Mathematics*, *173*(1): 39

Salmon, J. R., Liggett, J. A., & Gallagher, R. H. (1980). Dispersion analysis in homogeneous lakes. *International Journal for Numerical Methods in Engineering*, *15*(11): 1627–1642.

Singh, J., Kumar, D., & Sushila, D. (2011). Homotopy perturbation Sumudu transform method for nonlinear equations. *Advances in Theoretical and Applied Mechanics*, *4*(4): 165–175.

Tajadodi, H. (2020). A numerical approach of fractional advection-diffusion equation with Atangana–Baleanu derivative. *Chaos, Solitons & Fractals*, *130*: 109527.

Wazwaz, A. M. (2007). A comparison between the variational iteration method and Adomian decomposition method. *Journal of Computational and Applied Mathematics*, *207*(1): 129–136.

Zlatev, Z., Berkowicz, R., & Prahm, L. P. (1984). Implementation of a variable step-size variable formula method in the time-integration part of a code for treatment of long-range transport of air pollutants. *Journal of Computational Physics*, *55*(2): 278–301.

10 Differential Equation-Based Analytical Modeling of the Characteristics Parameters of the Junctionless MOSFET-Based Label-Free Biosensors

Manash Chanda
Meghnad Saha Institute of Technology

Papiya Debnath
Techno International New Town

Avtar Singh
ASTU

CONTENTS

DOI: 10.1201/9781003227847-10

10.1   INTRODUCTION

Recently Field Effect transistor (FET)-based biosensing applications have gained significant attention due to the demand for quick and accurate diagnosis of different enzymes, proteins, DNA, viruses, etc; cost-effective fabrication process; portability and better sensitivity and selectivity compared to the existing biosensors. FET (Chanda et al. 2016a; Das et al. 2014; Basak et al. 2021) is basically a three-terminal device with source, drain, and gate terminals. Basically, the gate terminal controls the current flow between the source and drain terminals. In FETs, first, a nanogap is created in the oxide layer or in the gate by etching adequate materials. When the biomolecules are trapped inside the nanocavity then the surface potentials change and also the threshold voltage varies. As a result, the output current also changes. Finally, by measuring the changes in the threshold voltage or the device current, one can easily detect the biomolecules easily. This process can offer a good amount of sensitivity and selectivity. These FET-based biosensors (Singh et al. 2020; Tripathi et al. 2019; Mendiratta et al. 2020; Bhattacharya et al. 2020a,b; Ajay et al. 2017; Kim et al. 2012; Goswami and Bhowmick 2019; Narang et al. 2015) are efficacious compared to the optical biosensors, resonant biosensors, thermal biosensors, electrochemical biosensors, and bioluminescence sensors. In biosensing label-free process is much more effective compared to the labeled biosensors because of the quick and more accurate diagnosis of the biomolecules.

Out of the FET, recently junctionless MOSFET (JL-MOSFET) (Tripathi et al. 2019; Mendiratta et al. 2020; Bhattacharya et al. 2020a,b; Kim et al. 2012; Goswami and Bhowmick 2019; Narang et al. 2015) has been reported as promising in terms of less short channel effects like improved drain induced barrier lowering (DIBL) and sub-threshold slope (SS), better on to off current, low-cost thermal budgeting etc. A plethora of JL-FET-based label-free biosensor architectures (Jana et al. 2021; Bhattacharya et al. 2020a,b, 2021; Kalra et al. 2016) have been reported in the literature with experimental data and simulated data. However, very few papers highlighted the analytical modeling (Bhattacharya et al. 2020a,b, 2021; Ajay et al. 2017) of the characteristic parameters of the label-free biosensors. In this chapter, we have detailed the analytical modeling of the surface potential, threshold voltage, and the device current by solving the differential equations and these methods will be efficacious to understand the physic in depth.

10.2   MODEL FORMULATION OF THE DEVICE

A cross-sectional view of the Junctionless label-free biosensor is depicted in Figure 10.1. Here,

T_{oxl} = oxide thickness of the silicon layer
T_{oxh} = oxide thickness of the HfO$_2$ layer
L_{ch} = Channel length
T_{si} = thickness of the silicon channel
N_{ch} = doping concentration of the channel

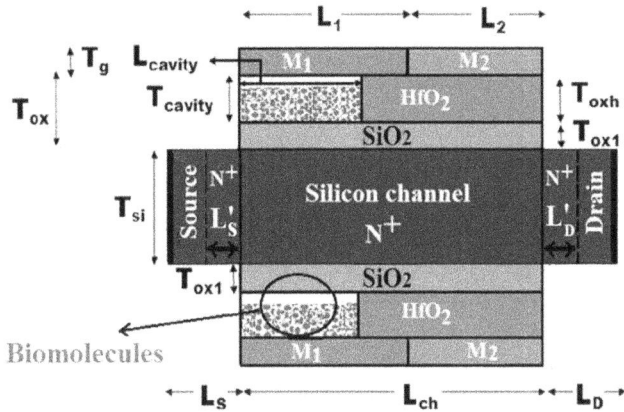

Figure 10.1 Cross-sectional view of the nanogap embedded junctionless MOSFET for label-free biosensing operations.

$N_{S/D}$ = doping concentration of the source/drain.
L_S = length of the source
L_D = Length of the Drain
WF_{M_1} and WF_{M_2} = work function of the gate metal 1 and 2

First, Poisson's equation (Ajay et al. 2017; Jana et al. 2019) for a 2-dimensional (2D) system can be written as,

$$\frac{\partial^2 \phi(x,z)}{\partial x^2} + \frac{\partial^2 \phi(x,z)}{\partial z^2} = \frac{qN_{ch}}{\varepsilon_{si}} \left[e^{\frac{\phi(x,z)-U(x)}{V_T}} - 1 \right] \qquad (10.1)$$

$$\text{for } 0 \leq x \leq L_{ch}; -T_{si} \leq z \leq 0$$

where, $\phi(x, z)$ = distribution of the surface potential in Si channel. Hence, X and Z directions are considered along the horizontal and vertical directions respectively throughout the modeling of the characteristic's parameters.

ε_{si} = permittivity of silicon channel;
N_{ch} = doping concentration of Si body;
$U(x)$ = quasi-fermi potential;
$V_T = KT/q$ (Thermal voltage)
T_{si} = thickness of the Si body
L_{ch} = length of Si channel

Considering the parabolic potential well, the solution of equation (10.1) can be represented as,

$$\phi(x,z) = \phi_s(x) + K_1(x)z + K_2(x)z^2; -T_{si} \leq z \leq 0 \qquad (10.2)$$

Here, surface potential, i.e., $\phi_s(x)$ of the device can be set by putting $z = 0$ in $\phi(x, z)$

$K_1(x)$ and $K_2(x)$ are the coefficients which depend on x only.

For the concept of dual metal, flat-band potentials V_{fb1} and V_{fb2}, correspondingly, related with gate material work functions WF_{M_1} and WF_{M_2} as:

$$V_{fb1} = WF_{M_1} - WF_{si} \tag{10.3}$$

$$\text{and } V_{fb2} = WF_{M_2} - WF_{si} \tag{10.4}$$

where WF_{si} is the work function of silicon. Potential distributions beneath dual gates are appeared as:

$$\phi_1(x,z) = \phi_{s1}(x) + K_{11}(x)z + K_{12}(x)z^2 \tag{10.5}$$

Considering the boundary conditions, $0 \leq x \leq L_1$ and $-T_{si}/2 \leq z \leq 0$,

$$\phi_2(x,z) = \phi_{s2}(x) + K_{21}(x)z + K_{22}(x)z^2$$
$$\text{for } L_1 \leq x \leq (L_1 + L_2); -T_{si}/2 \leq z \leq 0 \tag{10.6}$$

10.2.1   BOUNDARY CONDITIONS

With the subsequent boundary conditions, Poisson's equation for dual gates can be worked out:

1. At the interface of the gate oxide, electric flux (displacement) for dual metal gates is continuous, i.e.

$$\varepsilon_{si} \left[\frac{\partial \phi_1(x,z)}{\partial z} \right]_{z=0} = \frac{\varepsilon_{ox1}}{T_{ox1}} (V_{GS} - V_{fb1} - \phi_{s1}(x)) \quad \text{for gate metal } M_1 \tag{10.7}$$

$$\varepsilon_{si} \left[\frac{\partial \phi_2(x,y)}{\partial y} \right]_{y=0} = \frac{\varepsilon_{ox2}}{T_{oxh}} (V_{GS} - V_{fb2} - \psi_{s2}(x)) \quad \text{for gate metal } M_2 \tag{10.8}$$

where ε_{ox1} and ε_{ox2} are the permittivity of SiO_2 and HfO_2 layer respectively, T_{ox1} and T_{oxh} are their corresponding thickness; and, $V_{GS} = V_G - V_S =$ Gate potential with respect to the source node.

2. At the junction of the two different gate metals, the surface potential is continuous, i.e.

$$\phi_{s1}(L_1) = \phi_{s2}(L_1) \tag{10.9}$$

3. At the interface of two different gate metals, the electric field is continuous, i.e.

$$\phi'_{s1}(L_1) = \phi'_{s2}(L_2) \tag{10.10}$$

4. Assume the extension of depletion region toward source and drain ends are L'_S and L'_D respectively, hence we can estimate surface potential at the source side and drain side as:

$$\psi_{s1}(x=0) = V(x) - \frac{qN_{ch}(L'_S)^2}{2\varepsilon_{si}} \tag{10.11}$$

$$\psi_{s2}(x = L_1 + L_2) = V_{DS} + V(x) - \frac{qN_{ch}(L'_D)^2}{2\varepsilon_{si}} \tag{10.12}$$

where V_{DS} is the drain-to-source biasing potential.

5. Electric field can be estimated at both source and drain ends using boundary conditions given in equations (10.9) and (10.10) must be continuous, i.e.

$$\psi'_{s1}(x = 0) = -\frac{qN_{ch}L'_S}{\varepsilon_{si}} \tag{10.13}$$

$$\psi'_{s2}(x = L_1 + L_2) = \frac{qN_{ch}L'_D}{\varepsilon_{si}} \tag{10.14}$$

10.2.2 OPERATING MODES

The device operation can be explained by considering different modes of operations which mainly depend on the applied bias potentials at different terminals. In this case, the operating modes are given below for ease of understanding,

  i) Full depletion mode (when the applied gate potential, i.e., V_{GS} is below the threshold voltage, i.e., V_{th} of the device, and the device operates in the weak inversion regime mainly. Hence, the leakage current mainly flows in between the source and drain terminal which is very less in magnitude compared to the normal current).
  ii) Partially depleted mode: In this mode, the Si channel is depleted partially as the gate potential lies in between the threshold voltage and the flat-band voltage of the device.
  iii) Flat-band mode: In this case, the gate potential becomes equal with the flat-band voltage which forms almost flats the energy band diagram of the device.

Now to estimate the surface potential in different modes, (equation 10.1) has to be modified accordingly. These have been shown in detail here.

10.2.2.1 Full Depleted (FD) Mode

In this mode as mentioned above, V_{GS} ¡ V_{th}. Now, in full depletion mode (Zhengfan et al. 2007; Young 1989; Gnudi et al. 2013; Kumar and Chaudhary 2004; Colinge 1990; Chanda et al. 2016b), concentration of the mobile charges inside the channel can be ignored safely because of very small magnitude. So, $e^{\frac{\phi(x,z)-U(x)}{V_T}}$ is discarded. Finally, in full depletion mode, (equation 10.1) is rewritten as,

$$\frac{\partial^2 \phi(x,z)}{\partial x^2} + \frac{\partial^2 \phi(x,z)}{\partial z^2} = -\frac{qN_{ch}}{\varepsilon_{si}}; -T_{si}/2 \le y \le 0 \tag{10.15}$$

Due to the symmetricity of the device, the electric field vanishes or becomes zero at the mid-position of the device. Hence, it can be expressed as,

$$\left[\frac{\partial \psi (x, z = -T_{si}/2)}{\partial z}\right] = 0 \tag{10.16}$$

Considering the equations (10.7) and (10.8), and putting the values of the coefficients $K_1(x)$ and $K_2(x)$ mentioned in equation (10.2), $\psi (x, z)$ can be estimated. Putting $\phi (x, y)$ in equation (10.15) and assuming $z = 0$, following is obtained,

$$\frac{\partial^2 \phi_s(x)}{\partial x^2} - 2\frac{C_{ox}}{T_{si}\varepsilon_{si}} \phi_s(x) = b_n \tag{10.17}$$

Hence, assuming $a = 2\frac{C_{ox}}{T_{si}\varepsilon_{si}}$ and

$$b_n = -\left[\frac{qN_{ch}}{\varepsilon_{si}} + a(V_{GS} - WF_{M_n} + \phi_s)\right]$$

(b_n is related to metal gate 'n')

10.2.2.2 Partially Depleted (PD) Mode

Here, the Si channel is depleted when the functional gate voltage (V_{GS}) lies between threshold potential (V_{th}) and flat-band potential (V_{fb}). As a result, in the mid-channel region, a neutral zone is formed. Hence, equation (10.1) can be rewritten as,

$$\frac{\partial^2 \phi (x,z)}{\partial x^2} + \frac{\partial^2 \phi (x,z)}{\partial z^2} = -\frac{qN_{ch}}{\varepsilon_{si}}; -z_{dep} \leq z \leq 0 \tag{10.18}$$

Where z_{dep} measures the depth of the depletion region in Si channel. Assuming the uniformity of z_{dep} throughout the channel, and is derived from the 1-D surface potential model. If the center point of the device structure denotes the origin of surface potential, then the 1-D model is expressed as:

$$\phi (x,z) = \frac{-qN_{ch}}{2\varepsilon_{si}} (|z| - z_C)^2 - E_C (|z| - z_C) + U(x) - V_T \tag{10.19}$$
for $z_C \leq |z| \leq T_{si}/2$

Here, the depletion region starts from z_C with respect to the origin; Besides, at $z = z_C$, the electric field considered as E_C. This electric field can be estimated by applying the finite difference method (FDM) in $\phi (x, z_C) = U(x) - V_T$ i.e.

$$\frac{\partial \phi (x, z = z_C)}{\partial z} = E_C = \frac{-8V_T z_{dep}}{T_{si}^2} \tag{10.20}$$

$$z_{dep} = \frac{T_{si}}{2} - z_C \tag{10.21}$$

Utilizing equations (10.20) and (10.21) in equation (10.19), and then by utilizing equation (10.7) or (10.8), z_{dep} can be estimated as

$$z_{dep} = \frac{-\beta + \sqrt{\beta^2 - 4\alpha\gamma}}{2\alpha} \tag{10.22}$$

$$\text{where, } \beta = \frac{qN_{ch}}{C_{ox}} + \frac{8\varepsilon_{si}V_T}{C_{ox}T_{si}^2},$$

$$\alpha = \frac{qN_{ch}}{2\varepsilon_{si}} + \frac{8V_T}{T_{si}^2}$$

And $\gamma = V_T + V_{GS} - U(x) - V_{fb2}$

Using a simple parabolic function surface potential in a neutral area is developed and its estimated value at the center position of the device is considered as $U(x)$, i.e.

$$\phi\left(x, z = -\frac{T_{si}}{2}\right) = U(x) \tag{10.23}$$

The coefficients $K_{n1}(x)$ and $K_{n2}(x)$ are assumed from boundary conditions (10.7), (10.8) and (10.20). Put the constant values in equations (10.5) and (10.6) and subsequently in equation (10.19), the following is obtained,

$$\frac{\partial^2 \phi_{si}(x)}{\partial x^2} - a\phi_{si}(x) = b_n \tag{10.24}$$

$$\text{Where, } a = \frac{C_{ox}}{z_{dep}\varepsilon_{si}}$$

$$b_n + \left[\frac{qN_{ch}}{\varepsilon_{si}} + a\left(V_{GS} - WF_{M_n} + WF_{si}\right)\right] = 0$$

(b_n is related to metal gate 'n')

10.2.2.3 Next to Flat-Band (FB) Mode

Under the FB condition, the silicon channel has attained a doping concentration equal to N_{ch} ($\sim 10^{19}$/cm^3). Utilizing Taylor's series, equation (10.1) can be solved to estimate the potential distribution $\phi\,(x, z)$ in the Si channel.

$$\frac{\partial^2 \phi(x, z)}{\partial x^2} + \frac{\partial^2 \phi(x, z)}{\partial z^2} = \frac{qN_{ch}}{\varepsilon_{si}}\left(\frac{\phi(x, y) - U(x)}{V_T}\right); \quad \text{for } -T_{si}/2 \leq y \leq 0 \tag{10.25}$$

The coefficients $K_{n1}(x)$ and $K_{n2}(x)$ are assumed from boundary conditions (10.7), (10.8), and (10.16). Put the constant values in equations (10.5) and (10.6) and subsequently in equation (10.25), we get,

$$\frac{\partial^2 \phi_{si}(x)}{\partial x^2} - a\phi_{si}(x) = b_n \tag{10.26}$$

where $a = 2\dfrac{C_{ox}}{T_{si}\varepsilon_{si}} + \dfrac{qN_{ch}}{\varepsilon_{si}V_T}$;

$$b_n + \left[\frac{qN_{ch}}{\varepsilon_{si}V_T} + a\left(V_{GS} - WF_{M_n} + WF_{si}\right)\right] = 0$$

(b_n is related with metal gate 'n')

10.3 MODELING OF THE SURFACE POTENTIAL

With the variation of the applied gate to source biasing voltage, the channel regions underneath two gate metal regions will experience dissimilar operating modes. Thus, we need to find out the values of the surface potentials considering the dissimilar thicknesses of the depletion region (z_{dep1}, z_{dep2}) underneath two gate metals, which is more practical in the modeling of the surface potentials.

10.3.1 GATE METAL M_1 (FD MODE) AND GATE METAL M_2 (PD MODE)

Here, $V_{GS} < V_{th1}$, the channel region underneath gate metal M_1 is fully depleted and for $V_{th2} < V_{GS} < V_{fb2}$ that of M_2 is PD. Hence, the surface potential is characterized by utilizing equations (10.17) and (10.24) for gate metal M_1 and M_2 as:

$$\phi_{s1}(x) = M_1 e^{\sigma_1 x} + N_1 e^{-\sigma_1 x} - \frac{b_1}{a}; \text{ for } 0 \le x \le L_1 \tag{10.27}$$

$$\phi_{s2}(x) = M_2 e^{\sigma_2(x-L_1)} + N_2 e^{-\sigma_2(x-L_1)} - \frac{b_2}{a}; \text{ for } 0 \le x \le L_2 \tag{10.28}$$

where M_1, M_2, N_1, and N_2 are obtained by utilizing boundary conditions (10.9)–(10.14) as given in equation (10.34).

10.3.2 GATE METAL M_1 (PD MODE) AND GATE METAL M_2 (PD MODE)

When, $V_{GS} ¿ V_{th1}$, the channel region underneath gate metal M_1 is PD and for $V_{th2} < V_{GS} < V_{fb2}$ that of M_2 is also PD. Here, the surface potential is characterized by utilizing equation (10.24) for both gate metals M_1 and M_2 as:

$$\phi_{s1}(x) = N_1 e^{-\sigma_1 x} + M_1 e^{\sigma_1 x} - \frac{b_1}{a}; \text{ for } 0 \le x \le L_1 \tag{10.29}$$

$$\phi_{s2}(x) = N_2 e^{-\sigma_2(x-L_1)} + M_2 e^{\sigma_2(x-L_1)} - \frac{b_2}{a}; \text{ for } 0 \le x \le L_2 \tag{10.30}$$

where M_1, M_2, N_1, and N_2 are obtained by utilizing boundary conditions (10.9)–(10.14) as given in equation (10.34).

10.3.3 GATE METAL M_1 (PD MODE) AND GATE METAL M_2 (NEAR FB MODE)

Once $V_{th1} < V_{GS} < V_{fb1}$, gate metal M_1 drives the underneath channel into PD and for $V_{GS} ¿ V_{fb2}$ that of M_2 is near FB mode. Also, in the near FB mode, zero extension of the depletion region at the drain end is observed, i.e., ($L'_D = 0$). Here, the surface potential is characterized by utilizing equations (10.24) and (10.26) for gate metal M_1 and M_2 as:

$$\phi_{s1}(x) = M_1 e^{\sigma_1 x} + N_1 e^{-\sigma_1 x} - \frac{b_1}{a}; \text{ when } 0 \le x \le L_1 \tag{10.31}$$

$$\phi_{s2}(x) = M_2 e^{\sigma_2(x-L_1)} + N_2 e^{-\sigma_2(x-L_1)} - \frac{b_2}{a}; \text{ when } 0 \le x \le L_2 \tag{10.32}$$

where M_1, M_2, N_1, and N_2 are obtained by utilizing boundary conditions (10.9)–(10.13) as given in equation (10.34).

10.4 THRESHOLD POTENTIAL MODEL

Once $V_{GS} = V_{th}$, the depletion region dominates over the channel region. Now, putting $z_{dep} = -T_{si}/2$ to equation (10.22) and then solving for V_{GS}, we can obtain the threshold potential as (Figures 10.2–10.5):

$$V_{th} = V_{fb1} - \frac{qN_{ch}T_{si}^2}{8\varepsilon_{si}} - \frac{qN_{ch}T_{si}T_{ox}}{2\varepsilon_{ox}} + V(x) - 3V_T - \frac{4C_{si}}{C_{ox}}V_T \qquad (10.33)$$

In the dual metallic concept, the threshold voltage will be due to the gate metal having a higher work function. Here $WF_{M_1} > WF_{M_2}$. So, we get,

$$L'_S = \frac{-\beta_S + \sqrt{\beta_S^2 - 4\alpha_S\gamma_S}}{2\alpha_S};$$

$$L'_D = \frac{-\beta_D + \sqrt{\beta_D^2 - 4\alpha_D\gamma_D}}{2\alpha_D}$$

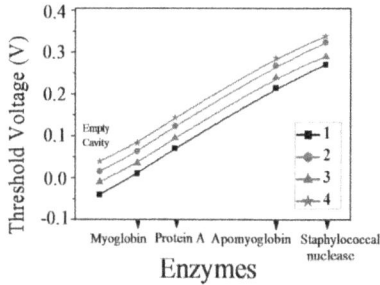

Figure 10.2 V_{th} variation has been measured for different dimensions of the cavity. Hence, line: model and symbol: simulation.

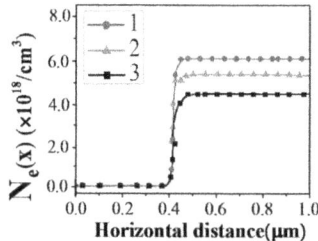

Figure 10.3 Electron concentration profile considering the charged biomolecules. Hence, line: model and symbol: simulation.

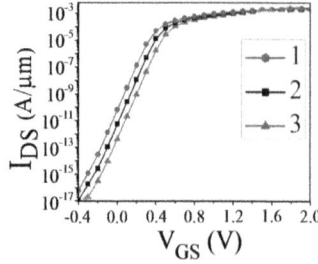

Figure 10.4 Variation of drain current for different gate to source potential. Hence, Channel length =1 μm, Silicon body thickness =8 nm, drain to source potential = 1V for the decreasing profile of the bio-molecules inside the cavity. Here, 1: Myoglobin, 2:Protein A, 3:Apomyoglobin, and 4: Staphylococcal nuclease 5: DNA and 6: Without Cavity. Hence, Line: Model and Symbol: Simulation. Very good agreement in between the model and the simulated data has been observed here.

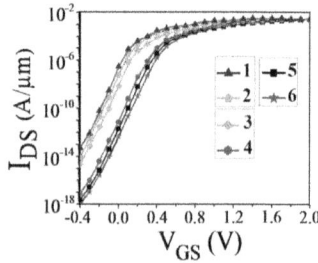

Figure 10.5 Variation of drain current for different gate to source potential. Hence, line: model and symbol: simulation.

$$\alpha_S = \frac{qN_{ch}}{2\varepsilon_{si}}; \quad \beta_S = \frac{qN_{ch}}{\varepsilon_{si}\sigma_1};$$

$$\gamma_S = \frac{-b_1}{a_1} - U(x)$$

$$\alpha_D = \frac{qN_{ch}}{2\varepsilon_{si}};$$

$$\beta_D = \frac{qN_{ch}}{\varepsilon_{si}\sigma_2};$$

$$\gamma_D = \frac{-b_2}{a_2} - U(x) - V_{dep}$$

$$\sigma_n = \sqrt{a_n};$$

$$N_1 = \left(\frac{b_1}{2a_1} + \frac{U(x)}{2} - \frac{(L'_S)}{2} \left(\frac{qN_{ch}}{2\varepsilon_{si}} (L'_S) + \frac{qN_{ch}}{\varepsilon_{si}\sigma_1} \right) \right)$$

$$M_2 = \left(\frac{b_2}{2a_2} + \frac{U(x)+V_{dep}}{2} - \frac{qN_{ch}}{4\varepsilon_{si}} (L'_D)^2 + \frac{qN_{ch}}{2\varepsilon_{si}\sigma_2} (L'_D) \right) e^{-\sigma_2 L_2}$$

for fully depleted modes

$$M_2 = \frac{\left[\left(V(x) + V_{\text{dep}} + \frac{b_2}{a_2}\right)(\sigma_1 + \sigma_2)e^{(\sigma_2 L_2)} + \left(\frac{b_1}{a_1} - \frac{b_2}{a_2}\right)\sigma_1 - 2\sigma_1 N_1 e^{-\sigma_1 L_1}\right]}{\sigma_1 - \sigma_2 + (\sigma_1 + \sigma_2)e^{2\sigma_2 L_2}}$$

and PD modes

$$M_2 = \frac{\left[\left(V(x) + V_{\text{dep}} + \frac{b_2}{a_2}\right)(\sigma_1 + \sigma_2)e^{(\sigma_2 L_2)} + \left(\frac{b_1}{a_1} - \frac{b_2}{a_2}\right)\sigma_1 - 2\sigma_1 N_1 e^{-\sigma_1 L_1}\right]}{\sigma_1 - \sigma_2 + (\sigma_1 + \sigma_2)e^{2\sigma_2 L_2}}$$

for next-to-FB mode

$$M_1 = \frac{\left[2M_2 - \left(1 - \frac{\sigma_1}{\sigma_2}\right)N_1 e^{-\sigma_1 L_1} + \frac{b_1}{a_1} - \frac{b_2}{a_2}\right]e^{-\sigma_1 L_1}}{1 + \frac{\sigma_1}{\sigma_2}}$$

$$N_2 = \left[\left(1 - \frac{\sigma_1}{\sigma_2}\right)M_1 e^{\sigma_1 L_1} + \left(1 + \frac{\sigma_1}{\sigma_2}\right)N_1 e^{-\sigma_1 L_1} - \frac{b_1}{a_1} + \frac{b_2}{a_2}\right]\frac{1}{2} \qquad (10.34)$$

According to different modes of operation shown by the particular gate metal, values of a_n, b_n, and σ_n are employed in all the above-mentioned equations.

In JL-MOSFET, sub-threshold current, I_S can be expressed using the EKV model (Chakraborty et al. 2013; Jain et al. 2013; Podder et al. 2017), as follows,

$$I_S = kT\mu \left(W/L_{\text{ch}}\right)n(0) \times \left[1 - \exp\left((-qV_{\text{DS}})/(kT)\right)\right] \qquad (10.35)$$

Here, K and $n(0)$ are Boltzmann's constant, the density of the electrons at near-source surface. Hence, $n(0)$ can be denoted as,

$$n(0) = N_{\text{ch}}q\exp\left[\left(\frac{qN_{\text{ch}}T_{\text{si}}^2}{8\varepsilon_{\text{si}}} + V_{\text{GS}} - V_{\text{fb}} + \frac{qN_{\text{ch}}T_{\text{si}}}{2C_{\text{eff},S}}\right)/KT\right]$$
$$\times \left(\frac{2\pi\varepsilon_{\text{si}}k_B T}{q^2 N_{\text{ch}}}\right)^{0.5} \times \text{erf}\left[\frac{T_{\text{si}}}{2} \times \left(\frac{q^2 N_{\text{ch}}}{2\varepsilon_{\text{si}}k_B T}\right)^{0.5}\right]$$

Considering, erf() = 1 because of high N_{ch}, we get,

$$I_{\text{sub}} = KT\mu\left(\frac{W}{L_{\text{ch}}}\right) \times N_{\text{ch}}q\exp\left[\left(\frac{qN_{\text{ch}}T_{\text{si}}^2}{8\varepsilon_{\text{si}}} + V_{\text{GS}} - V_{\text{fb}} + \frac{qN_{\text{ch}}T_{\text{si}}}{2C_{\text{eff},S}}\right)/KT\right]$$
$$\times \left(\frac{2\pi\varepsilon_{\text{si}}k_B T}{q^2 N_{\text{ch}}}\right)^{0.5} \times \left[1 - \exp\left(\frac{-V_{\text{DS}}q}{KT}\right)\right]$$

10.5 RESULTS AND SIMULATIONS

10.6 CONCLUSION

This chapter mainly shows the differential equation-based analytical modeling of the characteristic parameters of label-free biosensors. The differential equation-based approach simplifies the modeling of the sensor parameters like the surface potential, threshold voltage, and drain current. This helps to estimate the sensitivity of the sensors easily. Parameter modeling is very efficacious in the design and optimization of the sensor device for better efficiency. Hence, the model data have been obtained by extensive simulation of MATLAB whereas, the simulation data have been obtained by using the SILVACO ATLAS (ATLAS Device Simulation Software 2016) tool. Model data matches with simulated data very closely and so the efficiency of the model has been validated. This simple but effective approach to estimating the sensitivity of the label-free biosensor will definitely help the researchers in the design and implementation of optimized label-free biosensors for quick detection of the DNA, proteins, enzymes, etc.

REFERENCES

Ajay, S., Narang, R., Saxena, M. and Gupta, M. (2017). Modeling and simulation investigation of sensitivity of symmetric split gate junctionless FET for biosensing application, *IEEE Sensors Journal*, 17(15): 4853–4861.

ATLAS Device Simulation Software (2016). Silvaco Int., Santa Clara, CA, USA.

Basak, A., Chanda, M. and Sarkar, A. (2021). Drain current modelling of unipolar junction dual material double-gate MOSFET (UJDMDG) for SoC applications, *Journal of Microsystem Technologies*, 27(11): 3995–4005.

Bhattacharya, A., Chanda, M. and De, D. (2020a). Analysis of Partial hybridization and probe positioning on sensitivity of a dielectric modulated junctionless label free biosensor, *IEEE Transaction on Nanoelectronics*, 19(4): 719–727.

Bhattacharya, A., Chanda, M. and De, D. (2020b). Analysis of noise-immune dopingless heterojunction bio-TFET considering partial hybridization issue, *IEEE Transaction on Nanoelectronics*, 19: 769–777.

Bhattacharya, A., Chanda, M. and De, D. (2021). Dielectrically modulated Bio-FET for label-free detection of bio-molecules. In: Dutta, G., Biswas, A., and Chakrabarti, A. (Eds.), *Modern Techniques in Biosensors*, pp. 183–198. Springer Nature: Berlin.

Bhattacharya, A., Mukherjee, A., Chanda, M. and De, D. (2020). Advantages of charge plasma based double gate junctionless MOSFET over bulk MOSFET for label free biosensing, in *2020 IEEE VLSI Device Circuit and System (VLSI DCS)*. Meghnad Saha Institute of Technology (MSIT), Kolkata.

Chakraborty, A. S., Chanda, M. and Sarkar, C. K. (2013). Analysis of noise margin of CMOS inverter in sub-threshold regime, in *SCES*, Allahabad, pp. 1–5.

Chanda, M., Chakrabarty, A. S., and Sarkar, C. K. (2016a). Complete delay modeling of sub-threshold CMOS logic gates for low-power application, *International Journal of Numerical Modelling: Electronic Networks, Devices and Fields (Wiley Online Library)*, 29(2): 132–145.

Chanda, M., De, S. and Sarkar, C. K. (2016b). Modeling of characteristic parameters for nano-scale junctionless double gate MOSFET considering quantum mechanical effect, *Journal of Computational Electronics (Springer)*, 14(1): 262–269.

Colinge J. P. (1990). Conduction mechanisms in thin-film accumulation-mode SOI p-channel MOSFET's, *IEEE Transactions on Electron Devices*, 37(3): 718–723.

Das, D., De, S., Chanda, M. and Sarkar, C. K. (2014). Modelling of sub threshold surface potential for short channel double gate dual material double halo MOS-FET, *The IUP Journal of Electrical & Electronics Engineering, ICFAI University Press*, 7(4): 19–42.

Gnudi, A., Reggiani, S., Gnani, E. and Baccarani, G. (2013). Semi-analytical model of the subthreshold current in short-channel junctionless symmetric double-gate field-effect transistors, *IEEE Transactions on Electron Devices*, 60(4): 1342–1348.

Goswami, R. and Bhowmick, B. (2019). Comparative analyses of circular gate TFET and heterojunction TFET for dielectric-modulated label-free biosensing, *IEEE Sensors Journal*, 19(21): 9600–9609.

Jain, S., Chanda, M. and Sarkar, C. K. (2013). Comparative analysis of delay for sub-threshold CMOS logics, in *SCES*, Allahabad, pp. 1–4.

Jana, G., Debnath, P., and Chanda, M. (2021). A dielectric modulated MOS transistor for biosensing, in *IEEE 2021 Devices for Integrated Circuit (DevIC)*, Kalyani, pp. 484–488.

Jana, G., Sen, D. and Chanda, M. (2019). Junctionless double gate non-parabolic variable barrier height Si-MOSFET for energy efficient application, *Journal of Microsystem Technologies, Springer,* 27: 3987–3994.

Kalra, S., Kumar, M. J. and Dhawan, A. (2016). Dielectric modulated field effect transistor for DNA detection: Impact of DNA orientation, *IEEE Electron Device Letters*, 37(11): 1485–1488.

Kim, C. H., Ahn, J. H., Lee, K. B., Jung, C., Park, H. G. and Choi, Y. K. (2012). A new sensing metric to reduce data fluctuations in a nanogap-embedded field-effect transistor biosensor, *IEEE Transactions on Electron Devices*, 59(10): 2825–2831.

Kumar, M. J. and Chaudhary, A. (2004). Two-dimensional analytical modeling of fully depleted DMG SOI MOSFET and evidence for diminished SCEs, *IEEE Transactions on Electron Devices*, 51(4): 569–574.

Mendiratta, N., Tripathi, S. L., Padmanaban, S. and Hossain, E. (2020). Design and analysis of heavily doped n+ pocket asymmetrical junction-less double gate MOSFET for biomedical applications, *Applied Sciences, MDPI*, 10: 2499.

Narang, R., Saxena, M. and Gupta, M. (2015). Comparative analysis of dielectric-modulated FET and TFET-based biosensor, *IEEE Transactions on Nanotechnology*, 14(3): 427–435.

Podder, A., Mal, S., Chowdhury, A., Mondal, A. and Chanda, M. (2017). Design and analysis of adiabatic adder in near-threshold regime for low power application, in *Devices for Integrated Circuit (DevIC)*, Kalyani, pp. 670–675.

Singh, A., Chaudhury, S., Chanda, M. and Sarkar, C. K. (2020). Split gated silicon nanotube FET for bio-sensing applications, *IET Circuit, Device and System,* 14(8): 1289–1294.

Tripathi, S. L., Patel, R. and Agrawal, V. K. (2019). Low leakage pocket junction-less DGTFET with bio-sensing cavity region, *Turkish Journal of Electrical Engineering and Computer Sciences*, 27(4): 2466–2474.

Young, K. K. (1989). Analysis of conduction in fully depleted SOI MOSFETs, *IEEE Transactions on Electron Devices*, 36(3): 504–506.

Zhengfan, Z., Zhaoji, L., Kaizhou, T. and Jiabin, Z. (2007). Investigation into sub-threshold performance of double-gate accumulation-mode SOI PMOSFET, in *Proceedings of International Conference on ASIC (ASICON)*, Guilin, China, pp. 1150–1153.

11 Application of Differential Equation in Inventory Control

Arunava Majumder
Lovely Professional University

Tanmoy Chakraborty
SRM Institute of Science and Technology

Ruchi Chauhan and Richa Nandra
Lovely Professional University

CONTENTS

11.1 INTRODUCTION

Differential equations have been playing a versatile role in various aspects of mathematics, engineering, and marketing management. The application of differential equations comes into play when some variability exists in the phenomena. Basically,

DOI: 10.1201/9781003227847-11

in inventory control differential equation appears while dealing with the demand. The level of inventory decreases concerning time and at a demand rate, resulting in the formation of a boundary value problem. The rate of demand may be fixed or variable depending on which the governing differential equation changes. The formed differential equation is in general an ordinary differential equation (ODE) with two boundary conditions. On solving the differential equation, a relation between order quantity, time, and demand is obtained, which is further used for the optimization of the objective function (cost or profit).

The classical deterministic inventory model is of two types namely, economic order quantity (EOQ) and economic manufacturing quantity (EMQ) model. One of the most famous inventory decision-making models is EOQ which was pioneered by Harris (1990). After that, this model was extended with the EMQ model by Taft (1918). These two most commonly used inventory models were later modified with many attributes like lead time, shortages, fully or partially backlogging, etc. In EOQ and EMQ, the demand rate was considered constant and a simple governing differential equation (ordinary) was developed with two boundary conditions. Later on, the governing differential equation for demand was modified by introducing several variable quantities in demand or by considering inventory deterioration. A useful survey of numerous inventory models was studied by Goyal and Giri (2001). Depending on variation in demand function the following categories are shown (only deterministic demand functions are considered).

1. Uniform demand ("Deterministic demand")
2. Variation of demand with respect to time ("Time dependent demand")
3. Demand depending upon stock ("Stock dependent demand")
4. Demand depending upon price ("Price dependent demand")

Underneath are some of the fundamental forms of governing differential equations of inventory control.

11.1.1   THE CLASSICAL EOQ MODEL

Underneath the differential equation represents the level of inventory I at any time t in $[0, T]$ (D is the deterministic and constant demand rate)

$$\frac{dI}{dt} = -D$$

Subjected to the boundary constraints,

$$I = Q, t = 0 \qquad\qquad (11.1)$$

$$I = 0, t = t \qquad\qquad (11.2)$$

After solving these equations, we get

$$Q = Dt$$

11.1.2 EOQ MODEL WHERE DEMAND IS TIME-DEPENDENT i.e., *D* = *D*(*t*)

The following differential equation reflects the level of the inventory at any time t in $[0, T]$ is given by

$$\frac{dI}{dt} = -D(t)$$

Subjected to the boundary constraints,

$$I = Q, t = 0 \qquad (11.3)$$

$$I = 0, t = t$$
$$dI = -D(t)\,dt \qquad (11.4)$$

By integrating both sides we get,

$$I = -\int D(t)\,dt + C$$

11.1.3 EOQ MODEL WHERE DEMAND IS DEPENDING UPON STOCK i.e., *D* = *D*(*I*)

The level of inventory at any time t in $[0, T]$ is elucidated by the underneath differential equation

$$\frac{dI}{dt} = -D(I)$$

Subjected to the boundary conditions,

$$I = Q, t = 0 \qquad (11.5)$$

$$I = 0, t = t$$
$$-\frac{dI}{D(I)} = dt \qquad (11.6)$$

By integrating both sides we get,

$$t = -\int \frac{dI}{D(I)} + C$$

In this study, while considering ramp-type demand and deterioration in an inventory model, the application of a differential equation is represented. The variation concerning time "Weibull deterioration rate" is assumed in this model (Skouri et al., 2009).

Unlike others from the 1990s (Raaffat et al. 1991), extended a model for degenerating items employing a traditional computerized investigating procedure to evaluate its decision variables with fixed demand as well as determining replenishment rate. Goh et al. (1993) contemplated a system of two-stage perishable inventory, utilizing

FIFO policy with demand either for fresh or somewhat older items and compared two managerial strategies. Wee and Shum (1999) analyzed deteriorating inventory to obtain the best possible refill policy, deploying approaches of "the Wagner–Whitin, the Silver–Meal, and the least-unit-cost methods".

The concept of constant demand rate does not hold in all circumstances for certain goods with the possibility of change in demand either due to the introduction of competitive items or the age of inventory causing customer's loss of faith in items' quality. This encourages researchers to think of abating inventory models for demands that vary between continuous time and discrete time. Thereafter, $d(t) = a + bt$, $a > 0, b > $ or < 0 and $d(t) = Ae^{\alpha t}$, $A > 0, \alpha > $ or < 0 patterns are considered for continuous time inventory models.

Xu and Wang (1990) studied (Ti, Si) policy model for constant deterioration, unequal replenishment periods and linearly increasing/decreasing time-dependent demand. Hariga (1996) developed a model for continuous time-varying demand in log-concave fashion under three replenishment policies and allowed dearth. The author obtained results like minimal outlay policy and the finest service-level effectiveness. Chakraborty et al. (1998) had expressed time of deterioration in three-parameter Weibull distribution with the demand that changes with time and allowed shortages. Chang and Dye (1999) considered a replenishment policy which is having a constant deterioration rate and a time-continuous monotonic function of demand. In this paper, unlike others, they assumed backlogging as variable and dependent on the time between the next replenishment and defined it as $\frac{1}{1+\alpha(t_i - t)}$. Bhunia and Maiti (1998) incorporated a functional relation of replenishment, $R(t)$ $= \alpha - \beta Q(t) + rD(t)$, $\alpha > 0, r \geq 0, 0 \leq \beta < 1$, that depends on the available inventory and demand rate. They measured demand and deteriorating rate as linearly increasing time functions.

Hollier and Mak (1983) determined optimal replenishment policies for fixed and changeable intervals while analyzing exponentially declining demand. Andijani and AJ-Dajani (1998) solved numerically using constant, linear, and quadratic demand functions in its stated inventory problem. Aliyu and Boukas (1998) scrutinized discrete-time single as well as multi-item production inventory systems with arbitrary deterioration and obtain control policies using the LQ criterion.

It has been observed that stock tends to invite new clients. Thus, the thought of displaying more generates dilemmas like space allocation and investment inevitability. The deteriorating nature of inventory makes the situation more critical which originates the need of studying inventory models based on stock-dependent demand. This phenomenon has been studied by some including who derived two profit functions, with the backlog and without a backlog, for deteriorating items that had stock-dependent selling rates. Roy and Maiti (1998) analyzed multi-item, multi-objective inventory problems and determined to maximize profit and minimize wastage cost in the fuzzy environment using FNLP and FAGP and express all components of models by fuzzy membership functions as well as in fuzzy numbers. Kim et al. (1995) and Wee (1997) examined problems based on lot size and price in which products under consideration deteriorate constantly and varyingly respectively. Wee

(1995) took under consideration demand and inspected an inventory model of deteriorating items, finite planning horizon, and constant replenishment cycles while optimizing ordering policy and price. Abad (1996) explored the problem of fragile goods demanded by eager customers. The author used a new approach to illustrate the backlog phenomenon with dynamic pricing and lot size concerns stating it as a fraction of demand that too as a falling function of waiting time. Inventory deterioration is one of the most important aspects. Many researchers worked on the decay in inventory (linear deterioration, exponential deterioration, etc.) for many years. Chare and Schrader (1963) introduced the exponential decay in inventory initially. Later on, many researchers developed decaying inventory models such as "Weibull" and "Gamma" distribution deterioration (Covert and Philip 1973). The linear time-varying demand was first proposed by Donaldson (1977) and Resh et al. (1976). The time-varying deterioration became very popular and observed in many existing works of literature (Dave and Patel 1981; Datta and Pal 1988; Hariga 1996).

In a most practical situation, linear or exponential time-varying demand is not followed every time as the instantaneous increase in demand is not too idealistic. After a certain period, the increase in demand may cease and stabilize. Chung et al. (2000) took forward (Padmanabhan and Vrat's 1995) inventory problem by developing necessary and sufficient conditions for obtaining the best possible solutions for profit/time functions using the Newton-Raphson technique. Observing this scenario (Hill 1995), first introduces "ramp type" demand. Thereafter, the "ramp type" demand function was reinvestigated and modified with a "Generalized Weibull Distribution" deterioration rate (Wu 2001; Giri et al. 2003). It is quite the case that a merchant's ordered quantity relies on demand while the demand for a product depends on the price of the product.

This chapter deals with an inventory model with ramp-type demand. Deterioration is considered with Weibull distribution. Two sets of models are assumed. The first inventory model was formulated with no shortages at its initial. The second one considers shortages at the time of starting inventory. The optimal policy for both cases is derived and the minimum cost is obtained.

11.2 MATHEMATICAL MODEL

The following notations and assumptions are used to develop the model (Table 11.1).

Assumptions

1. The retailer places an order when the level of inventory reaches the reorder point R.
2. $\gamma(y)$ represents the rate at which shortages are backlogged, which is a non-increasing function of y ($\gamma'(y) \leq 0$) with $0 \leq \gamma(y) \leq 1, \gamma(0) = 1$ and y depicts the standby time till the succeeding is restored. Further, it is presumed that the relation $\gamma(y) + T\gamma'(y) \geq 0$ is satisfied by $\gamma(y)$, where $\gamma'(y)$ is a derivative of $\gamma(y)$. The cases with $\gamma(y) = 1$ (or 0) represents complete lost sales models (Skouri et al. 2009).

3. The time taken by item to deteriorate follows Weibull distribution (α, β), i.e., $\varnothing(t) = \alpha\beta t^{\beta-1} (\alpha > 0, \beta > 0, t > 0)$ represents deterioration rate. During the period T there is no repair of deteriorated units. For $\beta = 1, \varnothing(t) = $ constant representing exponentially decaying cases (Skouri et al. 2009).

Ramp-type demand function

The demand rate D(t) is given by

$$D(t) = \begin{cases} h(t), & t < \vartheta, \\ h(\vartheta), & t \geq \vartheta, \end{cases}$$

where $h(t)$ represents a positive, continuous function of $t \in (0, T]$.

- $\gamma(y)$ is a rate at which shortages are backlogged, which is a non-increasing function of y ($\gamma'(y) \leq 0$) with $0 \leq \gamma(y) \leq 1, \gamma(0) = 1$ and y is the standby time up to the succeeding refill. Further, it is presumed that $\gamma(y)$ satisfies the relation $\gamma(y) + T\gamma'(y) \geq 0$, where $\gamma'(y)$ is a derivative of $\gamma(y)$. Complete backlogging models are represented by $\gamma(y) = 1$ (or 0).
- The time required for deterioration of the item is following Weibull's distribution (α, β), i.e., the deterioration rate is $\varnothing(t) = \alpha\beta t^{\beta-1} (\alpha > 0, \beta > 0, t > 0)$. During the time period T there is no repair of deteriorated units. For $\beta = 1, \varnothing(t) = $ constant, depicting exponentially decaying cases.

The underneath model has been divided into two parts. In the first part, the inventory begins with no shortages and is followed by the inventory which starts with shortages.

Table 11.1

Details of the symbol used in the text

Notation

T	The uniform time period (cycle) (year)
t_1	The restoring time for the model initiating with a shortage
L	The maximal level of inventory during the time period (cycle)
a_1	The inventory holding cost per unit time ($ per unit per year)
a_2	The shortage cost per unit time ($ per unit per year)
a_3	The cost incurred from the deterioration of one unit ($ per unit)
a_4	The per unit opportunity cost due to the lost sales ($ per unit)
ϑ	The ramp-type demand function's (time point) parameter (year)
$I(t)$	The level of inventory at the time $t \in [0, T]$

11.2.1 FORMULATION OF MATHEMATICAL MODEL BEGINNING WITH NO SHORTAGE

In this section, the inventory model starting with no shortages is studied. The replenishment at the beginning of the cycle brings the inventory level up to L. Due to demand and deterioration, the inventory level gradually depletes during the period $(0, t_1)$ and falls to zero at $t = t_1$. Thereafter shortages occur during the period (t_1, T), which is partially backlogged. The backlogged demand is satisfied at the next replenishment. The inventory level, $I(t)$, $0 \leq t \leq T$ satisfies the following differential equations:

$$\frac{dI(t)}{dt} + \varnothing(t)I(t) = -D(t), \quad 0 \leq t \leq t_1, \quad I(t_1) = 0 \qquad (11.7)$$

$$\frac{dI(t)}{dt} = -D(t)\gamma(T-t), \quad t_1 \leq t \leq T, \quad I(t_1) = 0 \qquad (11.8)$$

The interconnection between t_1 and ϑ influences the results of differential equations through the demand rate function. Thus, for further analysis, upcoming cases must be considered and the accomplishment of inventory level is reflected in Figures 11.1 and 11.2.

11.2.1.1 Case I

$t_1 \leq \vartheta$

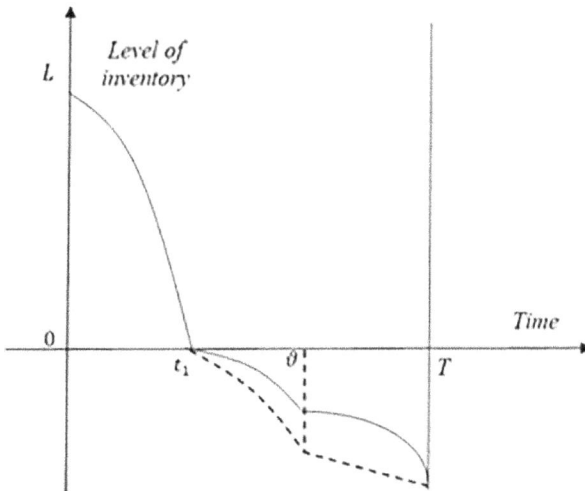

Figure 11.1 Representation of the inventory level for the case $t_1 \leq \vartheta$.

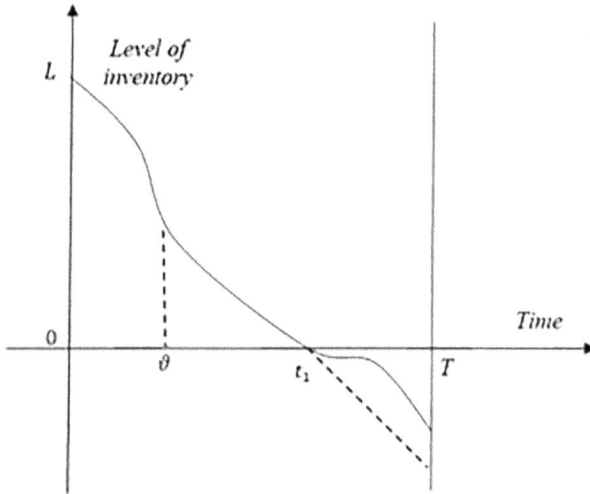

Figure 11.2 Representation of the inventory level for the case $t_1 > \vartheta$.

In this case, equation (11.7) becomes

$$\frac{dI(t)}{dt} + \alpha\beta t^{\beta-1} I(t) = -h(t), 0 \le t \le t_1, \; I(t_1) = 0 \tag{11.9}$$

Equation (11.8) advances to succeeding two:

$$\frac{dI(t)}{dt} = -h(t)\gamma(T-t), \;\; t_1 \le t \le \vartheta, \; I(t_1) = 0 \tag{11.10}$$

$$\frac{dI(t)}{dt} = -h(\vartheta)\gamma(T-t), \;\; \vartheta \le t \le T, \; I(\vartheta_-) = I(\vartheta_+) \tag{11.11}$$

The solutions of equations (11.9)–(11.11) are, respectively

$$I(t) = e^{-\alpha t^\beta} \int_t^{t_1} h(y) e^{\alpha y^\beta} dx, 0 \le t \le t_1, \tag{11.12}$$

$$I(t) = -\int_{t_1}^t h(y)\gamma(T-y)dy, \;\; t_1 \le t \le \vartheta \tag{11.13}$$

and

$$I(t) = -h(\vartheta)\int_\vartheta^t \gamma(T-y)dy - \int_{t_1}^\vartheta h(y)\gamma(T-y)dy, \;\; \vartheta \le t \le T \tag{11.14}$$

The total amount of deteriorated items in $[0, \; t_1]$ is

$$D = \int_0^{t_1} h(t) e^{\alpha t^\beta} dt - \int_0^{t_1} h(t) dt$$

Thus, in the interval $[0, T]$ the total cost is given by Skouri et al. (2009)

$$\text{Total cost} = \text{Holding cost} + \text{shortage cost} + \text{deterioration cost} + \text{opportunity cost}$$

$$TC_1(t_1) = a_1 I_1 + a_2 I_2 + a_3 D + a_4 L$$

$$= a_1 \left\{ \int_0^{t_1} e^{-\alpha t^\beta} \left[\int_t^{t_1} h(y) e^{\alpha y^\beta} dy \right] dt \right\} + a_3 \left\{ \int_0^{t_1} h(t) e^{\alpha t^\beta} dt - \int_0^{t_1} h(t) dt \right\}$$

$$+ a_2 \left\{ \int_{t_1}^{\vartheta} (\vartheta - t) h(t) \gamma(T - t) dt + h(\vartheta) \int_{\vartheta}^{T} \left[\int_{\vartheta}^{t} \gamma(T - y) dy \right] dt \right\}$$

$$+ a_2 \int_{\vartheta}^{T} \left[\int_{t_1}^{\vartheta} h(y) \gamma(T - y) dy \right] dt + a_4 \left\{ \int_{t_1}^{\vartheta} (1 - \gamma(T - t)) h(t) dt \right\}$$

$$+ a_4 h(\vartheta) \int_{\vartheta}^{T} (1 - \gamma(T - t)) dt \qquad (11.15)$$

11.2.1.2   Case II

$t_1 > \vartheta$

In this case, equation (11.7) gives the following two cases:

$$\frac{dI(t)}{dt} + \alpha \beta t^{\beta - 1} I(t) = -h(t), 0 \le t \le \vartheta, \ I(\vartheta^-) = I(\vartheta^+),$$

$$\frac{dI(t)}{dt} + \alpha \beta t^{\beta - 1} I(t) = -h(\vartheta), \ \vartheta \le t \le t_1, \ I(t_1) = 0.$$

Equation (11.8) becomes

$$\frac{dI(t)}{dt} = -h(\vartheta) \gamma(T - t), \ t_1 \le t \le T, \ I(t_1) = 0.$$

Their solutions are, respectively

$$I(t) = e^{-\alpha t^\beta} \left[\int_t^{\vartheta} h(y) e^{\alpha y^\beta} dy + h(\vartheta) \int_{\vartheta}^{t_1} e^{\alpha y^\beta} dy \right], 0 \le t \le \vartheta, \qquad (11.16)$$

$$I(t) = e^{-\alpha t^\beta} h(\vartheta) \int_t^{t_1} e^{\alpha y^\beta} dy, \ \vartheta \le t \le t_1, \qquad (11.17)$$

$$I(t) = -h(\vartheta) \int_{t_1}^{t} \gamma(T - y) dy, \ t_1 \le t \le T. \qquad (11.18)$$

The sum of deteriorated items in the interval $[0, t_1]$ is

$$D = I(0) - \int_0^{t_1} D(t) dt = \int_0^{\vartheta} h(t) e^{\alpha t^\beta} dt + h(\vartheta) \int_{\vartheta}^{t_1} e^{\alpha t^\beta} dt - \int_0^{\vartheta} h(t) dt - h(\vartheta)(t_1 - \vartheta).$$

Henceforth, the inventory cost for this case is given by

$$TC_2(t_1) = a_1 I_1 + a_2 I_2 + a_3 D + a_4 L$$

$$= a_1 \left\{ \int_0^\vartheta e^{-\alpha t^\beta} \left[\int_t^\vartheta h(y) e^{\alpha y^\beta} dy + h(\vartheta) \int_\vartheta^{t_1} e^{\alpha y^\beta} dy \right] dt \right\}$$

$$+ a_1 h(\vartheta) \int_\vartheta^{t_1} e^{-\alpha t^\beta} \left[\int_t^{t_1} e^{\alpha y^\beta} dy \right] dt + a_2 \left\{ h(\vartheta) \int_{t_1}^T (T-y)\gamma(T-y) dy \right\}$$

$$+ a_3 \left\{ \int_0^\vartheta h(t) e^{\alpha t^\beta} dt + h(\vartheta) \int_\vartheta^{t_1} e^{\alpha t^\beta} dt - \int_0^\vartheta h(t) dt - h(\vartheta)(t_1 - \vartheta) \right\}$$

$$+ a_4 \left\{ h(\vartheta) \int_{t_1}^T (1 - \gamma(T-t)) dt \right\}. \tag{11.19}$$

The aggregate cost function over the interval $[0, T]$ is given by

$$TC(t_1) = \begin{cases} TC_1(t_1), & \text{if } t_1 \leq \vartheta, \\ TC_2(t_1), & \text{if } \vartheta < t_1. \end{cases} \tag{11.20}$$

It's effortless to manifest that the aforementioned function is continuous ϑ. Minimization of these two sub-cases of function $TC(t_1)$ is the main hurdle. Thus, it requires a distinct examination of cases and then expressing the algorithm giving the optimal policy by amalgamating the results of these cases (Maity et al. (2020, 2021a, b) and Chakraborty et al. (2020)).

11.2.1.3 The Policy for Optimal Replenishment

In this segment, we will find a unique t_1, say t_1^*, which minimizes the total cost function for the model without shortages by using a classical approach.

The first and second-order derivatives of $TC_1(t_1)$ are, respectively

$$\frac{dTC_1(t_1)}{dt_1} = h(t_1) g(t_1),$$

$$\frac{d^2 TC_1(t_1)}{dt_1^2} = \frac{dh(t_1)}{dt_1} g(t_1) + h(t_1) \frac{dg(t_1)}{dt_1}, \tag{11.21}$$

Where,

$$g(t_1) = a_1 e^{\alpha t_1^\beta} \int_0^{t_1} e^{-\alpha t^\beta} dt + a_3 \left(e^{\alpha t_1^\beta} - 1 \right) - a_2 (T - t_1)\gamma(T - t_1) - a_4 (1 - \gamma(T - t_1)). \tag{11.22}$$

For this t_1^*, we have

$$\frac{d^2 TC_1(t_1^*)}{dt_1^{*2}} = \frac{dh(t_1^*)}{dt_1^*} g(t_1^*) + h(t_1^*) \frac{dg(t_1^*)}{dt_1^*} = h(t_1^*) \frac{dg(t_1^*)}{dt_1^*} > 0$$

where t_1^* tie in with the non-restricted global minimum of $TC_1(t_1)$.

If t_1^* is feasible, i.e., $t_1^* \leq \vartheta$, the optimal value of the order level, $L = I(0)$, is

$$L^* = \int_0^{t_1^*} h(t) e^{\alpha t^\beta} dt \tag{11.23}$$

Thus, Q^* symbolizing the optimal order quantity is given by

$$Q^* = L^* + \int_{t_1^*}^{\vartheta} \gamma(T-t) h(t) dt + h(\vartheta) \int_{\vartheta}^{T} \gamma(T-t) dt. \tag{11.24}$$

For the case $\gamma(y) = 0$, $\frac{dg(t_1)}{dt_1} > 0$ and $g(0) < 0$, nevertheless, $g(T)$ might be lesser or greater to zero. If $g(T) > 0$ then t_1^* tie in with the non-restricted global minimum of $TC_1(t_1)$. If $g(T) < 0$ then $TC_1(t_1)$ attains minimum value at ϑ by stringently declining to decrease.

The first and second-order derivatives of $TC_1(t_1)$ are

$$\frac{dTC_2(t_1)}{dt_1} = h(\vartheta) g(t_1), \tag{11.25}$$

$$\frac{d^2TC_2(t_1)}{dt_1^2} = h(\vartheta) \frac{dg(t_1)}{dt_1} > 0, \tag{11.26}$$

where the function $g(t_1)$ is given by equation (11.22).

If t_1^* is feasible, i.e., $t_1^* \leq \vartheta$, the optimal value of the order level, $L = I(0)$, is

$$L^* = \int_0^{\vartheta} h(t) e^{\alpha t^\beta} dt + h(\vartheta) \int_{\vartheta}^{t_1^*} e^{\alpha t^\beta} dt \tag{11.27}$$

The optimal order quantity Q^* in $TC_1(t_1)$ is given by

$$Q^* = L^* + h(\vartheta) \int_{t_1^*}^{T} \gamma(T-t) dt. \tag{11.28}$$

11.2.2   FORMULATION OF MATHEMATICAL MODEL STARTING WITH SHORTAGES

Underneath segment examines the inventory model starting with shortages. The model starts with shortages, which materialize in the course of the period $[0, \ t_1]$ and are partially backlogged. During this time t_1 a refilling brings the level of inventory to L. The inventory level falls to zero at $t = T$ in the time period $[t_1, T]$ because of demand and deterioration of the items. Again, the following two cases must be examined and the accomplishment of inventory level is reflected in Figures 11.3 and 11.4.

11.2.2.1   Case I

$t_1 < \vartheta$

The level of inventory $I(t), 0 \leq t \leq T$ satisfies the underneath differential equations:

$$\frac{dI(t)}{dt} = -h(t) \gamma(t_1 - t), 0 \leq t \leq t_1, I(0) = 0, \tag{11.29}$$

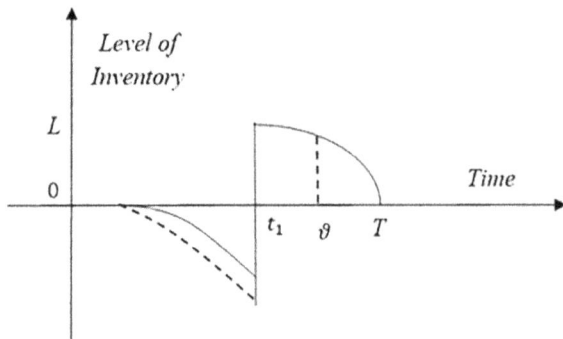

Figure 11.3 Representation of the inventory level for the case $t_1 < \vartheta$.

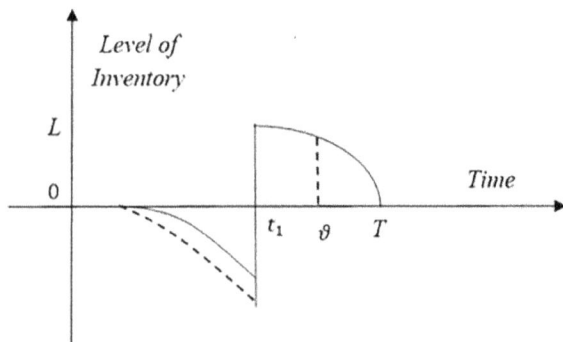

Figure 11.4 Representation of the inventory level for the case $t_1 > \vartheta$.

$$\frac{dI(t)}{dt} + \alpha\beta t^{\beta-1} I(t) = -h(t), \quad t_1 \le t \le \vartheta, \, I(\vartheta_-) = I(\vartheta_+), \tag{11.30}$$

$$\frac{dI(t)}{dt} + \alpha\beta t^{\beta-1} I(t) = -h(\vartheta), \quad \vartheta \le t \le T, \, I(T) = 0. \tag{11.31}$$

The results of equations (11.29)–(11.31) are respectively

$$I(t) = -\int_0^t h(y)\,\gamma(t_1 - t), 0 \le t \le t_1, \tag{11.32}$$

$$I(t) = e^{-\alpha t^\beta} \left[\int_t^\vartheta e^{\alpha y^\beta} h(y)\,dy + h(\vartheta) \int_\vartheta^T e^{\alpha y^\beta} dy \right], \quad t_1 \le t \le \vartheta \tag{11.33}$$

and

$$I(t) = e^{-\alpha t^\beta} h(\vartheta) \int_t^T e^{\alpha y^\beta} dy, \quad \vartheta \le t \le T. \tag{11.34}$$

The total amount of deteriorated units during $[t_1, \, T]$ is

$$D = e^{-\alpha t_1^\beta} \left[\int_{t_1}^\vartheta h(t)\,e^{\alpha t^\beta} + h(\vartheta) \int_\vartheta^T e^{\alpha t^\beta} dt \right] - \int_{t_1}^\vartheta h(t)\,dt - h(\vartheta)(T - \vartheta).$$

The inventory cost during the time interval $[0, T]$ is given by

Inventory cost = Holding cost + shortage cost + deterioration cost + opportunity cost

$$TC_1(t_1) = a_1 I_1 + a_2 I_2 + a_3 D + a_4 L$$

$$= a_1 \left\{ \int_{t_1}^{\vartheta} e^{-\alpha t^\beta} \left[\int_t^{\vartheta} h(y) e^{\alpha y^\beta} dx \right. \right.$$

$$+ h(\vartheta) \int_{\vartheta}^{T} e^{\alpha y^\beta} dy \right] dt + h(\vartheta) \int_{\vartheta}^{T} e^{-\alpha t^\beta} \left[\int_t^{T} e^{\alpha y^\beta} dy \right] dt \right\}$$

$$+ \int_0^{t_1} [a_2 (t_1 - t) \gamma(t_1 - t) + a_4 (1 - \gamma(t_1 - t))] h(t) dt + a_3$$

$$\left\{ e^{-\alpha t_1^\beta} \left[\int_{t_1}^{\vartheta} h(t) e^{\alpha t^\beta} dt + h(\vartheta) \int_{\vartheta}^{T} \int_{\vartheta}^{T} e^{\alpha t^\beta} dt \right] \right.$$

$$\left. - \int_{t_1}^{\vartheta} h(t) dt - h(\vartheta)(T - \vartheta) \right\} \tag{11.35}$$

11.2.2.2 Case II

$t_1 > \vartheta$

The level of inventory $I(t), 0 \le t \le T$ satisfies the underneath differential equations

$$\frac{dI(t)}{dt} = -h(t)\gamma(t_1 - t), 0 \le t \le \vartheta, I(0) = 0 \tag{11.36}$$

$$\frac{dI(t)}{dt} = -h(\vartheta)\gamma(t_1 - t), \ \vartheta \le t \le t_1, I(\vartheta^-) = I(\vartheta^+), \tag{11.37}$$

$$\frac{dI(t)}{dt} + \alpha\beta t^{\beta-1} I(t) = -h(\vartheta), \ t_1 \le t \le T, I(T) = 0. \tag{11.38}$$

Thus, the respective results are,

$$I(t) = -\int_0^t h(y)\gamma(t_1 - y) dy, 0 \le t \le \vartheta, \tag{11.39}$$

$$I(t) = -\int_0^{\vartheta} h(y)\gamma(t_1 - y) dy - h(\vartheta)\int_{\vartheta}^t \gamma(t_1 - y) dy \ \vartheta \le t \le t_1, \tag{11.40}$$

$$I(t) = -e^{\alpha t^\beta} h(\vartheta) \int_t^T e^{\alpha y^\beta} dy, \ t_1 \le t \le T. \tag{11.41}$$

The number of deteriorated items during $[t_1, T]$ are,

$$D = e^{-\alpha t_1^\beta} h(\vartheta) \int_{t_1}^T e^{\alpha t^\beta} dt - h(\vartheta)(T - t_1)$$

The total cost for this case is

$$
\begin{aligned}
TC_2\,(t_1) =\,& a_1 I_1 + a_2 I_2 + a_3 D + a_4 L \\
=\,& a_1 h\,(\vartheta) \int_{t_1}^{T} e^{-\alpha t^{\beta}} \left[\int_{t}^{T} e^{\alpha y^{\beta}} dy \right] dt \\
& + a_2 \left\{ \int_{0}^{\vartheta} \left[\int_{0}^{t} h\,(y)\,\gamma(t_1 - y)\,dy \right] dt \right. \\
& \left. + \int_{\vartheta}^{t_1} \left[\int_{0}^{\vartheta} h\,(y)\,\gamma(t_1 - y)\,dy + h\,(\vartheta) \int_{\vartheta}^{t} \gamma(t_1 - y)\,dy \right] dt \right\} \\
& + a_3 \left\{ e^{-\alpha t_1^{\beta}} h\,(\vartheta) \int_{t_1}^{T} e^{\alpha t^{\beta}} dt - h\,(\vartheta)\,(T - t_1) \right\} \\
& + a_4 \left\{ \int_{0}^{\vartheta} [1 - \gamma(t_1 - t)]\,h\,(t)\,dt + h\,(\vartheta) \int_{\vartheta}^{t_1} [1 - \gamma(t_1 - t)]\,dt \right\} \quad (11.42)
\end{aligned}
$$

The aggregate cost function over $[0, T]$ is given by

$$
TC\,(t_1) = \begin{cases} TC_1\,(t_1), & \text{if } t_1 \leq \vartheta, \\ TC_2\,(t_1), & \text{if } \vartheta < t_1. \end{cases} \quad (11.43)
$$

It is effortless to examine the continuity of the function at ϑ. Minimization of these two cases in the function $TC\,(t_1)$ is the main hurdle now.

11.2.2.3 The Optimal Replenishment Policy

This section attains the optimal replenishment policy, i.e., we minimize the total cost function using the classical approach by finding the value, say t_1^* which minimizes it. Substituting the first-order derivative of $TC_1\,(t_1)$ to zero gives

$$
\begin{aligned}
& - \left(a_1 + a_3 \alpha \beta t_1^{\beta - 1} \right) e^{-\alpha t_1^{\beta}} \left[\int_{t_1}^{\vartheta} e^{\alpha x^{\beta}} h\,(y)\,dy + h\,(\vartheta) \int_{\vartheta}^{T} e^{\alpha y^{\beta}} dy \right] \\
& + \int_{0}^{t_1} [a_2 \gamma(t_1 - t) + a_2\,(t_1 - t)\,\gamma'\,(t_1 - t) - a_4 \gamma'\,(t_1 - t)]\,h\,(t)\,dt = 0
\end{aligned} \quad (11.44)
$$

So, the ordering quantity is

$$
Q^* = \int_{0}^{t_1^*} h\,(y)\,\gamma(t_1^* - y)\,dy + L^* \quad (11.45)
$$

and the total cost is $TC_1\,(t_1^*)$.

Similarly, for $TC_2\,(t_1)$, Q^* is given by

$$
Q^* = \int_{0}^{\vartheta} h\,(y)\,\gamma(t_1^* - y)\,dy + h\,(\vartheta) \int_{\vartheta}^{t_1^*} \gamma(t_1^* - y)\,dy + L^* \quad (11.46)
$$

and the total cost is $TC_2\,(t_1^*)$.

11.3 NUMERICAL EXAMPLES

Underneath example exemplifies the results of the aforementioned model.

Example 11.1

This is taken from (Skouri et al. 2009) and is modified according to our models. The input parameters are: $a_1 = \$4$ per unit per year, $a_2 = \$14$ per unit per year, $a_3 = \$6$ per unit per year, $a_4 = \$19$ per unit per year, $\vartheta = 0.12$ year, $\alpha = 0.001$, $\beta = 2$, $T = 1$ year, $h(t) = 33e^{4.5t}$ and $\gamma(x) = e^{-0.2x}$.

Model starting with no shortages:

Using equations (11.23) or (11.27) the optimal value of t_1 is $t_1^* = 0.8603 > \vartheta$. The optimal ordering quantity is $Q^* = 4.988$ (from equation (11.28)) and the minimum cost is $TC(t_1^*) = 6.6431$ (from equation (11.19)).

Model starting with shortages:

$$t_1^* = 0.1632\,(\vartheta < t_1^*),\ Q^* = 4.998\ (\text{from equation (11.46)}),$$
$$\text{and } TC(t_1^*) = 6.3776\,(\text{from equation (11.42)}).$$

Example 11.2

This example is executed on the same lines as Example 11.1, the only difference is that $\vartheta = 0.95$ year.

Model starting with no shortages:

$$t_1^* = 0.8509\ (\vartheta > t_1^*),\ Q^* = 54.6211\ (\text{from equation (11.24)})$$
$$\text{and } TC(t_1^*) = 91.9102\,(\text{from equation (11.15)}).$$

Model starting with shortages:

$$t_1^* = 0.5574\,(\text{from equation (11.44)})\,(\vartheta > t_1^*),\ Q^* = 54.60\ (\text{from equation (11.45)})$$
$$\text{and } TC(t_1^*) = 62.991\,(\text{from equation (11.35)}).$$

From the aforementioned examples, it can be observed that the total cost for the model with no shortages is more in comparison to the total cost for the model with shortages.

11.4 CONCLUDING REMARKS

The chapter includes various applications of differential equations in inventory control. The formation of governing differential equation to establish the relationship between inventory level with demand and other factors was studied. In this particular study, we have shown several forms of the differential equation under ramp-type demand and inventory deterioration. The two-parameter Weibull distribution was considered for deterioration. Two separate inventory models have been considered and their governing differential equations were shown. The models included no shortages at the initial inventory position and shortages at the initial position. Finally, the solution methodology for solving the differential equations and the optimization model was illustrated to obtain the optimal EOQ. For model validation, the numerical solution was also obtained for both cases. The study consists of only a single echelon inventory model with only a single retailer. But practically, in a large industrial scenario, multi-retailer inventory management is observed. Recently, multiple parties are considered an integrated inventory model which makes a higher profit than a single-party inventory model. Thus, the current study can be extended to multi-retailer along with several parties associated with it.

REFERENCES

Abad, P. L. (1996). Optimal pricing and lot-sizing under conditions of perishability and partial backordering, *Management Science*, 42(8): 1093–1104.

Aliyu, M. D. S., and Boukas, E. K. (1998). Discrete-time inventory models with deteriorating items, *International Journal of Systems Science*, 29(9): 1007–1014.

Andijani, A., and Al-Dajani, M. (1998). Analysis of deteriorating inventory/production systems using a linear quadratic regulator, *European Journal of Operational Research*, 106(1): 82–89.

Bhunia, A. K., and Maiti, M. (1998). Deterministic inventory model for deteriorating items with finite rate of replenishment dependent on inventory level, *Computers & Operations Research*, 25(11): 997–1006.

Chakraborty, A., Maity, S., Jain, S., Mondal, S. P., and Alam, S. (2020). Hexagonal fuzzy number and its distinctive representation, ranking, defuzzification technique and application in production inventory management problem, *Granular Computing*, 1–15: 507–521.

Chakraborty, T., Giri, B. C., and Chaudhuri, K. S. (1998). An EOQ model for items with Weibull distribution deterioration, shortages and trended demand: An extension of Philip's model, *Computers & Operations Research*, 25(7–8): 649–657.

Chang, H. J., and Dye, C. Y. (1999). An EOQ model for deteriorating items with time varying demand and partial backlogging, *Journal of the Operational Research Society*, 50(11): 1176–1182.

Chare, P., and Schrader, G. (1963). A model for exponentially decaying inventories, *Journal of Industrial Engineering*, 15: 238–243.

Chung, K. J., Chu, P., and Lan, S. P. (2000). A note on EOQ models for deteriorating items under stock dependent selling rate, *European Journal of Operational Research*, 124(3): 550–559.

Covert, R. P., and Philip, G. C. (1973). An EOQ model for items with Weibull distribution deterioration, *AIIE Transactions*, 5(4): 323–326.

Datta, T. A., and Pal, A. K. (1988). Order level inventory system with power demand pattern for items with variable rate of deterioration, *Indian Journal of Pure and Applied Mathematics*, 19(11): 1043–1053.

Dave, U., and Patel, L. K. (1981). (T, Si) policy inventory model for deteriorating items with time proportional demand, *Journal of the Operational Research Society*, 32(2): 137–142.

Donaldson, W. A. (1977). Inventory replenishment policy for a linear trend in demand: An analytical solution, *Journal of the Operational Research Society*, 28(3): 663–670.

Giri, B. C., Jalan, A. K., and Chaudhuri, K. S. (2003). Economic order quantity model with Weibull deterioration distribution, shortage and ramp-type demand, *International Journal of Systems Science*, 34(4): 237–243.

Goh, C. H., Greenberg, B. S., and Matsuo, H. (1993). Two-stage perishable inventory models, *Management Science*, 39(5): 633–649.

Goyal, S. K., and Giri, B. C. (2001). Recent trends in modeling of deteriorating inventory, *European Journal of Operational Research*, 134(1): 1–16.

Hariga, M. (1996). Optimal EOQ models for deteriorating items with time-varying demand, *Journal of the Operational Research Society*, 47(10): 1228–1246.

Harris, F. W. (1990). How many parts to make at once, *Operations Research*, 38(6): 947–950.

Hill, R. M. (1995). Inventory models for increasing demand followed by level demand, *Journal of the Operational Research Society*, 46(10): 1250–1259.

Hollier, R. H., and Mak, K. L. (1983). Inventory replenishment policies for deteriorating items in a declining market, *International Journal of Production Research*, 21(6): 813–836.

Kim, J., Hwang, H., and Shinn, S. (1995). An optimal credit policy to increase supplier's profits with price-dependent demand functions, *Production Planning & Control*, 6(1): 45–50.

Maity, S., Chakraborty, A., De, S. K., Mondal, S. P., and Alam, S. (2020). A comprehensive study of a backlogging EOQ model with nonlinear heptagonal dense fuzzy environment, *RAIRO-Operations Research*, 54(1): 267–286.

Maity, S., De, S. K., Pal, M., and Mondal, S. P. (2021a). A study of an EOQ model with public-screened discounted items under cloudy Fuzzy demand rate, *Journal of Intelligent & Fuzzy System*, 41(6): 6923–6934.

Maity, S., De, S. K., Pal, M., and Mondal, S. P. (2021b). A study of an EOQ model of growing items with parabolic dense Fuzzy lock demand rate, *Applied System Innovation*, 4(4): 81.

Raafat, F. F., Wolfe, P. M., and Eldin, H. K. (1991). An inventory model for deteriorating items, *Computers & Industrial Engineering*, 20(1): 89–94.

Resh, M., Friedman, M., and Barbosa, L. C. (1976). On a general solution of the deterministic lot size problem with time-proportional demand, *Operations Research*, 24(4): 718–725.

Roy, T. K. and Maiti, M. (1998). Multi-objective inventory models of deteriorating items with some constraints in a fuzzy environment, *Computers & Operations Research*, 25(12): 1085–1095.

Skouri, K., Konstantaras, I., Papachristos, S., and Ganas, I. (2009). Inventory models with ramp type demand rate, partial backlogging and Weibull deterioration rate, *European Journal of Operational Research*, 192(1): 79–92.

Taft, E. W. (1918). The most economical production lot, *The Iron Age*, 101: 1410–1412.

Wee, H. M. (1995). Joint pricing and replenishment policy for deteriorating inventory with declining market, *International Journal of Production Economics*, 40(2–3): 163–171.

Wee, H. M. (1997). A replenishment policy for items with a price-dependent demand and a varying rate of deterioration, *Production Planning & Control*, 8(5): 494–499.

Wee, H. M., and Shum, Y. S. (1999). Model development for deteriorating inventory in material requirement planning systems, *Computers & Industrial Engineering*, 36(1): 219–225.

Wu, K. S. (2001). An EOQ inventory model for items with Weibull distribution deterioration, ramp type demand rate and partial backlogging, *Production Planning & Control*, 12(8): 787–793.

Xu, H., and Wang, H. P. B. (1990). An economic ordering policy model for deteriorating items with time proportional demand, *European Journal of Operational Research*, 46(1): 21–27.

12 The Behavior of Interacting Faults under Increasing Tectonic Forces

Papiya Debnath
Techno International New Town

Mitali Sarkar
Chung-Ang University

R. T. Goswami
Techno International New Town

CONTENTS

12.1 INTRODUCTION

There are certain regions on the earth which are found to be seismically active. Earthquakes occur in these regions at regular intervals. The phase in the middle of two seismic events is known as the aseismic period and it is observed through regular geodetic surveys that there are small-scale ground deformations for the duration of such an aseismic period. Chinnery (1961) and Dreger et al. (2011) developed some mathematical models to simulate the nature of such small ground deformation during the aseismic period. When an aseismic state is established in the vicinity of an active seismic fault after a seismic event, usually there is a relative residual surface displacement across the fault as compared to the pre-seismic state. This displacement is visible on the terrestrial surface in case of surface-breaking faults. Geodetic surveys conducted in the region around an earthquake-generating fault before and after the occurrence of an earthquake normally show that there is a residual displacement on the free surface, both across and near the fault, after the fault movement. This feature is associated not only with surface-breaking faults but also with buried faults.

During the last few decades, numerous theoretical models have been developed with a view to investigating the residual surface displacements by Cohen (1980a,b). In some theoretical models' authors contemplated an infinite, vertical strike-slip fault located in a viscoelastic layer, composed of a material of the Maxwell-type, free to slide over the material beneath it. In most of these models, faults located in static, elastically, or viscoelastically isotropic half spaces, either homogeneous or layered, have been considered by Sato (1971), Rybicki (1973), Hearn and Burgmann (2005), Sen and Karmakar (2013), Debnath and Sen (2015c), Mondal and Sen (2016), Mondal and Sen (2017), Manna and Sen (2017), Debnath and Sen (2020), and Mondal and Debnath (2021). Theoretical models with long vertical strike-slip faults situated in a viscoelastic half-space were considered by Sen et al. (1993), Lovely et al. (2009), Mondal et al. (2019), and Manna and Sen (2019). Some of the theoretical models have been developed for stress accumulation patterns due to a fault movement in seismically active regions of the lithosphere-asthenosphere system during the aseismic period by Mukhopadhyay et al. (1979) and Mondal et al. (2018a,b).

The lithosphere-asthenosphere system of the earth was represented in various ways, such as Maxwell-type of elastic or viscoelastic half-space or layered medium, elastic or viscoelastic resting on elastic or viscoelastic half-space. During the last few years, some theoretical models have been introduced by Debnath and Sen (2014a,b) and Debnath and Sen (2015a,b) working with time-dependent aseismic stresses, strains and displacements elaborating ways of obtaining it in a natural way. These models consider locked or creeping faults, normally characterised by Volterra-type dislocations, located in a viscoelastic layer or half-space, or in an elastic layer which is in welded contact along with a viscoelastic half-space, or in a layered viscoelastic half-space.

During mantle convection, heat is transported from the deep interior of the earth to the surface. In seismically active regions, this convection current produces the tectonic force which also helps to accumulate the stress during the aseismic period. The general form of the tectonic force is $\left(\tau_\infty(t) - \tau_\infty(0)\right)/\tau_\infty(0) = kt$, where k is a constant which is set by the model parameters. Sarkar et al. (2012) explain the interacting effect between the surrounding faults. Considering, $\tau_\infty(t) = $ constand, the impact of the increasing value of $\tau_\infty(t)$ has been investigated here. In the present study, we have introduced Maxwell-type and linearly viscoelastic-type materials for the half-space.

12.2 FORMULATION OF MATHEMATICAL MODEL

Here we established the main theory of the mathematical model based on lithosphere-asthenosphere structure. Two interacting inclined strike slip faults (SSFs) are considered in a viscoelastic half-space with tectonic forces rising with time.

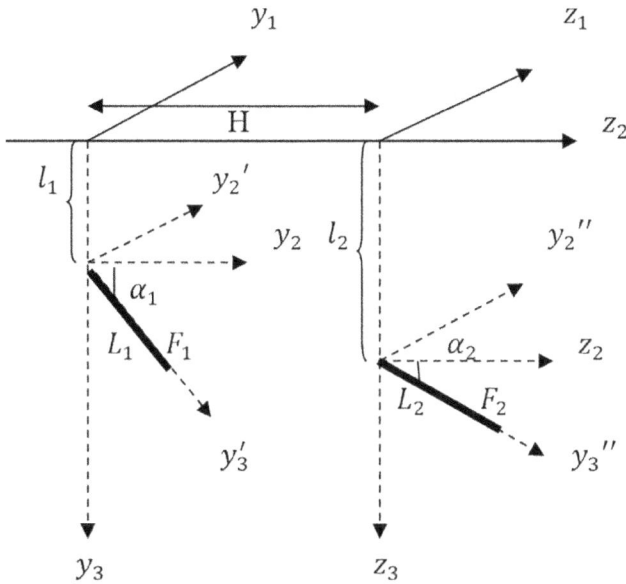

Figure 12.1 Sectional representation of the model by $y_1 = 0$.

In Figure (12.1), F_1 and F_2 represent long buried SSF, having horizontal lower and upper edges, in a half-space inclined to the horizontal α_1 and α_2 respectively. Let l_1 and l_2 represents upper edges depths of the faults and H is the space calculated horizontally. Where L_i $(i = 1,2)$ and α_i $(i = 1,2)$ are the length and angle of F_i $(i = 1,2)$ with the horizontal. Introducing rectangular Cartesian coordinates (o, y_1, y_2, y_3), (o', y'_1, y'_2, y'_3), $\left(o'', y''_1, y''_2, y''_3\right)$ as given in the figure. The relationships are given by:

$$\left.\begin{array}{l} y'_1 - y_1 = 0 \\ y'_2 - y_2 \sin\alpha_1 + (y_3 - l_1)\cos\alpha_1 = 0 \\ y'_3 = y_2 \cos\alpha_1 + (y_3 - l_1)\sin\alpha_1 \\ \quad \text{and} \\ y''_1 = z_1 \\ y''_2 = z_2 \sin\alpha_2 - z_3 \cos\alpha_2 \\ y''_3 = z_2 \cos\alpha_2 + z_3 \sin\alpha_2 \end{array}\right\}$$

where $z_2 = y_2 - H$, $z_3 = y_3 - l_2$.

Assuming y_1 axes along the strike of the first fault F_1 and the fault lengths $\gg$ fault widths. The time-dependent relation between stress and strain can explain by the constitutive laws. Therefore, for a long fault, the displacement, stresses, and strain become independent of y_1 and finally, we can have a 2D problem as a function of t and y_i $(i = 1,2)$.

Let u, $\tau_{12}, \tau_{13}, \tau_{23}, \tau_{11}, \tau_{22}$ and e_{ij} $(i, j = 1, 2, 3)$ are the components of stress, strain and displacement (SSD). As these are independent of y_1 and are functions of y_2, y_3 and t, the components of SSD are separated into two independent groups. One group containing u in y_1 direction and two related stress components (τ_{12}, τ_{13}) and the strain components (e_{12}, e_{13}) along the SSF movement. Hence, only the movement of SSF is considered here.

The u and (τ_{12}, τ_{13}) associated with the SSF movement are connected by the following stress-strain relation:

$$\left. \begin{array}{l} \left(\frac{1}{\eta} + \frac{1}{\mu} \frac{\partial}{\partial t} \right) \tau_{12} = \frac{\partial}{\partial t} (e_{12}) = \frac{\partial^2 u}{\partial t \partial y_2} \\[2mm] \left(\frac{1}{\eta} + \frac{1}{\mu} \frac{\partial}{\partial t} \right) \tau_{13} = \frac{\partial}{\partial t} (e_{13}) = \frac{\partial^2 u}{\partial t \partial y_3} \end{array} \right\} \tag{12.1}$$

where the effective viscosity is η and the effective rigidity of the material is μ.

The equilibrium equation of momentum is $\frac{\partial \tau_{ij}}{\partial y_j} + \rho f_i = \rho \frac{\partial V}{\partial t}$, $(i, j = 1, 2, 3)$.

In small deformation, $\frac{\partial V}{\partial t} = \frac{\partial^2 u}{\partial t^2}$, where $V = du/dt$. So, the non-quasi-static deformation will not be violated. f_i and ρ are the body force and density of the system. Besides, the acceleration can be safely ignored to assume the quasi-static equilibrium during the motion with the governing equation, $\frac{\partial \tau_{ij}}{\partial y_j} + \rho f_i = 0$. In the absence of f_i, considering $\frac{\partial \tau_{ij}}{\partial y_j} = 0$,

$$\left. \begin{array}{l} \frac{\partial}{\partial y_2} (\tau_{12}) + \frac{\partial}{\partial y_3} (\tau_{13}) = 0, \\[2mm] y_2 \in (-\infty, \infty),\, t \geq 0, y_3 \geq 0 \end{array} \right\} \tag{12.2}$$

In this model, we consider initial conditions at the time $t = 0$, $(u)_0$, $(\tau_{12})_0$, $(\tau_{13})_0$, $(e_{12})_0$ are the values of u, τ_{12}, τ_{13}, e_{12} individually. [$t = 0$ signifying a time while the model is in an aseismic state]

The boundary conditions are taken to be

$$\left. \begin{array}{l} \tau_{13} = 0, \text{ on the free surface } y_3 = 0 \\[2mm] \tau_{13} \to 0 \text{ when } y_3 \to \infty \end{array} \right\} \tag{12.3}$$

Where, $y_2 \in (-\infty, \infty)$ and $t \geq 0$.
Assuming $\tau_\infty(t)$ increasing linearly with time,

$$\left. \begin{array}{l} \tau_{12} \text{ approaches to } \tau_\infty(t), \text{ where } \tau_\infty(t) / \tau_\infty(0) = (1 + kt), \\[2mm] \text{as } |y_2| \to \infty, \text{ for } y_3 \geq 0, t \geq 0 \end{array} \right\} \tag{12.4}$$

$$\left. \begin{array}{l} \tau_\infty(t) = \tau_\infty(0) \\[2mm] \tau_{12}(0) \to \tau_\infty(0) \text{ as } |y_2| \to \infty, \text{ for } t = 0 \end{array} \right\} \tag{12.5}$$

Solving equations (12.1) and (12.2) we get,

$$\nabla^2 u(y_2, y_3,\, t) = 0 \tag{12.6}$$

12.3 SOLUTION OF SSD IN ABSENCE OF FAULT MOVEMENT

SSD is continuous when no fault acts. The exact expressions for SSD are obtained as:

$$
\left.
\begin{aligned}
u &= (u)_0 + y_2 \tau_\infty (0) \left[\frac{kt}{\mu} + \frac{t}{\eta} + \frac{kt^2}{2\eta} \right] \\
e_{12} &= (e_{12})_0 + \tau_\infty (0) \left[\frac{kt}{\mu} + \frac{t}{\eta} + \frac{kt^2}{2\eta} \right] \quad \tau_{12} = (\tau_{12})_0 e^{-\frac{\mu t}{\eta}} + \tau_\infty (0) \left(1 + kt - e^{-\frac{\mu t}{\eta}} \right) \\
\tau_{13} &= (\tau_{13})_0 e^{-\frac{\mu t}{\eta}} \quad \tau_{1'2'} = \text{The stresses across the fault } F_1 = \tau_{12} \sin \alpha_1 - \tau_{13} \cos \alpha_1 \\
&= (\tau_{1'2'})_0 e^{-\frac{\mu t}{\eta}} + \tau_\infty (0) \left(1 + kt - e^{-\frac{\mu t}{\eta}} \right) \sin \alpha_1, t \geq 0 \ \tau_{1''2''} \\
&= \text{The stresses across the fault } F_2 = \tau_{12} \sin \alpha_2 - \tau_{13} \cos \alpha_2 \\
&= (\tau_{1''2''})_0 e^{-\frac{\mu t}{\eta}} + \tau_\infty (0) \left(1 + kt - e^{-\frac{\mu t}{\eta}} \right) \sin \alpha_2, t \geq 0
\end{aligned}
\right\}
$$

$$(12.7)$$

Here, with t, $\tau_{1'2'}$ and $\tau_{1''2''}$ increases as $\tau_\infty (0) \left(1 + kt - e^{-\frac{\mu t}{\eta}} \right) \sin \alpha_1$ and $\tau_\infty (0) \left(1 + kt - e^{-\frac{\mu t}{\eta}} \right) \sin \alpha_2$ respectively, noting that $e^{-\frac{\mu t}{\eta}} \to 0$ as $t \to \infty$.

For distinct values of k, in the case of before fault movement, this has been shown in Figure 12.2.

Stress values are very less. For $k = 10^{-9}$ stress is increasing with respect to time.

Assuming, the faults F_1 and F_2, can endure $(\tau_c)_1 = 200 \, \text{bar}$, $(\tau_c)_2 = 250 \, \text{bar}$ respectively. Additionally, we assume $\tau_\infty (0) = 50 \, \text{bar}$. To reach these critical values,

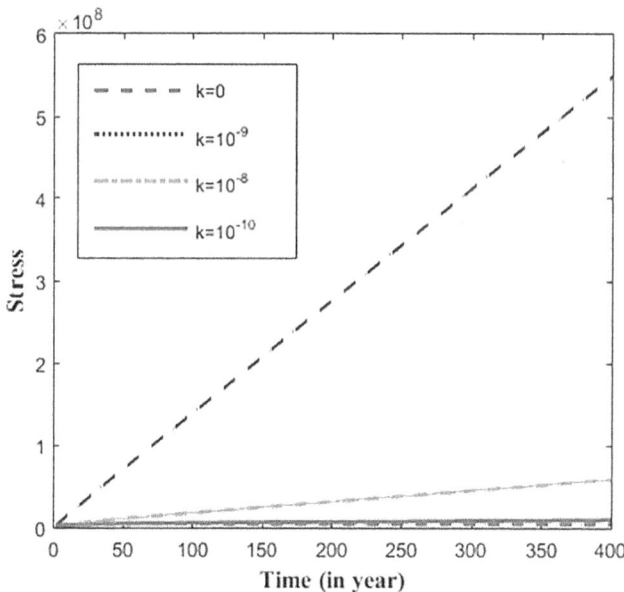

Figure 12.2 Stress with respect to time for different increasing values of k.

it is found that, the time taken is $T_1 = 117$ years and $T_2 = 152$ years (assuming $\alpha_1 = \alpha_2 = 60°$).

After time T_1 years, when $\tau_{1'2'}$ attains the value $(\tau_c)_1$, F_1 begins creeping with a velocity v_1 cm/year. The creeping dislocation is characterised by

$$[u]_{F_1} = U_1(t_1) f(y'_3) H(T - t_1) \tag{12.8}$$

12.4 PROBLEM SOLUTION AFTER THE CREEPING MOVEMENT ACROSS F_1 FOR $0 < T \leq T_1 < T_2$

For $t = T_1 \geq 0$, solutions become,

$$\left.\begin{array}{l} u = (u)_1 + (u)_2 \\ e_{12} = (e_{12})_1 + (e_{12})_2 \\ \tau_i = (\tau_i)_1 + (\tau_i)_2 \end{array}\right\} \tag{12.9}$$

Here, $i = 12, 13, 1'2', 1''2''$
with replacing the equation $\tau_{12} \to \tau_\infty(t)$ by

$$\tau_{12} \to 0 \text{ as } |y_2| \to \infty, \text{ for } y_3 \geq 0, t \geq 0. \tag{12.10}$$

and an added dislocation condition

$$[u]_{F_1} = U_1(t_1) f(y'_3) H(t - T_1)$$

where $[u]$ represents the discontinuity in displacement u through F_1 and $H(t - T_1)$ is the Heaviside unit step function.

When there was no fault movement, we got the solutions for $(u)_1$, $(e_{12})_1$, ... as

$$u = (u)_0 + y_2 \tau_\infty(0) \left[\frac{kt}{\mu} + \frac{t}{\eta} + \frac{kt^2}{2\eta}\right] e_{12}$$

$$= (e_{12})_0 + \tau_\infty(0) \left[\frac{kt}{\mu} + \frac{t}{\eta} + \frac{kt^2}{2\eta}\right] \tau_{12}$$

$$= (\tau_{12})_0 e^{-\frac{\mu t}{\eta}} + \tau_\infty(0) \left(1 + kt - e^{-\frac{\mu t}{\eta}}\right) \tau_{13} = (\tau_{13})_0 e^{-\frac{\mu t}{\eta}} \quad \tau_{1'2'}$$

$$= \tau_{12} \sin\alpha_1 - \tau_{13} \cos\alpha_1$$

$$= (\tau_{1'2'})_0 e^{-\frac{\mu t}{\eta}} + \tau_\infty(0) \left(1 + kt - e^{-\frac{\mu t}{\eta}}\right) \sin\alpha_1, t \geq 0 \quad \tau_{1''2''}$$

$$= \tau_{12} \sin\alpha_2 - \tau_{13} \cos\alpha_2 = (\tau_{1''2''})_0 e^{-\frac{\mu t}{\eta}} + \tau_\infty(0) \left(1 + kt - e^{-\frac{\mu t}{\eta}}\right) \sin\alpha_2, t \geq 0 \tag{12.11}$$

This is a Boundary Value Problem (BVP) that satisfies equations (12.1)–(12.6), (12.8), and (12.10) which is including the function of y_2, y_3 and t.

$$\left. \begin{aligned} \left(\frac{1}{\eta}+\frac{1}{\mu}\frac{\partial}{\partial t_1}\right)(\tau_{12})_2 &= \frac{\partial^2(u)_2}{\partial t_1 \partial y_2} \\ \left(\frac{1}{\eta}+\frac{1}{\mu}\frac{\partial}{\partial t_1}\right)(\tau_{13})_2 &= \frac{\partial^2(u)_2}{\partial t_1 \partial y_3} \end{aligned} \right\} y_2 \in (-\infty,\infty),\ y_3 \geq 0,\ t \geq T_1 \qquad (12.12)$$

$$\frac{\partial}{\partial y_2}(\tau_{12})_2 + \frac{\partial}{\partial y_3}(\tau_{13})_2 = 0$$

Boundary conditions become

$$\left. \begin{aligned} (\tau_{13})_2 &= 0 \text{ on } y_3 = 0, y_2 \in (-\infty,\infty),\ t \geq T_1 \\ (\tau_{13})_2 &\to 0 \text{ as } y_3 \to \infty,\ y_2 \in (-\infty,\infty),\ t \geq T_1 \\ (\tau_{12})_2 &\to 0 \text{ as } |y_2| \to \infty,\ y_3 \geq 0, t \geq T_1 \end{aligned} \right\} \qquad (12.13)$$

Solving we get,

$$(\nabla^2 u)_2 = 0 \qquad (12.14)$$

$$[u]_2 = U_1(t_1) f(y'_3) H(t-T_1) \qquad (12.15)$$

Across the first fault F_1 $(y'_2 = 0, 0 \leq y'_3 \leq H)$ with $U_1(t_1) = 0$ for $t_1 \leq 0$

Here, we are using Laplace transform (LT) of equation (12.11)–(12.15) with respect to $(t_1 = t - T_1)$ and achieve a BVP including $(\underline{u})_2$, $(\underline{e_{12}})_2$, $(\underline{\tau_{12}})_2$, $(\underline{\tau_{13}})_2$ which are represented by the LT of $(u)_2$, $(e_{12})_2$, $(\tau_{12})_2$, $(\tau_{13})_2$ respectively with respect to t_1 and are defined as

$$\{(\underline{u})_2, (\underline{u})_3, (\underline{\tau_{12}})_2\} = \int_0^\infty \{(u)_2, (u)_3, (\tau_{12})_2\}\, e^{-pt_1}\, dt_1$$

where LT variable is p.

The subsequent BVP is described by the relations given by

$$\left. \begin{aligned} (\underline{\tau_{12}})_2 &= \frac{p}{\left(\frac{p}{\mu}+\frac{1}{\eta}\right)}\frac{\partial(\underline{u})_2}{\partial y_2} \\ (\underline{\tau_{13}})_2 &= \frac{p}{\left(\frac{p}{\mu}+\frac{1}{\eta}\right)}\frac{\partial(\underline{u})_2}{\partial y_3} \\ (y_2 \in (-\infty,\infty),\ y_3 \geq 0\,) \end{aligned} \right\} \qquad (12.16)$$

$$\frac{\partial}{\partial y_2}(\underline{\tau_{12}})_2 + \frac{\partial}{\partial y_3}(\underline{\tau_{13}})_2 = 0 \qquad (12.17)$$

$$\left. \begin{aligned} (\underline{\tau_{13}})_2 &= 0 \text{ on } y_3 = 0, y_2 \in (-\infty,\infty) \text{ and } t \geq 0 \\ (\underline{\tau_{13}})_2 &\to 0 \text{ as } y_3 \to \infty,\ y_2 \in (-\infty,\infty) \text{ and } t \geq 0 \\ (\underline{\tau_{12}})_2 &\to 0 \text{ as } |y_2| \to \infty,\ \text{ for } y_3 \geq 0 \end{aligned} \right\} \qquad (12.18)$$

$$\nabla^2(\underline{u})_2 = 0 \qquad (12.19)$$

And $[(\underline{u})_2] = \underline{U_1}(p) f(y'_3)$

$$\text{Through } F_1: y'_2 = 0, y'_3 \in [0,H] \qquad (12.20)$$

where $\underline{U_1}(p)$ is the L.T of $U_1(t_1)$ with respect to t_1,

$$\underline{U_1}(p) = \int_0^\infty U_1(t_1)\ e^{-pt_1}\ dt_1$$

We solve this BVP by using (Maruyama 1966; Rybicki 1971),

$$(\underline{u})_2(Q) = \int (\underline{u})_2(P)\left\{G^1{}_{13}(Q,P)\,d\xi_2 - G^1{}_{12}(Q,P)\,d\xi_3\right\} \qquad (12.21)$$

Along the fault F_1, integration is carried out and in the half-space, $Q(y_1,y_2,y_3)$ is the field point, not on the fault. Now $G^1_{13}(Q,P)$ and $G^1_{12}(Q,P)$ are given by:

$$G^1_{13}(Q,P) = \frac{1}{2\pi}\left[\frac{y_3 - \xi_3}{L^2} - \frac{y_3 + \xi_3}{M^2}\right]$$

and

$$G^1_{12}(Q,P) = \frac{1}{2\pi}\left[\frac{y_2 - \xi_2}{L^2} + \frac{y_2 - \xi_2}{M^2}\right]$$

where,

$$L^2 = (y_2 - \xi_2)^2 + (y_3 - \xi_3)^2,\ M^2 = (y_2 - \xi_2)^2 + (y_3 + \xi_3)^2$$

Now on the fault F_1, $P(\xi_1, \xi_2, \xi_3)$ be any point, $\xi_2 \in [0, L_1 \cos \alpha_1]$, $\xi_3 \in [0, L_1 \cos \alpha_1]$ and $\xi_2 = \xi_3 \cot \alpha_1$. Hence the coordinate axes have been changed from (ξ_1, ξ_2, ξ_3) to (ξ'_1, ξ'_2, ξ'_3), given by the following,

$$\left.\begin{array}{l} \xi_1 - \xi'_1 = 0 \\ \xi_2 - \xi'_2 \sin \alpha_1 - \xi'_3 \cos \alpha_1 = 0 \\ \xi_3 - l_1 + \xi'_2 \cos \alpha_1 - \xi'_3 \sin \alpha_1 = 0 \end{array}\right\}\ \xi'_2 = 0,\ \xi'_3 \in [0, H] \text{ on } F_1.$$

Then from (12.20) using (12.21) we have

$$(\underline{u})_2(Q) = \frac{U_1(p)}{2\pi}\int_0^{L_1} f\left(\xi'_3\right)\left[\frac{y_2 \sin \alpha_1 - (y_3 - l_1)\cos \alpha_1}{\xi'^2_3 - 2\xi'_3\{y_2 \cos \alpha_1 + (y_3 - l_1)\sin \alpha_1\} + y_2^2 + (y_3 - l_1)^2}\right.$$
$$\left. + \frac{y_2 \sin \alpha_1 + (y_3 + l_1)\cos \alpha_1}{\xi'^2_3 - 2\xi'_3\{y_2 \cos \alpha_1 - (y_3 + l_1)\sin \alpha_1\} + y_2^2 + (y_3 + l_1)^2}\right]\,d\xi'_3$$

or,

$$(\underline{u})_2(Q) = \frac{U_1(p)}{2\pi}\psi_1(y_2, y_3)$$

where,

$$\psi_1(y_2, y_3) = \int_0^{L_1} f\left(\xi'_3\right)\left[\frac{y_2 \sin \alpha_1 - (y_3 - l_1)\cos \alpha_1}{\xi'^2_3 - 2\xi'_3\{y_2 \cos \alpha_1 + (y_3 - l_1)\sin \alpha_1\} + y_2^2 + (y_3 - l_1)^2}\right.$$
$$\left. + \frac{y_2 \sin \alpha_1 + (y_3 + d_1)\cos \alpha_1}{\xi'^2_3 - 2\xi'_3\{y_2 \cos \alpha_1 - (y_3 + l_1)\sin \alpha_1\} + y_2^2 + (y_3 + l_1)^2}\right]\,d\xi'_3$$

Now, $(\underline{u})_2 = 0$ for $t_1 \le 0$ and taking inverse L.T. w.r.t t_1

$$(u)_2 = \frac{U_1(t_1)}{2\pi}\psi_1(y_2, y_3)\,H(t - T_1)$$

Now from (12.16),

$$(\tau_{12})_2 = \frac{p}{\left(\frac{p}{\mu}+\frac{1}{\eta}\right)} \frac{\partial (u)_2}{\partial y_2}$$

$$= \frac{p}{\left(\frac{p}{\mu}+\frac{1}{\eta}\right)} \frac{U_1(p)}{2\pi} \psi_2(y_2,y_3)$$

where,

$$\psi_2(y_2,y_3) = \frac{\partial}{\partial y_2}\{\psi_1(y_2,y_3)\}$$

$$= \int_0^{L_1} f\left(\xi'_3\right)\left[\frac{\xi'^2_3 \sin\alpha_1 - 2\xi'_3(y_3-l_1)-\{y_2^2-(y_3-l_1)^2\}\sin\alpha_1+2y_2(y_3-l_1)\cos\alpha_1}{\left[\xi'^2_3-2\xi'_3\{y_2\cos\alpha_1+(y_3-l_1)\sin\alpha_1\}+y_2^2+(y_3-l_1)^2\right]^2}\right.$$

$$\left.+\frac{\xi'^2_3\sin+2\xi'_3(y_3+l_1)-\{y_2^2-(y_3+l_1)^2\}\sin\alpha_1-2y_2(y_3+l_1)\cos\alpha_1}{\left[\xi'^2_3-2\xi'_3\{y_2\cos\alpha_1-(y_3+l_1)\sin\alpha_1\}+y_2^2+(y_3+l_1)^2\right]^2}\right] d\xi'_3$$

Again, $(\tau_{12})_2 = 0$ for $t_1 \leq 0$ and taking inverse L.T. we get

$$(\tau_{12})_2 = \frac{\mu}{2\pi}H(t-T_1)\left(U_1(t_1)-\frac{\mu}{\eta}\int_0^{t-T_1}U_1(\tau)e^{\frac{-\mu(t-T_1-\tau)}{\eta}}d\tau\right)\psi_2(y_2,y_3)$$

$$(\tau_{13})_2 = \frac{\mu}{2\pi}H(t-T_1)\left(U_1(t_1)-\frac{\mu}{\eta}\int_0^{t-T_1}U_1(\tau)e^{\frac{-\mu(t-T_1-\tau)}{\eta}}d\tau\right)\psi_3(y_2,y_3)$$

where,

$$\psi_3(y_2,y_3) = -\int_0^{L_1} f\left(\xi'_3\right)\left[\frac{\xi'^2_3\cos\alpha_1-2\xi'_3y_2+\{y_2^2-(y_3-l_1)^2\}\cos\alpha_1+2y_2(y_3-l_1)\sin\alpha_1}{\left[\xi'^2_3-2\xi'_3\{y_2\cos\alpha_1+(y_3-l_1)\sin\alpha_1\}+y_2^2+(y_3-l_1)^2\right]^2}\right.$$

$$\left.-\frac{\xi'^2_3\cos\alpha_1-2\xi'_3y_2+\{y_2^2-(y_3+l_1)^2\}\cos\alpha_1-2y_2(y_3+l_1)\sin\alpha_1}{\left[\xi'^2_3-2\xi'_3\{y_2\cos\alpha_1-(y_3+l_1)\sin\alpha_1\}+y_2^2+(y_3+l_1)^2\right]^2}\right] d\xi'_3$$

We are assuming, $U_1(t_1) = v_1t_1$

$$(u)_2 = H(t-T_1)\frac{v_1t_1}{2\pi}\psi_1(y_2,y_3)$$

$$(e_{12})_2 = H(t-T_1)\frac{v_1t_1}{2\pi}\psi_2(y_2,y_3)$$

$$(\tau_{12})_2 = H(t-T_1)\frac{v_1\eta}{2\pi}\left(1-e^{\frac{-\mu(t-T_1)}{\eta}}\right)\psi_2(y_2,y_3)$$

$$(\tau_{13})_2 = H(t-T_1)\frac{v_1\eta}{2\pi}\left(1-e^{\frac{-\mu(t-T_1)}{\eta}}\right)\psi_3(y_2,y_3)$$

$$(\tau_{1'2'})_2 = H(t-T_1)\frac{v_1\eta}{2\pi}\left(1-e^{\frac{-\mu(t-T_1)}{\eta}}\right)(\psi_2\sin\alpha_1 - \psi_3\cos\alpha_1)$$

$$(\tau_{1''2''})_2 = H(t-T_1)\frac{v_1\eta}{2\pi}\left(1-e^{\frac{-\mu(t-T_1)}{\eta}}\right)(\psi_2\sin\alpha_2 - \psi_3\cos\alpha_2)$$

12.5 AFTER COMMENCEMENT OF THE FAULT CREEP THROUGH SECOND FAULT F_2, RESULT FOR SSD

Near the second fault F_2, if the accumulated stress crosses $(\tau_c)_2$, a creeping move-ment starts across F_2 and starts with a velocity v_2 cm/year.

Following the same approach, the final solution after time $t > T_2$ is given by,

$$
\begin{aligned}
u &= (u)_0 + y_2 \tau_\infty (0) \left[\tfrac{kt}{\mu} + \tfrac{t}{\eta} + \tfrac{kt^2}{2\eta} \right] + H(t - T_1) \tfrac{v_1 t_1}{2\pi} \psi_1 (y_2, y_3) \\
&\quad + H(t - T_2) \tfrac{v_2 t_2}{2\pi} \phi_1 (y_2, y_3) \\
e_{12} &= (e_{12})_0 + \tau_\infty (0) \left[\tfrac{kt}{\mu} + \tfrac{t}{\eta} + \tfrac{kt^2}{2\eta} \right] + H(t - T_1) \tfrac{v_1 t_1}{2\pi} \psi_2 (y_2, y_3) \\
&\quad + H(t - T_2) \tfrac{v_2 t_2}{2\pi} \phi_2 (y_2, y_3) \\
\tau_{12} &= (\tau_{12})_0 e^{-\tfrac{\mu t}{\eta}} + \tau_\infty (0) \left(1 + kt - e^{-\tfrac{\mu t}{\eta}} \right) \\
&\quad + H(t - T_1) \tfrac{v_1 \eta}{2\pi} \left(1 - e^{\tfrac{-\mu(t-T_1)}{\eta}} \right) \psi_2 (y_2, y_3) \\
&\quad + H(t - T_2) \tfrac{v_2 \eta}{2\pi} \left(1 - e^{\tfrac{-\mu(t-T_2)}{\eta}} \right) \phi_2 (y_2, y_3) \\
\tau_{13} &= (\tau_{13})_0 e^{-\tfrac{\mu t}{\eta}} + H(t - T_1) \tfrac{v_1 \eta}{2\pi} \left(1 - e^{\tfrac{-\mu(t-T_1)}{\eta}} \right) \psi_3 (y_2, y_3) \\
&\quad + H(t - T_2) \tfrac{v_2 \eta}{2\pi} \left(1 - e^{\tfrac{-\mu(t-T_2)}{\eta}} \right) \phi_3 (y_2, y_3) \\
\tau_{1'2'} &= (\tau_{1'2'})_0 e^{-\tfrac{\mu t}{\eta}} + \tau_\infty (0) \left(1 + kt - e^{-\tfrac{\mu t}{\eta}} \right) \sin \alpha_1 \\
&\quad + H(t - T_1) \tfrac{v_1 \eta}{2\pi} \left(1 - e^{\tfrac{-\mu(t-T_1)}{\eta}} \right) \\
&(\psi_2 \sin \alpha_1 - \psi_3 \cos \alpha_1) + H(t - T_2) \tfrac{v_2 \eta}{2\pi} \left(1 - e^{\tfrac{-\mu(t-T_2)}{\eta}} \right) (\phi_2 \sin \alpha_1 - \phi_3 \cos \alpha_1) \\
\tau_{1''2''} &= (\tau_{1''2''})_0 e^{-\tfrac{\mu t}{\eta}} + \tau_\infty (0) \left(1 + kt - e^{-\tfrac{\mu t}{\eta}} \right) \sin \alpha_2 \\
&\quad + H(t - T_1) \tfrac{v_1 \eta}{2\pi} \left(1 - e^{\tfrac{-\mu(t-T_1)}{\eta}} \right) \\
&(\psi_2 \sin \alpha_2 - \psi_3 \cos \alpha_2) + H(t - T_2) \tfrac{v_2 \eta}{2\pi} \left(1 - e^{\tfrac{-\mu(t-T_2)}{\eta}} \right) (\phi_3 \sin \alpha_2 - \phi_3 \cos \alpha_2)
\end{aligned}
$$

$$(12.22)$$

where,

$$
\phi_1 (y_2, y_3) = \int_0^{L_2} f_2 (\eta'_3) \left[\frac{(y_2 - H) \sin \alpha_2 - (y_3 - l_2) \cos \alpha_2}{\eta'^2_3 - 2\eta'_3 \{(y_2 - H) \cos \alpha_2 + (y_3 - l_2) \sin \alpha_2\} + (y_2 - H)^2 + (y_3 - l_2)^2} \right.
$$
$$
\left. + \frac{(y_2 - H) \sin \alpha_2 + (y_3 - l_2) \cos \alpha_2}{\eta'^2_3 - 2\eta'_3 \{(y_2 - H) \cos \alpha_2 - (y_3 - l_2) \sin \alpha_2\} + (y_2 - H)^2 + (y_3 - l_2)^2} \right] d\eta'_3
$$

$$\phi_2(y_2, y_3) =$$

$$\int_0^{L_2} f_2(\eta'_3) \left[\frac{\eta'^2_3 \sin\alpha_2 - 2\eta'_3(y_3-l_2) - \{(y_2-H)^2 - (y_3-l_2)^2\}\sin\alpha_2 + 2(y_2-H)(y_3-l_2)\cos\alpha_2}{[\eta'^2_3 - 2\eta'_3\{(y_2-H)\cos\alpha_2 + (y_3-l_2)\sin\alpha_2\} + (y_2-H)^2 + (y_3-l_2)^2]^2} \right.$$

$$\left. + \frac{\eta'^2_3 \sin\alpha_2 + 2\eta'_3(y_3+l_2) - \{(y_2-H)^2 - (y_3+l_2)^2\}\sin\alpha_2 - 2(y_2-H)(y_3+l_2)\cos\alpha_2}{[\eta'^2_3 - 2\eta'_3\{(y_2-H)\cos\alpha_2 + (y_3+l_2)\sin\alpha_2\} + (y_2-H)^2 + (y_3+l_2)^2]^2} \right] d\eta'_3$$

$$\phi_3(y_2, y_3) =$$

$$-\int_0^{L_2} f_2(\eta'_3) \left[\frac{\eta'^2_3 \cos\alpha_2 - 2\eta'_3(y_2-H) + \{(y_2-H)^2 - (y_3-l_2)^2\}\cos\alpha_2 + 2(y_2-H)(y_3-l_2)\sin\alpha_2}{[\eta'^2_3 - 2\eta'_3\{(y_2-H)\cos\alpha_2 + (y_3-l_2)\sin\alpha_2\} + (y_2-H)^2 + (y_3-l_2)^2]^2} \right.$$

$$\left. - \frac{\eta'^2_3 \cos\alpha_2 - 2\eta'_3(y_2-H) + \{(y_2-H)^2 - (y_3+l_2)^2\}\cos\alpha_2 - 2(y_2-H)(y_3+l_2)\sin\alpha_2}{[\eta'^2_3 - 2\eta'_3\{(y_2-H)\cos\alpha_2 + (y_3+l_2)\sin\alpha_2\} + (y_2-H)^2 + (y_3+l_2)^2]^2} \right] d\eta'_3$$

Hence, the functions f_1 and f_2 also fulfil the subsequent sufficient conditions (Rybicki 1973):

I. $f(y_3)$ and $f'(y_3)$ are both continuous and $y_3 \in [0, L_1]$
II. Either (i) $f''(y_3)$ is continuous with $y_3 \in [0, L_1]$
or (i) $f''(y_3)$ is continuous with $y_3 \in [0, L_1]$, excluding for a finite number of points with finite discontinuity and $y_3 \in [0, L_1]$,
or (c) $f''(y_3)$ is continuous with $y_3 \in [0, L_1]$, excluding possibly for a finite number of points of finite discontinuity and for the ends points of $(0, L_1)$, there exist real constants $m < 1$ and $n < 1$ such that $y_3^m f''(y_3) \to 0$ or to a finite limit as $y_3 \to 0 + 0$ and $(L_1 - y_3)^n f''(y_3) \to 0$ or to a finite limit as $y_3 \to L_1 - 0$ and
III. $f(L_1) = 0 = f'(L_1)$, $f'(0) = 0$,
The above-mentioned conditions are adequate for SSD considering all finite (y_2, y_3, t).
We calculate the integrals if $f(y_3)$ is given as,

$$f(y'_3) = \frac{y'^2_3 (y'_3 - L_1)^2}{\left(\frac{L_1}{2}\right)^4}$$

Similar conditions are applicable for the second fault F_2.

12.6 NUMERICAL COMPUTATIONS

Assuming, $f_1(\xi'_3)$ to be

$$f_1(\xi'_3) = \frac{\xi'^2_3 (\xi'_3 - L_1)^2}{\left(\frac{L_1}{2}\right)^4}$$

Hence, the values of the model parameters (based on the rheologic performance of the crust and upper mantle (Mondal and Debnath 2021, Aki and Richards (1980), Catlhes (1975), Chift et al. (2002), and Karato (2010)) are given as (Table 12.1),

Table 12.1

Values of the Model Parameters

Parameter	Symbol	Value
Rigidity	μ	3.5×10^{10} N/m^2
Viscosity	η	3×10^{19} Pa s
Depth of the first fault	l_1	5×10^3 m
Depth of the second fault	l_2	10×10^3 m
The gap between the upper edges of the fault (along the horizontal direction)	H	10×10^3 m
Length of the first fault	L_1	8×10^3 m
Length of the second fault	L_2	10×10^3 m
Initial stress	$\tau_\infty(0)$	50×10^5 N/m^2
Initial stress	$(\tau_{12})_0$	50×10^5 N/m^2
Initial stress	$(\tau_{13})_0$	50×10^5 N/m^2
Critical stress for the first fault	$(\tau_c)_1$	2×10^7 N/m^2
Critical stress for the second fault	$(\tau_c)_2$	2.5×10^7 N/m^2
Inclination of the fault F to the free surface	α_1, α_2	$60°, 45°$
The time taken to reach the first critical value	T_1	$T_1 = 117$ years, when $\alpha_1 = 60°$
The time taken to reach the second critical value	T_2	$T_2 = 152$ years, when $\alpha_2 = 60°$

12.7   RESULT AND DISCUSSION

Surface share strain along with time is represented by the following expression:

$$E_{12} = e_{12} - (e_{12})_0 = \tau_\infty(0)\left[\frac{kt}{\mu} + \frac{t}{\eta} + \frac{kt^2}{2\eta}\right] + H(t - T_1)\frac{v_1 t_1}{2\pi}\psi_2(y_2, y_3)$$
$$+ H(t - T_2)\frac{v_2 t_2}{2\pi}\phi_2(y_2, y_3)$$

Figure (12.3) shows the rate of change of shear strain with respect to the time. It has been observed that the rate of change is increasing with time because of the incremental tectonic forces. The approximate magnitude is observed in the order 10^{-3}.

Share stress with respect to depth along a line L is calculated for the different inclination of the fault $60°$ and $90°$ with $y_2 = 5$ km. The deviation of $\tau_{1'2'}$ versus depth along a line L is shown in Figure (12.4). From the figure, stress accumulation and stress release region are clearly visible. Stress magnitude becomes negligibly small for the depth from 20 km. Hence, an almost similar nature of dependency has been observed in the case of stress with respect to the depth and inclination of the fault.

Total stress for various creeping velocities v_1 and v_2, at a point $y_2 = 15$ km, $y_3 = 15$ km, versus time is shown in Figure (12.5). Stress $\tau_{1''2''}$ is calculated with respect to time for various creep velocities given by $v_1 = v_2 = 0, 15, 20$ cm/year. Up to $T_1 = 117$ year, stress is steadily rising. With the increasing creep velocities, the stress accumulation rate increases up to $T_2 = 152$ year, earlier the second fault movement. If the creep velocities across F_2 increase, the magnitude of the stress falls down. Though, in a few decades, the reduced stresses again start increasing due to the increase in $\tau_\infty(t)$ to make up for the stress drop.

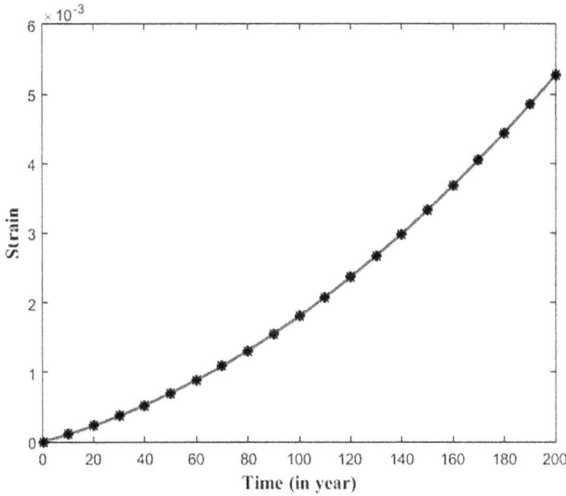

Figure 12.3   Shear strain due to fault movement at the surface.

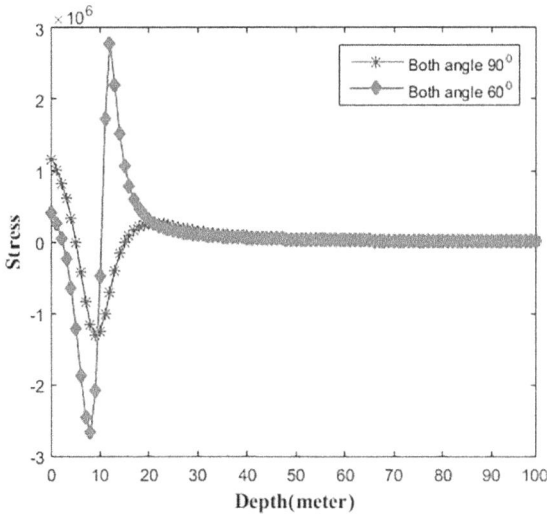

Figure 12.4   Stress representation due to depth.

The contour map of stress accumulation and reduction region due to the fault movement across the fault F_1 only is shown in Figure (12.6) as $t = T_1 + 1$ year. Figure (12.7) shows the stress accumulation and reduction region across F_2 due to the fault movement at time $t = T_2 + 1$ year year. Due to the fault movement across both the faults F_1 and F_2, the stress accumulation and release region is shown in Figure (12.8) one year after T_2.

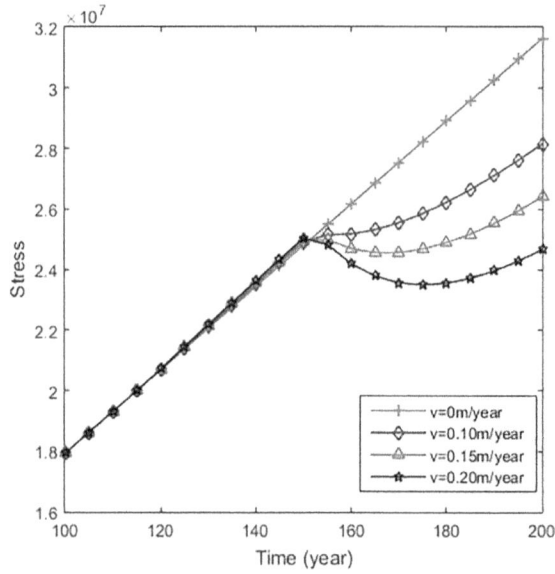

Figure 12.5   Stress representation due to time with several creep velocities.

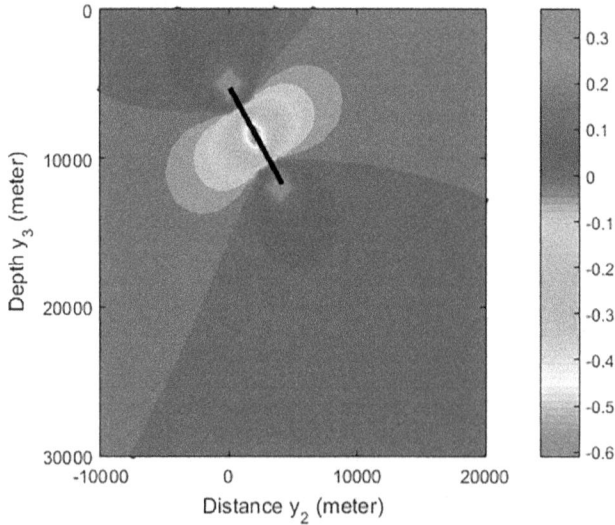

Figure 12.6   The contour representation of shear stress in viscoelastic half-space for the first fault.

12.8   CONCLUSION

In this chapter, mainly the behavior of the two interacting inclined SSF in a viscoelastic half-space has been modeled properly and also has been validated using

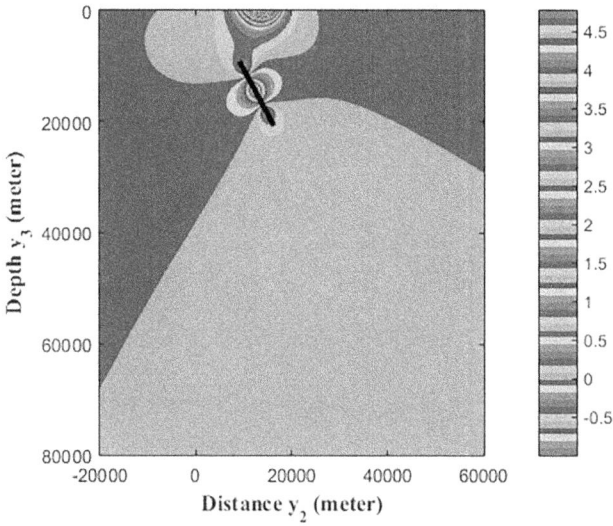

Figure 12.7   The contour representation of shear stress in viscoelastic half-space for the second fault.

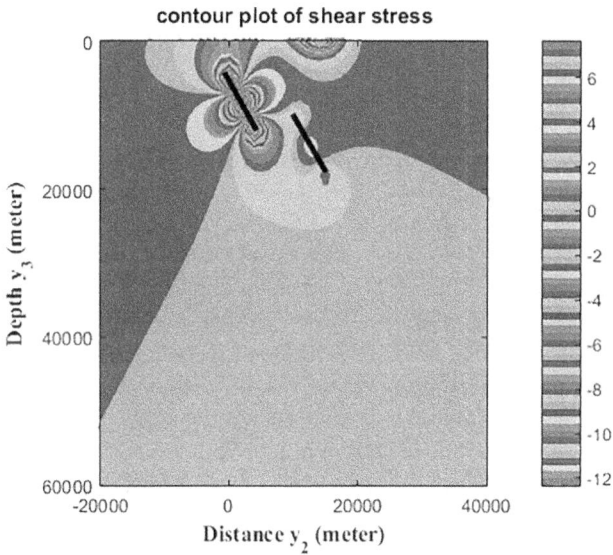

Figure 12.8   The contour representation of shear stress in viscoelastic half-space for both the faults.

the MATLAB simulation. In this viscoelastic half-space, the stresses rise gradually and an earthquake occurs when it crosses the threshold limit. This interaction depends upon the relative position of the faults with respect to one another and also on the inclination of the faults. Some finite faults, Dip-slip faults, and layered models can be applied in future work to identify the stress-strain relation more exclusively in the lithosphere-aesthenosphere fault system. Hence the modelling of SSD in the presence or absence of the fault has been detailed also. Adequate graphs have been detailed which validate the efficacy of the proposed models. These formulations can be used by researchers for further analysis of earthquakes and associated geological phenomena.

REFERENCES

Aki, K., and Richards, P.G. (1980). *Quantitative Seismology: Theory and Methods*. W.H. Freeman, San Francisco, CA.

Catlhes III, L.M. (1975). *The Viscoelasticity of the Earth's Mantle*. Princeton University Press, Princeton, N.J.

Chift, P., Lin, J., and Barcktiausen, U. (2002). Evidence of low flexural rigidity and low viscosity lower continental crust during continental break-up in the South China Sea. *Marine and Petroleum Geology*, 19:951–970.

Chinnery, M.A. (1961). The deformation of the ground around surface faults. *Bulletin of the Seismological Society of America*, 51:355–372.

Cohen, S.C. (1980a). Post-seismic viscoelastic surface deformations and stress, 1, theoretical considerations: Displacement and strain calculations. *Journal of Geophysical Research*, 85(B6):3131–3150.

Cohen, S.C. (1980b). Post-seismic viscoelastic surface deformations and stress, 2, stress theory and computation: Dependence of displacement, strain and stress on fault parameters. *Journal of Geophysical Research*, 85(B6):3151–3158.

Debnath, P., and Sen, S. (2014a). Creeping movement across a long strike-slip fault in a half space of linear viscoelastic material representing the lithosphere-asthenosphere system. *Frontiers in Science*, 4(2):21–28.

Debnath, P., and Sen, S. (2014b). Two neighbouring strike slip faults and their interaction. *IOSR Journal of Applied Geology and Geophysics (IOSR-JAGG)*, 2(6): 44–56.

Debnath, P., and Sen, S. (2015a). A vertical creeping strike slip fault in a viscoelastic half space under the action of tectonic forces varying with time. *IOSR Journal of Mathematics (IOSR-JM)*, 11(3):105–114.

Debnath, P., and Sen, S. (2015b). Two interacting strike slip faults in a viscoelastic half space under increasing tectonic forces. *International Journal of Scientific and Engineering Research*, 6(5):378–386.

Debnath, P., and Sen, S. (2015c). A finite rectangular strike slip fault in a linear viscoelastic half space creeping under tectonic forces. *International Journal of Current Research*, 7(7):18365–18373.

Debnath, P., and Sen, S. (2020). Interaction among neighbouring rectangular finite strike slip faults in a linear viscoelastic half space representing lithosphere-asthenosphere system. *Mausam*, 71(4):699–708.

Dreger, D., et al. (2011). Near-field across-fault seismic ground motions. *Bulletin of the Seismological Society of America*, 101:202–221.

Hearn, E.H., and Burgmann, R. (2005). The effect of elastic layering of inversion of GPS data for coseismic slip and resulting stress changes: Strike-slip earthquakes. *Bulletin of the Seismological Society of America*, 95(5):1637–1653.

Karato, S. (2010). Rheology of the Earth's mantle: A historical review. *Elsevier*, 18(1):17–45.

Lovely, P.J., et al. (2009). Regions of reduced static stress drop near fault tips for large strike-slip earthquake. *Bulletin of the Seismological Society of America*, 99(3):1691–1704.

Manna, K., and Sen, S. (2017). Interacting inclined strike-slip faults in a layered medium. *Mausam*, 68(3):487–498.

Manna, K., and Sen, S. (2019). Stress-strain accumulation due to interactions of finite and long strike-slip faults in a viscoelastic half-space. *Bulletin of the Calcutta Mathematical Society*, 111(2):173–198.

Maruyama, T. (1966). On two dimensional dislocation in an infinite and semi-infinite medium. *Bulletin of the Earthquake Research Institute, University of Tokyo*, 44(3):811–871.

Mondal, B., and Sen, S. (2016). Long vertical strike-slip fault in a multi-layered elastic media. *Geosciences*, 6(2): 29–40.

Mondal, B., and Sen, S. (2017). Pattern of stress accumulation due to a sudden movement across a long vertical strike-slip fault in a three-layered elastic/viscoelastic model. *IOSR Journal of Applied Geology and Geophysics (IOSR-JAGG)*, 5(1): 34–39.

Mondal, D., and Debnath, P. (2021). An application of fractional calculus to geophysics: Effect of a strike-slip fault on displacement, stresses and strains in a fractional order Maxwell type visco-elastic half space. *International Journal of Applied Mathematics*, 34(5):873–888.

Mondal, S., Sen, S., and Debsarma, S. (2018a). A mathematical model to study the stress distribution due to a strike slip fault creeping with a reducing velocity. *Bulletin of the Calcutta Mathematical Society*, 110(4):265–280.

Mondal, S., Sen, S., and Debsarma, S. (2018b). A numerical approach for solution of aseismic ground deformation problem. *Journal of Geoscience and Geomatics*, 6(1):27–34.

Mondal, S., Sen, S., and Debsarma, S. (2019). A mathematical model for analyzing the ground deformation due to a creeping movement across a strike slip fault. *International Journal on Geomathematics*, 10:16.

Mukhopadhyay, A., et.al. (1979). On stress accumulation near finite fault. *Mousam*, 30:347–352.

Rybicki, K. (1971). The elastic residual filed of a very long strike-slip fault in the presence of a discontinuity. *Bulletin of the Seismological Society of America*, 61:79–92.

Rybicki, K. (1973). Static deformation of a multi-layered half space by very long strike slip fault. *Pure and Applied Geophysics*, 110:1955–1966.

Sarkar, S., et al. (2012). Interaction among neighbouring strike slip faults inclined to the vertical in a viscoelastic half space. *American Journal of mathematics and Mathematical Sciences*, 1(2):129–138.

Sato, R. (1971). Crustal due to dislocation in a multi-layered medium. *Journal of Physics of the Earth*, 19(1):31–46.

Sen, S., and Karmakar, A. (2013). The nature of stress pattern due to a sudden movement across a nonplanar buried strike-slip fault in a layered medium. *European Journal of Mathematical Sciences*, 2(2):178–208.

Sen, S., et al. (1993). A creeping and surface breaking strike-slip fault inclined to the vertical in a viscoelastic half space. *Mausam*, 44(4):365–372.

13 Applications of Fixed-Point Theory in Differential Equations

Deepak Kumar
Lovely Professional University

CONTENTS

13.1 INTRODUCTION

Fixed-point theory is the most dynamic area of research, with numerous applications both in pure and applied mathematics, along with physical, chemical and social sciences. This theory is useful as it describes the ways to establish the existence and uniqueness of solutions of algebraic, differential and integral equations. Also, the theory has a number of applications in game theory, control theory, eigenvalue problems, boundary value problems and best approximation problems. The theory has gained a remarkable scope of research in nonlinear analysis also. Many nonlinear problems can be described in a unitary manner by the following scheme. For a given object T, find another object x satisfying two conditions:

  i) The object x belongs to a given class X of objects;
  ii) The object x is in a certain relation R to the object T.

An object x satisfying these conditions will be called a solution to the given problem. This problem can be described as $\{x \in X : xRT\}$."

In turn, any problem of the above form can be written equivalently as $x = Tx$, where $T : X \to X$ is a mapping defined on a non-empty space X that allows us to use constructive tools for obtaining the desired solution. The solutions of the equation $x = Tx$ are called the fixed points of the mapping T. The theory of fixed points is of

DOI: 10.1201/9781003227847-13

interest in itself as it provides ways to check the existence of a solution to a set of equations. For example, in theoretical economics, while dealing with the problems in general equilibrium theory, a situation may arise where one needs to check whether the solution for a given system of equations exists or under which conditions we can obtain the solution. The mathematical analysis of these problems can be dealt with by using fixed-point theory.

The origin of fixed-point theory lies in the method of successive approximations used to establish the existence and uniqueness of solutions of differential equations by Liouville (1837) and Picard (1890) independently. The formal theoretic approach of the fixed-point originated from the work of Picard. However, it was the Polish mathematician Banach (1922), who underlined the idea into an abstract framework and provides a constructive tool to establish the fixed points of a mapping. From the historical point of view, the three turning points in fixed-point theorem for nonlinear analysis are: Brouwer's fixed theorem (Brouwer 1912), Banach fixed-point theorem (1922) in 1922, and Browder's fixed-point theorem (Browder 1968).

A significant contribution to fixed-point theory was given by Brouwer (1912). He stated that

Under a continuous map of the unit cube into itself which displace every point less than half a unit, the image has an interior point.

In this theorem, Brouwer established that under a continuous mapping of an object to itself there is atleast one point, where the object inevitably confronts itself. He proved this theorem as: "If X denotes a unit closed ball with center at origin in R^n and T be a continuous self-mapping of X into itself then T has a fixed-point in X, or $Tx = x$ has a solution." This theorem was extended by Schauder (1927) to the case in which X is a compact convex set in normed vector space. Schauder stated, "Any non-empty compact convex subset X of a Banach space has the topological fixed point property". The compactness condition of the subset X is a stronger one. But, in analysis many problems do not have a compact setting. Thus modification of the theorem by relaxing the condition of compactness was required. Schauder (1930) also proved a theorem which is known as Schauder second form of the above theorem and it can be stated as: "Every compact self-mapping of a closed bounded convex subset of a Banach space has atleast one fixed point". In order to understand the application of the Schauder fixed-point theorem let us consider different types of mathematical models.

13.2 MATHEMATICAL MODEL 1

Consider the mathematical model of the steady flow of a "Casson fluid" within the two horizontal coaxial rotating cylinders under the effect of the magnetic field in the cylindrical coordinate system under generalized boundary conditions (Kumar et al. 2018a):

$$\left(1 + \frac{1}{\beta}\right)\left(\frac{d^2 v}{dr^2} + \frac{1}{r}\frac{dv}{dr} - \frac{v}{r^2}\right) - B_1 v = 0, \tag{13.1}$$

with the generalized boundary condition of the form

$$v = \begin{cases} a, & \text{at } r = 1. \\ b, & \text{at } r = R. \end{cases}$$ (13.2)

In order to verify the existence of a solution to equation (13.1) under the boundary conditions given by equation (13.2), we take

$$v_\theta = n(r-1) + a, 1 \leq r \leq r,$$ (13.3)

where $n = (b-a)/(R-1)$.

By defining $p(r)$ as follow

$$p(r) = v(r) - v_\theta(r).$$ (13.4)

Using the equations (13.3) and (13.4), we get

$$p'(r) = v'(r) - n \text{ and } p''(r) = v''(r).$$ (13.5)

Substituting the equations (13.3) and (13.4) in the equation (13.1), the obtained equation can be written as

$$p''(r) + \frac{1}{r}p'(r) - \frac{1}{r^2}\left(1 + \frac{\beta}{1+\beta}B_1 r^2\right)p = g(r),$$ (13.6)

with boundary conditions given by

$$p(1) = 0 \text{ and } p(R) = 0,$$ (13.7)

where, $g(r) = \left(v_\theta + (\beta/(1+\beta))B_1 r^2 v_\theta - mr\right)/r^2$.

Now, letting $p = zr$ and substituting its derivatives in the equations (13.6) and (13.7), we get

$$z'' + \frac{3}{r}z' - \frac{\beta}{1+\beta}B_1 z = g_1(r),$$ (13.8)

$$z(1) = z(R) = 0,$$ (13.9)

in which,

$$g_1(r) = \frac{g(r)}{r}.$$ (13.10)

Now, consider a function $y(r)$ as:

$$y(r) = -\exp\left(-\int_1^r \frac{3}{\eta}d\eta\right)\int_1^r \frac{\beta B_1 y(\eta)}{1+\beta}\exp\left(\int_1^\eta \frac{3}{\xi}d\xi\right)d\eta.$$ (13.11)

Introducing $k(r) = z'(r) + y(r)$ in equation (13.8), we get

$$k'(r) + \frac{3}{r}k(r) - \left(y'(r) + \frac{3}{r}y(r) + \frac{\beta B_1 z}{1+\beta}\right) = g_1(r),$$ (13.12)

From equation (13.11), we have

$$y'(r) + \frac{3}{r}y(r) + \frac{\beta B_1 z}{1+\beta} = 0. \tag{13.13}$$

From equations (13.12) and (13.13), we have

$$k'(r) + N(r)k(r) = g_1(r), \tag{13.14}$$

where, $N(r) = 3/r$.

On solving the equation (13.14), we get

$$k(r) = A_0 \exp\left(-\int_1^r N(\eta)\,d\eta\right) + \int_1^r g_1(\eta)\exp\left(-\int_\eta^r N(\xi)\,d\xi\right)d\eta. \tag{13.15}$$

Using equations (13.11) and (13.15), in $k(r) = z'(r) + y(r)$, the obtained equation can be rewritten as

$$z'(r) = A_0 \exp\left(-\int_1^r N(\eta)\,d\eta\right) + \int_1^r g_1(\eta)\exp\left(-\int_\eta^r N(\xi)\,d\xi\right)d\eta$$
$$+ \exp\left(-\int_1^r N(\eta)\,d\eta\right)\int_1^r \frac{\beta}{1+\beta}B_1 z(\eta)\exp\left(\int_1^\eta N(\xi)\,d\xi\right)d\eta. \tag{13.16}$$

Introducing the equation (13.16) using the boundary conditions, we get

$$A_0 = -\frac{A_2}{A_1}, \tag{13.17}$$

where,

$$A_1 = \int_1^R \exp\left(-\int_1^\eta N(\xi)\,d\xi\right)d\eta, \tag{13.18}$$

$$A_2 = \int_1^R \left(\int_1^\mu g_1(\eta)\exp\left(-\int_\eta^\mu N(\xi)\,d\xi\right)d\eta\right)d\mu$$
$$+ \int_1^R \left(\exp\left(-\int_1^\mu N(\eta)\,d\eta\right)\int_1^\mu \frac{\beta}{1+\beta}B_1 z(\eta)\exp\left(\int_1^\eta N(\xi)\,d\xi\right)d\eta\right)d\mu, \tag{13.19}$$

and also, we have

$$z(r) = \int_1^r z'(\eta)\,d\eta. \tag{13.20}$$

The existence of a solution of the equation (13.1) can be claimed on the basis of the existence of a solution of the equation (13.16). To check, we have applied Schauder's fixed-point theorem (SFPT) for $T: S \to S$, where S is a subset of a Banach space "C" of a continuous function φ on the closed $[1,R]$ which becomes zero at 1 and R with the norm.

$$\|\varphi\| = \sup_{1 \le r \le R} |\varphi(r)|, \tag{13.21}$$

where, $(T\varphi)(r)$ is equivalent to the RHS of the equation (13.16).

Now, from the equation (13.10), we have

$$g_1(r) = \left(\frac{v_\theta}{r^3} - \frac{n}{r^2}\right) + \frac{\beta B_1 v_\theta}{r(1+\beta)}. \tag{13.22}$$

Hence, we obtain

$$|g_1(r)| \le \left|\frac{v_\theta}{r^3} - \frac{n}{r^3}\right| + \frac{\beta B_1}{1+\beta}\frac{|v_\theta|}{|r|},$$

$$\le |v_\theta - r| + \frac{\beta B_1}{1+\beta}|v_\theta|,$$

$$\le \max\{|a|,|b|\} + \frac{\beta B_1}{1+\beta}\max\{|a|,|b|\},$$

$$\le \left(1 + \frac{\beta B_1}{(1+\beta)}\right)\max\{|a|,|b|\} = O_1 \text{ (say)}, \tag{13.23}$$

This is because of the fact that $|v_\theta| \le \max\{|a|,|b|\}$ and $|v_\theta - n| \le \max\{|a|,|b|\}$. On integrating $N(r) = 3/r$ the range η to r, we obtain

$$\int_\eta^r N(r)\,dr = \int_\eta^r \frac{3}{r}\,dr = 3\log\left(\frac{r}{\eta}\right). \tag{13.24}$$

Now $1 \le r \le R$, therefore

$$\log\left(\eta^{-3}\right) \le \int_\eta^r N(r)\,dr \le \log\left(\frac{R}{\eta}\right)^3, \tag{13.25}$$

$$\eta^{-3} \le \exp\left(\int_\eta^r N(r)\,dr\right) \le \left(\frac{R}{\eta}\right)^3, \tag{13.26}$$

hence,

$$1 \le \exp\left(\int_1^r N(r)\,dr\right) \le R^3, \tag{13.27}$$

and

$$\left(\frac{\eta}{R}\right)^3 \le \exp\left(-\int_\eta^r N(r)\,dr\right) \le \eta^3. \tag{13.28}$$

Now, for the function $\beta B_1 z/(1+\beta)$, we have

$$\left|\frac{\beta B_1 z}{1+\beta}\right| = \frac{\beta}{1+\beta}|B_1||z|. \tag{13.29}$$

For $z = p/r$, we obtain

$$\frac{|r|}{R} \leq |z| \leq |r|. \tag{13.30}$$

Also,

$$|p| \leq |v_\theta| + |v| = \max\{|a|, |b|\} + |v| \tag{13.31}$$

Now, in order to check the boundedness of velocity distribution consider the magnetic field as zero, thus equation (13.16) becomes

$$z'_1(r) = A_0 \exp\left(-\int_1^r N(\eta)\, d\eta\right) + \int_1^r g_1(\eta) \exp\left(-\int_\eta^r N(\xi)\, d\xi\right) d\eta \tag{13.32}$$

From the equations (13.23), (13.27), (13.28) and (13.32), we get

$$|z'_1(r)| \leq |A_0| + \int_1^r O_1 \eta^3 d\eta \leq |A_0| + \frac{O_1}{4}(R^4 - 1). \tag{13.33}$$

$$|z_1(r)| \leq \int_1^r |z'(\eta)|\, d\eta \leq \left(|A_0| + \frac{O_1}{4}(R^4 - 1)\right)(R - 1). \tag{13.34}$$

Regarding equation (13.30), we get

$$|p_1| \leq R|z_1| \tag{13.35}$$

From the equation (13.4), we have

$$|v_1| = |p_1 + v_{\theta_1}| \leq |p_1| + |v_{\theta_1}| \tag{13.36}$$

Using the equation (13.35), we get

$$|v_1| \leq R|z_1| + |v_{\theta_1}| \leq R\left(|A_0| + \frac{O_1}{4}(R^4 - 1)\right)(R - 1) + \max\{|a|, |b|\} = O_2(A_0)\ (\text{say}). \tag{13.37}$$

If the magnetic field acts on the fluid element, then the maximum increment in energy because of the magnetic field will be the same as that of the energy of the magnetic field and it is written as

$$\frac{1}{2}m_0\left[v^2(r) - v_1^2(r)\right] \leq W_B, \tag{13.38}$$

where m_0 is the "mass of the fluid element", r is the "distance between the element and the cylinder axis" and W_B is the "energy of the magnetic field in a cross section of the flow of fluid". The energy of the magnetic field with the supposition of the constant magnetic field B is given by Farooq et al. (2015),

$$W_B = \frac{1}{2}HlB_a, \tag{13.39}$$

in which, the intensity of the magnetic field is represented by $H(=B/\mu_m)$, the characteristic length by l and the area crossing the magnetic field flow with unit depth by a, these two parameters $l\&a$ can be written as

$$l = 2\pi \left(\frac{r_1 + r_2}{2} \right) = \pi (r_1 + r_2), \tag{13.40}$$

$$a = (r_2 - r_1) \times 1 = r_2 - r_1. \tag{13.41}$$

Using the equations (13.40) and (13.41) in the equation (13.39), we get

$$W_B = \frac{B_0^2}{2\mu_m} \pi \left(r_2^2 - r_1^2 \right). \tag{13.42}$$

The energy of the magnetic field in a cross section of the fluid flow is

$$w_B = \frac{W_B}{2\pi} = \frac{B_0^2}{4\mu_m} \left(r_2^2 - r_1^2 \right). \tag{13.43}$$

From the equations (13.37), (13.38) and (13.43), we get

$$v^2 \leq \frac{B_0^2 r_1^2}{2m_0 \mu_m} \left(R^2 - 1 \right) + v_1^2 \tag{13.44}$$

Thus,

$$|v| \leq \left(\frac{B_0^2 r_1^2}{2m_0 \mu_m} \left(R^2 - 1 \right) + O_2^2 (A_0) \right)^{\frac{1}{2}} = O_3 (A_0) \, (\text{say}) \tag{13.45}$$

From the equations (13.30), (13.31) and (13.45), we get

$$|z| \leq |p| \leq \max \{|a|, |b|\} + |v| \leq \max \{|a|, |b|\} + O_3 (A_0) \tag{13.46}$$

From equation (13.29), we get

$$\left| \frac{\beta B_1 z}{1 + \beta} \right| = \frac{\beta}{1 + \beta} |B_1| |z| \leq \frac{\beta}{1 + \beta} |B_1| (\max \{|a|, |b|\} + O_3 (A_0)) = O_4 (A_0) \, (\text{say}) \tag{13.47}$$

Using the equations (13.23) and (13.28), we have

$$\left| \int_1^r g_1 (\eta) \exp \left(- \int_\eta^r N(\xi) d\xi \right) d\eta \right| \leq \int_1^r g_1 (\eta) \left| \exp \left(- \int_\eta^r N(\xi) d\xi \right) \right| d\eta$$

$$\leq \int_1^r O_1 \eta^3 d\eta = \frac{O_1}{4} \left(R^4 - 1 \right) = O_5 \, (\text{say}) \tag{13.48}$$

From the equations (13.27) and (13.47), we have

$$\left| \int_1^r \frac{\beta}{1+\beta} B_1 z(\eta) \exp \left(\int_1^\eta N(\xi) d\xi \right) d\eta \right| \leq \int_1^r \left| \frac{\beta}{1+\beta} B_1 z(\eta) \right| R^3 d\eta$$

$$\leq \int_1^r O_4(A_0) R^3 d\eta$$

$$\leq O_4(A_0) R^3 (R-1) = O_6(A_0).$$

$$(13.49)$$

From the equations (13.18) and (13.27), we have

$$A_1 \geq R^{-3}(R-1), \qquad (13.50)$$

$$|A_2| \leq O_5(R-1) + O_6(R-1) = (O_5 + O_6)(R-1). \qquad (13.51)$$

Therefore,

$$|A_0| \leq (O_5 + O_6) R^3, \qquad (13.52)$$

Using the equations (13.23), (13.32), (13.45), (13.47), (13.48) and (13.49), we get

$$|A_0| \leq C_1/4\sqrt{2m_0\mu_m} \left(1 - |O_1| R^7 (R-1)^2 \beta/(1+\beta) \right) = O_7 \text{ (say)}, \qquad (13.53)$$

where,

$$C_1 = 8\sqrt{2m_0\mu_m} |B_1| R^6 (R-1) \max\{|a|,|b|\} \beta/(1+\beta)$$

$$+ 4B_0 |B_1| r_1 R^6 (R+1)^{\frac{1}{2}} (R-1)^{\frac{3}{2}} \beta/(1+\beta)$$

$$+ \sqrt{2m_0\mu_m} O_1 |B_1| R^7 (R+1)(R-1)^3 \beta/(1+\beta)$$

$$+ \sqrt{2m_0\mu_m} O_1 R^3 (R^4 - 1).$$

Therefore,

$$|T(\varphi)(r)| \leq O_8, \text{ where } O_8 = O_5 + O_6 + O_7. \qquad (13.54)$$

Since $O_5, O_5, \& O_7$ are independent of φ therefore, all the mappings $(T\varphi)$ are uniformly bounded. The function can be considered from Banach space N to S, where

$$S = \{\varphi \in N: \|\varphi\| \leq O_8\}, \qquad (13.55)$$

and S is a closed convex subset of N.

Now, we define a mapping $T: S \to S$, and take the derivative of $(T\varphi)(r)$

$$T'\varphi(r) = -N(r) T\varphi(r) + g_1(r) + \frac{\beta B_1 z(r)}{1+\beta} + N(r) \int_1^r g_1(\eta) \exp \left(-\int_\eta^r N(\xi) d\xi \right) d\eta.$$

$$|T'\varphi(r)| \leq |N(r)| |T\varphi(r)| + |g_1(r)| + \frac{\beta |B_1| \|z(r)\|}{1+\beta}$$

$$+ |N(r)| \left| \int_1^r g_1(\eta) \exp \left(-\int_\eta^r N(\xi) d\xi \right) d\eta \right|, \qquad (13.56)$$

$$|T'\varphi(r)| \leq 3O_8 + O_1 + O_4 + 3O_5 = O_9 \text{ (say)}.$$

Since O_9 is independent of φ, therefore the set of the functions $(T\varphi)$, $\varphi \in S$ is equicontinuous. Consider the mappings $T: S \to S_N = T(S)$, where

$$S_N = \left\{ \varphi \in N: \ \|\varphi\| \leq O_8, \ \|\varphi'\| \leq O_9 \right\}, \tag{13.57}$$

consists of uniformly bounded and equicontinuous functions. Then via the Ascoli-Azerla theorem S_N is a relatively compact subset of S. To complete the theorem it is sufficient to show that $T: S \to S$ is continuous mapping. By the chain rule, it can be shown that there exists a constant $O_{10} = O_{10}(R, O_1, O_2, \ldots, O_9)$ such that $\frac{\partial T(\varphi)}{\partial \varphi} \leq O_{10}$, Consequently

$$\|T\varphi_1 - T\varphi_2\| \leq \|\varphi_1 - \varphi_2\|.$$

This implies the continuity of the mapping $T: S \to S$. Therefore, by SFPT it is claimed that the exist at least one solution of the equation (13.1).

13.3 MATHEMATICAL MODEL 2

Consider a mathematical model for the flow of an incompressible "Powell-Eyring fluid" between two concentric rotating cylinders (Kumar et al. 2018b).

$$(1+\in)\left(\frac{d^2v_\phi}{dr^2} + \frac{1}{r}\frac{dv_\phi}{dr} - \frac{v_\phi}{r^2}\right) - \beta\left(\frac{dv_\phi}{dr} - \frac{v_\phi}{r}\right)^2\left(3\frac{d^2v_\phi}{dr^2} - \frac{1}{r}\frac{dv_\phi}{dr} + \frac{v_\phi}{r^2}\right) = 0,$$
$$\tag{13.58}$$

with generalized boundary conditions

$$v_\phi = \begin{cases} a, & \text{at } r = 1. \\ b, & \text{at } r = R. \end{cases} \tag{13.59}$$

Introducing equation (13.60) in equation (13.58), we get

$$\frac{dv_\phi}{dr} - \frac{v_\phi}{r} = \omega, \tag{13.60}$$

$$\frac{d^2v_\phi}{dr^2} + \frac{g(\omega)}{r}\frac{dv_\phi}{dr} - g(\omega)\frac{v_\phi}{r^2} = 0, \tag{13.61}$$

where $g(\omega) = (1 + \in + \beta\omega^2)/(1 + \in - 3\beta\omega^2)$.

To check the existence of a solution of the equation (13.61) using boundary conditions given by equation (13.59). Assume that

$$\mu(r) = n(r-1) + a, 1 \leq r \leq R, \tag{13.62}$$

where $n = (b-a)/(R-1)$.

We define $y(r)$ as follows

$$y(r) = v_\phi(r) - \mu(r). \tag{13.63}$$

On differentiating equation (13.63) and using equation (13.62), we get

$$y'(r) = v'_\phi(r) - n \ and \ y''(r) = v''_\phi(r). \tag{13.64}$$

By substituting equations (13.63) and (13.64) in equation (13.61), we get

$$y''(r) + \frac{1}{r}y'(r) - \frac{1}{r^2}g(\omega)y(r) = h_1(r), \tag{13.65}$$

where,

$$h_1(r) = \left(\frac{\mu(r)}{r^2}g(\omega) - \frac{n}{r}g(\omega)\right) \tag{13.66}$$

where the corresponding boundary conditions are

$$y(1) = 0 \ and \ y(R) = 0. \tag{13.67}$$

From equations (13.60), (13.63) and (13.64), we get

$$\omega = v'_\phi - \frac{v_\phi}{r} = \left(y' - \frac{y}{r}\right) + \left(n - \frac{\mu}{r}\right). \tag{13.68}$$

Now, using $y = zr$ and its derivatives in the equation (13.65), we get

$$z'' + \left(\frac{2 + g(\omega)}{r}\right)z' = h(r), \tag{13.69}$$

$$z(1) = z(R) = 0, \tag{13.70}$$

where,

$$h(r) = \frac{h_1(r)}{r}. \tag{13.71}$$

Now we define a function $q(r)$ as follows

$$q(r) = \exp\left(-\int_1^r \frac{1}{\sigma}(2 + g(\omega(\sigma)))d\sigma\right). \tag{13.72}$$

Introducing $k(r) = z'(r) + q(r)$ in equation (13.69), we get

$$k'(r) + \left(\frac{2 + g(\omega)}{r}\right)k(r) - \left(q'(r) + \frac{2 + g(\omega)}{r}q(r)\right) = h(r), \tag{13.73}$$

but from the equation (13.72), we have

$$q'(r) + \left(\frac{2 + g(\omega)}{r}\right)q(r) = 0 \tag{13.74}$$

From the equations (13.73) and (13.74), we get

$$k'(r) + A(r)k(r) = h(r), \tag{13.75}$$

in which, $A(r) = (2 + g(\omega))/r$.

The solution of the equation (13.75) is obtained as

$$k(r) = A_0 \exp\left(-\int_1^r A(\sigma)\,d\sigma\right) + \int_1^r h(\sigma)\exp\left(-\int_\sigma^r A(\xi)\,d\xi\right)d\sigma. \quad (13.76)$$

Now, using equations (13.72) and (13.74), in the expression $z'(r) = k(r) - q(r)$, the obtained equation can be rewritten as

$$z'(r) = A_0 \exp\left(-\int_1^r A(\sigma)\,d\sigma\right) + \int_1^r h(\sigma)\exp\left(-\int_\sigma^r A(\xi)\,d\xi\right)d\sigma$$

$$- \exp\left(-\int_1^r A(\sigma)\,d\sigma\right). \quad (13.77)$$

Integrating equation (13.77) and using the boundary conditions, we get

$$A_0 = \frac{\int_1^R \left\{\int_1^\rho h(\sigma)\exp\left(-\int_\sigma^\rho A(\xi)\,d\xi\right)d\sigma\right\}d\rho - \int_1^R \left\{\exp\left(-\int_1^\rho A(\sigma)\,d\sigma\right)\right\}d\rho}{\int_1^R \exp\left(-\int_1^\sigma A(\xi)\,d\xi\right)d\sigma}, \quad (13.78)$$

and also, we have

$$z(r) = \int_1^r z'(\sigma)\,d\sigma. \quad (13.79)$$

To check the existence of a solution to the equation (13.58) we can check the existence of a solution to equation (13.77). To check we apply, SFPT for the mapping $T: S \rightarrow S$, where S is a subset of Banach space "C" of a continuous function φ on $[1, R]$ which becomes zero at 1 and R with the norm.

$$\|\varphi\| = \sup_{1 \leq r \leq R} |\varphi(r)|, \quad (13.80)$$

where, $(T\varphi)(r)$ is equivalent to the RHS of the equation (13.77) with $\omega(r)$ replaced with,

$$\omega(r) = rz' + \left(n - \frac{\mu}{r}\right) = r\varphi(r) + \left(n - \frac{\mu}{r}\right). \quad (13.81)$$

From $g(\omega) = (1 + \in + \beta\omega^2)(1 + \in -3\beta\omega^2)$, we have.

$$-\frac{1}{3} \leq g(\omega) \leq 1. \quad (13.82)$$

Therefore,

$$\frac{5}{3r} \leq A(r) \leq \frac{3}{r}. \quad (13.83)$$

Also, we can assume that

$$|g'(\omega)| < \gamma(\omega), -\infty < \omega < \infty, \quad (13.84)$$

where γ is a positive continuous mapping of ω. From equations (13.66) and (13.71), we obtain

$$h(r) = \left(\frac{\mu}{r} - \frac{n}{r^2}\right)g(\omega). \quad (13.85)$$

Hence, we have

$$|h(r)| \leq \left| \left(\frac{\mu}{r^3} - \frac{n}{r^2} \right) g(\omega) \right| \leq |\mu - n| |g(\omega)| \leq \max\{|a|, |b|\} = O_1 \text{ (say)}. \quad (13.86)$$

This is because of the fact that, we have $|\mu - n| \leq \max\{|a|, |b|\}$.
On integrating equation (13.83) between the range σ to r, we obtain

$$\log \left(\frac{r}{\sigma} \right)^{\frac{5}{3}} \leq \int_\sigma^r A(r) \, dr \leq \log \left(\frac{r}{\sigma} \right)^3. \quad (13.87)$$

Since, $1 \leq r \leq R$, therefore we have

$$\log \left(\sigma^{\left(-\frac{5}{3}\right)} \right) \leq \int_\sigma^r A(r) \, dr \leq \log \left(\frac{R}{\sigma} \right)^3, \quad (13.88)$$

hence,

$$1 \leq \exp \left(\int_1^r A(r) \, dr \right) \leq R^3, \quad (13.89)$$

and

$$\left(\frac{\sigma}{R} \right)^3 \leq \exp \left(-\int_\sigma^r A(r) \, dr \right) \leq \sigma^{\frac{5}{3}}. \quad (13.90)$$

Now, from equation (13.90), we have

$$\int_1^R \frac{1}{R^3} \, dr \leq \int_1^R \exp \left(-\int_1^r A(r) \, dr \right) dr \leq \int_1^R dr, \quad (13.91)$$

On solving

$$(R-1) \geq \left| \int_1^R \exp \left(-\int_1^r A(r) \, dr \right) dr \right| \geq R^{-3}(R-1). \quad (13.92)$$

Using equations (13.86) and (13.90), we have

$$\left| \int_1^\rho h(\sigma) \exp \left(-\int_\sigma^\rho A(\xi) \, d\xi \right) d\sigma \right| \leq \int_1^r |h(\sigma)| \left| \exp \left(-\int_\sigma^\rho A(\xi) \, d\xi \right) \right| d\sigma$$

$$\leq \int_1^r O_1 \sigma^{\frac{5}{3}} \, d\sigma = \frac{3O_1}{8} \left(R^{\frac{8}{3}} - 1 \right) = O_2 \text{ (say)}.$$

$$(13.93)$$

Using equations (13.89) and (13.93), we have

$$\left| \int_1^R \left\{ \int_1^\rho h(\sigma) \exp \left(-\int_\sigma^\rho A(\xi) \, d\xi \right) d\sigma \right\} d\rho \right.$$

$$\left. - \int_1^R \left\{ \exp \left(-\int_1^\rho A(\sigma) \, d\sigma \right) \right\} d\rho \right| \leq (O_2 + 1)(R - 1). \quad (13.94)$$

Therefore,

$$|A_0| \leq \frac{(O_2+1)(R-1)}{R^{-3}(R-1)} = R^3(O_2+1) = O_3 \text{ (say)}, \tag{13.95}$$

Therefore,

$$|T(\varphi)(r)| \leq O_4, \text{ where } O_4 = O_3 + O_2 + 1. \tag{13.96}$$

Since $O_3, O_2 \& 1$ are independent of φ, therefore, all the mappings $(T\varphi)$ are uniformly bounded. The function can be considered from Banach space N to S, where

$$S = \{\varphi \in N : \|\varphi\| \leq O_4\}, \tag{13.97}$$

and S is a closed convex subset of N.

Now, we define a mapping $T : S \rightarrow S$, and take the derivative of $(T\varphi)(r)$

$$T'\varphi(r) = -A(r)T\varphi(r) + h(r) + A(r)\int_1^r h(\sigma) \exp\left(-\int_\sigma^r A(\xi)d\xi\right)d\sigma,$$
$$|T'\varphi(r)| \leq |A(r)||T\varphi(r)| + |h(r)| + |A(r)|\left|\int_1^r h(\sigma) \exp\left(-\int_\sigma^r A(\xi)d\xi\right)d\sigma\right|,$$
$$|T'\varphi(r)| \leq 3O_4 + 3O_2 + O_1 = O_5 \text{ (say)}$$
$$\tag{13.98}$$

Since O_5 is independent of φ, thus the set of functions $(T\varphi), \varphi \in S$ is equicontinuous. Consider the mappings $T : S \rightarrow S_N = T(S)$, where

$$S_N = \{\varphi \in N : \|\varphi\| \leq O_4, \|\varphi'\| \leq O_5\}, \tag{13.99}$$

consists of uniformly bounded and equicontinuous functions. Then via the Ascoli-Azerla theorem S_N is a relatively compact subset of S. In order to complete the theorem it is sufficient to show that $T : S \rightarrow S$ is continuous mapping. Regarding the discussed notices, $\omega = r\varphi(r) + (n - \mu/r)$ and on the basis of the fact that

$$|g'(\omega)| \leq O_6, \tag{13.100}$$

where,

$$O_6 = \max_{\{|\omega| \leq O_7\}} |\gamma(\omega)| \tag{13.101}$$

and

$$O_7 = RO_4 + |n| + \max\{|a|,|b|\}. \tag{13.102}$$

By the chain rule, it can be shown that there exists a constant $O_8 = O_8(R, O_1, O_2, \ldots, O_7)$ such that $\frac{\partial T(\varphi)}{\partial \varphi} \leq O_8$, Consequently

$$\|T\varphi_1 - T\varphi_2\| \leq O_8 \|\varphi_1 - \varphi_2\|.$$

This implies the continuity of the mapping $T : S \rightarrow S$. Therefore, by SFPT it is claimed that the exist at least one solution of the equation (13.61).

13.4 CONCLUSIONS

In the present chapter:

1. The first model is based on the steady flow of a Casson fluid within two horizontal coaxial rotating cylinders under the effect of the magnetic field. The inner and outer cylinders are rotating at a constant angular speed around the axis. The problem is formulated in a cylindrical coordinate system. The obtained mathematical model is linear
2. The second model is formulated for the flow of an incompressible Powell-Eyring fluid between concentric cylinders. The problem is formulated in a cylindrical coordinate system. The obtained mathematical model is nonlinear.

The existence of the solution in both models linear as well as nonlinear is proved using SFPT. On the basis of the above study for models 1 and 2, it can be concluded that Sahauder's fixed-point theorem can be used to verify the existence of linear as well nonlinear mathematical models.

REFERENCES

Banach, S. (1922). Sur les opérationsdans les ensembles abstraits et leur application auxéquationsintégrales. *Fundamenta Mathematicae*, 3: 133–181.

Brouwer, L.E.J. (1912). Überabbildung von mannigfaltigkiten. *Mathematische Annalen*, 71: 97–115.

Browder, F.E. (1968). The fixed-point theory of multivalued mappings in topological vector spaces. *Mathematische Annalen*, 177: 283–301.

Farooq, M., Gull, N., Alsaedi, A., Hayat, T. (2015). MHD flow of Jeffrey fluid with Newtonian heating. *Journal of Mechanics*, 31(3): 319–329.

Kumar, D., Katta, R., Chandok, S. (2018a). Influence of an external magnetic field on the flow of a Casson fluid in micro-annulus between two concentric cylinders. *Bulletin of the Brazilian Mathematical Society, New Series*, 50: 515–531.

Kumar, D., Katta, R., Chandok, S. (2018b). Mathematical modeling and simulation for the flow of magneto-Powell-Eyring fluid in an annulus with concentric rotating cylinders. *Chinese Journal of Physics*, 65: 187–197.

Liouville, J. (1837). Second mémoire sur le development des fonctionsou parties de fonctionsensériesdont divers termessontassujettis à satisfaire a uneméme equation differentielle du second ordrecontenant un parameter variable. *Journal de MathématiquesPures et Appliquées*, 2: 16–35.

Picard, E. (1890). Memoire sur la theorie des equations aux deriveespartielles et la method des approximations successives. *Journal de MathématiquesPures et Appliquées*, 6: 145–210.

Schauder, J. (1927). Zur Theoriestetiger Abbildungen in Funktionalräumen. *Mathematische Zeitschrift*, 26: 47–65.

Schauder, J. (1930). Der fix punktsatz in functional raumen. *Studia Mathematica*, 2: 171–180.

14 Differential Equation-Based Compact 2-D Modeling of Asymmetric Gate Oxide Heterojunction Tunnel FET

Sudipta Ghosh
Jadavpur University

Arghyadeep Sarkar
Macmaster University

CONTENTS

14.1 INTRODUCTION

Tunnel Field Effect Transistor (TFET) has emerged as an effective alternative device to replace MOSFET for a few decades. The major drawbacks of MOSFET devices are the short-channel effects, due to which the leakage current increases with a decrease in device dimension (Chanda et al. 2018). Besides, because of tunneling-based current conduction, TFETs can mitigate the short-channel effects (SCE). So, scaling down TFET is more efficacious than that of MOSFETs. Sub-threshold swing (SS) is another advantageous characteristic of TFET devices for high-speed digital applications. In TFETs the subthreshold swing (SS) could be well below 60 mV/decade

DOI: 10.1201/9781003227847-14

(Dutta and Sarkar 2019; Kumar et al. 2018), which is the thermal limit for MOS-FET devices and therefore makes it more suitable than MOSFET for faster switching applications. It is observed from the literature studies that the performances of the TFET devices have been explored thoroughly by using 2-D TCAD simulation (Mallik and Chattopadhyay 2011; Bhattacharyya et al. 2020a, 2022) but an analytical model is always essential to understand the physical behavior of the device and the physics behind this; which facilitates further, the analysis of the device performances at circuit level as and when implemented. Several analytical approaches have been explored to date to model the TFET structure (Dutta and Sarkar 2019; Kumar et al. 2018; Mallik and Chattopadhyay 2011; Bhattacharyya et al. 2020a,b; Mojumder and Roy 2009; Safa et al. 2017; Saha and Sarkar 2019; Bagga and Sarkar 2015; Liang and Taur 2004), some of which are 1-D modeling, pseudo-2-D modeling, and 2-D compact modeling. 1-D modeling does not include the influence of drain bias to derive the drain current (Mojumder and Roy 2009; Safa et al. 2017). Pseudo-2-D modeling, based on Young's parabolic approximation of surface potential model (Saha and Sarkar 2019; Bagga and Sarkar 2015; Bhattacharyya et al. 2020b), suffers inaccuracy in predicting drain current in many cases as it considers a constant electric field distribution over both directions of the channel region (lateral and longitudinal). Therefore compact 2-D modeling has come into the picture (Liang and Taur 2004; Liu et al. 2012), which is a very useful mathematical tool to derive electrical characteristics of semiconductor devices, like MOSFETs and TFETs.

Here, in this chapter, a 2-D compact modeling approach for an asymmetric gate oxide dual metal double gate heterojunction TFET (A-DMDG-HTFET) device has been proposed. Variable separation method has been adopted in this work for modeling the asymmetric dual metal heterojunction TFET (A-DMDG-HTFET) and a comparative performance study is carried out with a single metal double gate heterojunction TFET (SMDG-HTFET) structure (Cui et al. 2013; Lee and Choi 2011) and therefore results are documented. Several design parameters and semiconductor materials have been varied and the corresponding changes in outputs are compared with SILVACO TCAD simulation results to validate the proposed analytical model. The organization of this article is as follows. In Section 14.2, the proposed structure of the TFET device and the physical parameters are discussed in detail and tabulated. The changes in design parameters are also mentioned in the given table. The analytical model is thoroughly discussed in Section 14.3, followed by results and discussions in Section 14.4. In this section, the findings are demonstrated, analyzed, and explained thoroughly with the help of device physics. Finally, the conclusion is made in Section 14.5, followed by an "Appendix", in which detailed calculations of some characteristic parameters are furnished.

14.2 DESCRIPTION OF THE DEVICE PARAMETERS

Figure 14.1 depicts the cross-sectional view of the asymmetric gate oxide dual metal double gate heterojunction TFET (A-DMDG-HTFET) with the gate length and the body thickness are considered as 60 and 10 nm, respectively. The source or drain length is taken as 40 nm each. Asymmetric oxide thickness has been assumed here

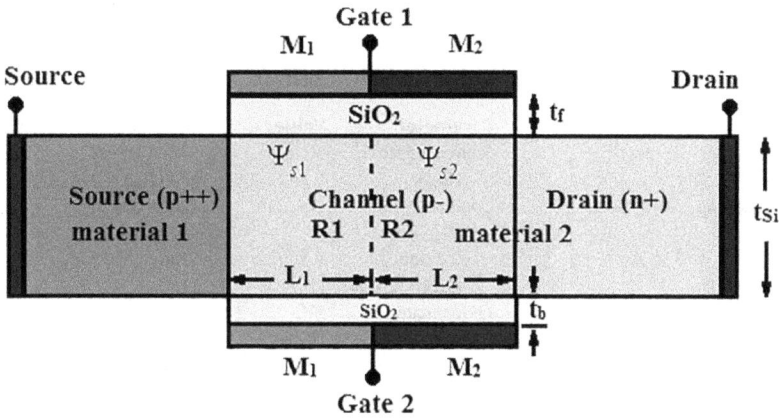

Figure 14.1 The 2D cross-sectional view of asymmetric gate oxide dual metal double gate heterojunction TFET (A-DMDG-HTFET).

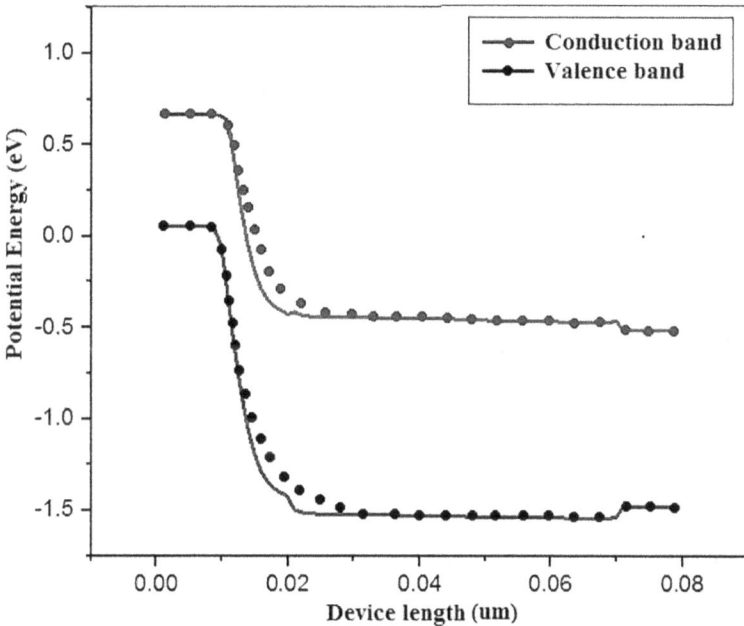

Figure 14.2 Comparison of band diagrams of A-DMDG-HTFET with A-SMDG-HTFET at the ON state of the devices.

for desired and efficient electrostatic controls over the channel region. For tunneling and auxiliary gates, two gate metals with work functions of 4.2 and 4.6 eV have been considered here respectively. The other dimensions of the device are as follows. Hence, L_g = gate length; L_i is the length of the ith region. According to Figure 14.1,

Table 14.1

Device Parameters Used in Model and Simulation

Parameter	Value	Parameter	Value	Parameter	Value
Channel length (L_g)	60 nm	Source length (L_{src})	30 nm	Drain length (L_{dr})	30 nm
Channel doping (N_{ch})	$10^{16}/cm^3$	Source doping (N_{src})	$10^{20}/cm^3$	Drain doping (N_{dr})	$5 \times 10^{18}/cm^3$
Front oxide thickness (t_f)	2 nm	Back oxide thickness (t_b)	1.5/2/2.5 nm	Body thickness (t_{si})	10 nm
Heterojunction systems	GaSb-InAs Ge-Si GaAsSb-InGaAs	Oxide materials	SiO_2 HfO_2 Si_3N_4	Φ_{M_1} Φ_{M_2} Φ_M	4.2 eV 4.6 eV 4.4 eV

$L_1 + L_2 = L_g \cdot N_{ch}$, N_{src}, and N_{dr} are the channel, source, and drain doping respectively. The thickness of the front gate oxide and back-gate oxide layers are t_f and t_b. The design parameters of the proposed device, used for analytical modeling and simulation are provided in Table 14.1. The said table also includes the changes in design parameters for which the experimental studies are carried out in detail.

The energy band diagram is shown in Figure 14.2. The dotted lines represent the energy bands of the SMDG structure and the solid lines represent the same for the DMDG structure. The diagram indicates the impacts of the work function engineering over the band bending of the structures at their ON state, which finally makes a huge difference in the drive currents of the respective devices.

14.3   MODEL DERIVATION

14.3.1   SURFACE POTENTIAL MODELING

The junction depletion region has not been considered in model derivation due to high and abrupt source/drain doping (Sarkhel et al. 2016). The influence of the mobile charges could be neglected for the small value of V_{GS}. Hence Poisson's equation at the channel region with depletion charges is written as follows,

$$\frac{\partial^2 \Psi(x,y)}{\partial x^2} + \frac{\partial^2 \Psi(x,y)}{\partial y^2} = \frac{qN_{ch}}{\varepsilon_{si}} \tag{14.1}$$

Where $\Psi(x, y)$ is the electrostatic potential. Using the superposition theorem of electrostatic potential, the $\Psi(x, y)$ of Region i ($i = 1, 2$) are written as follows (Mamidala et al. 2016).

$$\Psi_{si}(x,y) = v_i(y) + u_{Li}(x,y) + u_{Ri}(x,y), \quad i = 1,2. \tag{14.2}$$

Here $i = 1, 2$ represent Reg1 and Reg2 respectively, where $v_i(y)$ are the solutions to 1D Poisson's Equation and $u_{Li}(x, y)$ and $u_{Ri}(x, y)$ are the solutions to 2D Laplace

Equations for corresponding regions. The 1D Poisson's equation in channel & gate oxide is given as follows,

$$\frac{\partial^2 v_i(y)}{\partial y^2} = \frac{qN_{ch}}{\varepsilon_{si}} \text{ in Silicon body } \& \frac{\partial^2 v_i(y)}{\partial y^2} = 0 \text{ in Gate Oxide.} \tag{14.3}$$

Solutions of the set of equation (14.3) are,

$$v_i(y) = \frac{qN_{ch}}{\varepsilon_{si}}y^2 + \alpha_{1i}y + \beta_{1i} \text{ in Silicon body } \& \; v_i(y) = \alpha_{2i}y + \beta_{2i} \text{ in Gate Oxide.}$$

The constants α_{1i}, β_{1i}, α_{2i} and β_{2i} are determined by suitable boundary conditions given as, $v_i(-t_{ox}) = V_{GS,i}$, $v_i(t_{si}/2) = 0$ and the continuity of $v_i(y)$ and $dv_i(y)/dy$ at the gate oxide and silicon body interface. Figure 14.3 shows the double gate symmetric TFET splits into two single gate TFET structures with boundary conditions for calculation purposes superposition of electrostatic potentials of these two substructures resulted in the potential solution of the 1D Poisson's equations. The concept with boundary conditions is illustrated in Figure 14.3. The derived constants are given below for Reg1 and Reg2 respectively (Figure 14.4),

$$\begin{aligned}
\alpha_{11} &= \left(2\varepsilon_f V_{gs1} + \tfrac{qN_{ch}\varepsilon_f t_{si}^2}{4\varepsilon_{si}}\right) / \left(2\varepsilon_{si}t_f - \varepsilon_f t_{si}\right), \\
\beta_{11} &= \left(\tfrac{qN_{ch}t_{si}^2}{4\varepsilon_{si}} - t_{si}\varepsilon_f V_{gs1}\right) / \left(\varepsilon_{si}t_f - \varepsilon_f t_{si}/2\right) \\
\alpha_{12} &= \left(2\varepsilon_f V_{gs2} + \tfrac{qN_{ch}\varepsilon_b t_{si}^2}{4\varepsilon_{si}}\right) / \left(2\varepsilon_{si}t_b - \varepsilon_b t_{si}\right), \\
\beta_{12} &= \left(\tfrac{qN_{ch}t_{si}^2}{4\varepsilon_{si}} - t_{si}\varepsilon_b V_{gs2}\right) / \left(\varepsilon_{si}t_b - \varepsilon_b t_{si}/2\right)
\end{aligned} \tag{14.4}$$

The general solutions of 2D Laplace equations are given below (Gholizadeh and Hosseini 2014),

$$\begin{aligned}
u_{L1}(x,y) &= \sum_{i=1}^{n} c_{n1} \frac{\sinh((l_1-x)/\lambda_n)}{\sinh(l_1/\lambda_n)} \sin\left((y+A)/\lambda_n\right) \\
u_{R1}(x,y) &= \sum_{i=1}^{n} d_{n1} \frac{\sinh(x/\lambda_n)}{\sinh(l_1/\lambda_n)} \sin\left((y+A)/\lambda_n\right), \quad (0 \le x \le l_1), \text{Reg1}
\end{aligned} \tag{14.5}$$

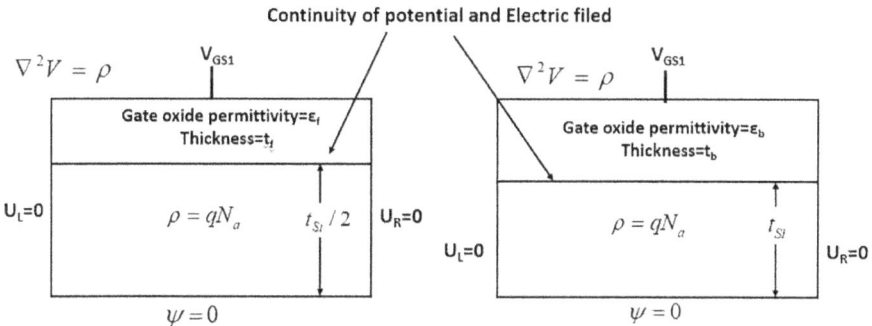

Figure 14.3 Double gate asymmetric TFET splits into two single gate TFET structures with boundary conditions for calculation purposes.

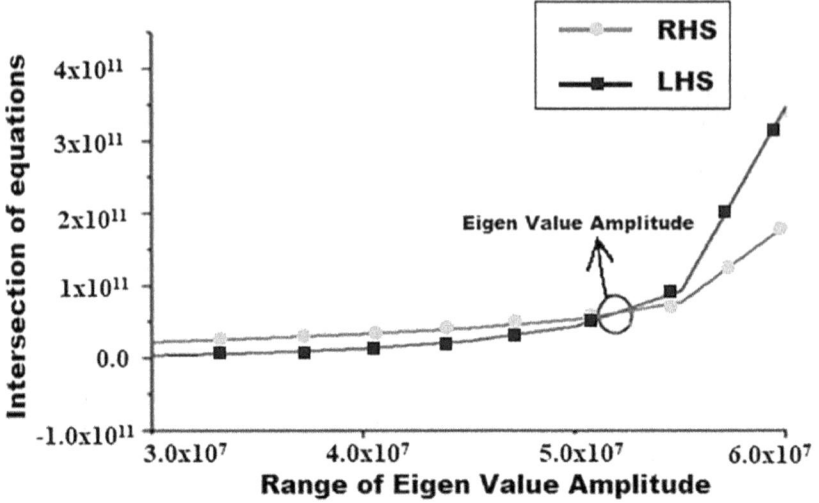

Figure 14.4 Solution of the first-order Eigenvalue.

$$u_{L2}(x,y) = \sum_{i=1}^{n} c_{n2} \frac{\sinh((l_1+l_2-x)/\lambda_n)}{\sinh(l_2/\lambda_n)} \sin((y+A)/\lambda_n)$$

$$u_{R2}(x,y) = \sum_{i=1}^{n} d_{n2} \frac{\sinh((x-l_1)/\lambda_n)}{\sinh(l_2/\lambda_n)} \sin((y+A)/\lambda_n), \quad (l_1 \le x \le l_1 + l_2 (= l_g)), \text{ Reg2}$$

$$(14.6)$$

Where $A = \pm n\pi/2\lambda_1$ considering first-order Eigenvalue. The constants c_{n1}, d_{n1}, c_{n2}, and d_{n2} are derived by using suitable boundary conditions given below. The expressions of the constants are given in Appendix.

$$u_{L1}(x=0,y) = V_s - V(y) \tag{14.7}$$

$$u_{R2}(x=l_2,y) = V_d - V(y) \tag{14.8}$$

$$\Psi_{s1}(x=l_1,y) = \Psi_{s2}(x=0,y) \tag{14.9}$$

$$\frac{d\Psi_{s1}(x,y)}{dx}\Big|_{x=l_1,y} = \frac{d\Psi_{s2}(x,y)}{dx}\Big|_{x=0,y} \tag{14.10}$$

Where $V_s = V_{bis}$ and $V_d = V_{bid} + V_{DS}$; V_{bis} and V_{bid} are the source and drain sided built-in potential

Hence, $V_{bis} = \frac{\chi_1-\chi_2}{2} + \frac{E_{g1}-E_{g2}}{2} + \frac{kT}{q}\ln\left(\frac{N_{ch}}{n_i}\right)$ considering the heterojunction and $V_{bid} = \frac{kT}{q}\ln\left(\frac{N_{ch}N_{dr}}{n_i^2}\right)$.

χ_1, χ_2 are the electron affinities; E_{g1}, E_{g2} are the energy gaps of the source and channel material respectively.

14.3.2   EIGENFUNCTION & EIGENVALUE

For the given asymmetric structure (Figure 14.1), the equation of Eigen function (Frank et al. 1998) is given by,

$$\frac{\varepsilon_{si}}{\varepsilon_f \varepsilon_b} \tan\left(\frac{t_f}{\lambda_n}\right) \tan\left(\frac{t_{si}}{\lambda_n}\right) \tan\left(\frac{t_b}{\lambda_n}\right) = \frac{1}{\varepsilon_f} \tan\left(\frac{t_f}{\lambda_n}\right) + \frac{1}{\varepsilon_{si}} \tan\left(\frac{t_{si}}{\lambda_n}\right) + \frac{1}{\varepsilon_b} \tan\left(\frac{t_b}{\lambda_n}\right),$$

$$n = \text{odd integers} \tag{14.11}$$

It is reported (that λ_n is the scaling length of the device and the maximum limit of scaling length (λ_n) could be $(t_{si} + t_f + t_b)$ (Das et al. 2022; Wu et al. 2015). Hence the expressions of the odd number of Eigenvalues are $\lambda_n = (t_{si} + t_f + t_b)$, $(t_{si} + t_f + t_b)/3$, $(t_{si} + t_f + t_b)/5$, and so on. In this work the equation (14.11) is graphically solved, given in Figure 14.5, for λ_1 (first-order Eigen value). Therefore the higher degree Eigenvalue e.g. λ_3, λ_5, λ_7 can be easily derived in a similar way. In our mathematical model, the first-order Eigenvalue (λ_1) is sufficient to calculate the surface potential due to the existence of "Sine Hyperbolic" function in u_L and u_R components of electrostatic potential and it has been also validated in results and discussions section. Figure 14.5 depicts the characteristics of the 'Sine Hyperbolic' function having an argument of λ_n value which shows that the function decays very fast for the higher-order Eigenvalues. The amplitudes of the hyperbolic function for higher-order lambda values are several decades lesser than that obtained from the first-order *lambda*, as shown in the inset of Figure 14.3.

Therefore, the expressions of the surface potential for A-DMDG-HTFET for Reg1 and Reg2 are given as follows,

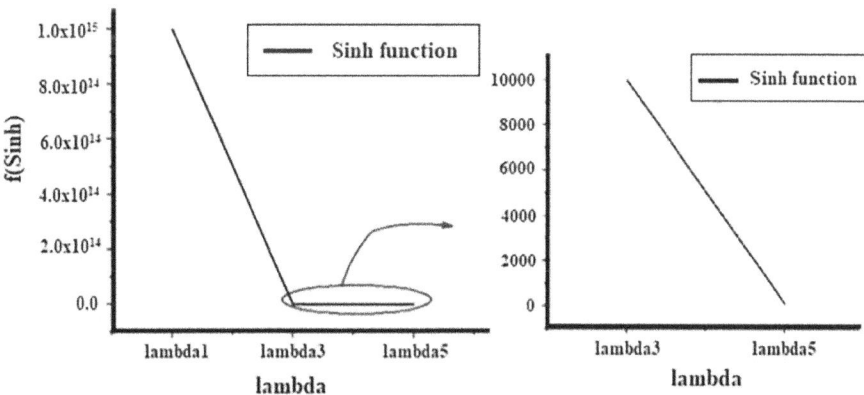

Figure 14.5   Steep variation of the "Sinh" functions corresponding to u_L and u_R expressions for odd no of Eigenvalues.

$$\Psi_{s1}(x,y) = v_1(y) + \sum_{i=1}^{n} c_{n1} \frac{\sinh((l_1-x)/\lambda_n)}{\sinh(l_1/\lambda_n)} \sin((y+A)/\lambda_n)$$

$$+ \sum_{i=1}^{n} d_{n1} \frac{\sinh(x/\lambda_n)}{\sinh(l_1/\lambda_n)} \sin((y+A)/\lambda_n)$$

$$\Psi_{s2}(x,y) = v_2(y) + \sum_{i=1}^{n} c_{n2} \frac{\sinh((l_1+l_2-x)/\lambda_n)}{\sinh(l_2/\lambda_n)} \sin((y+A)/\lambda_n)$$

$$+ \sum_{i=1}^{n} d_{n2} \frac{\sinh((x-l_1)/\lambda_n)}{\sinh(l_2/\lambda_n)} \sin((y+A)/\lambda_n)$$

(14.12)

14.3.3 ELECTRIC FIELD MODELING

The expression of lateral Electric Fields for Reg1 and Reg2 are derived in equations (14.13) and (14.14) respectively.

$$E_{x1}(x,y)|_{\text{Reg1}} = -\partial\Psi_{s1}(x,y)/\partial x$$

$$= -\left[\sum_{i=1}^{\infty} c_{n1} \frac{\cosh((l_1-x)/\lambda_1)}{\lambda_n \sinh(l_1/\lambda_n)} \sin((y+A)/\lambda_n)\right.$$

$$\left. + \sum_{i=1}^{\infty} d_{n1} \frac{\cosh(x/\lambda_n)}{\lambda_n \sinh(l_1/\lambda_n)} \sin((y+A)/\lambda_n)\right], \quad n = 1,3,5\ldots \quad (14.13)$$

$$E_{x2}(x,y)|_{\text{Reg2}} = -\partial\Psi_{s2}(x,y)/\partial x$$

$$= -\left[\sum_{i=1}^{\infty} c_{n2} \frac{\cosh((l_1+l_2-x)/\lambda_1)}{\lambda_n \sinh(l_2/\lambda_n)} \sin((y+A)/\lambda_n)\right.$$

$$\left. + \sum_{i=1}^{\infty} d_{n2} \frac{\cosh(x/\lambda_n)}{\lambda_n \sinh(l_2/\lambda_n)} \sin((y+A)/\lambda_n)\right], \quad n = 1,3,5\ldots \quad (14.14)$$

14.3.4 DRAIN CURRENT MODELING

As mentioned above, (14.13) and (14.14) give the equations of the lateral electric field under the tunneling gate and auxiliary gate respectively and it is sufficient to consider only this field to compute the drain current (Kumar et al. 2016). The longitudinal electric field has no significant effect on drain current and is hence neglected. Here both equations contain two components. The first component of equation (14.13) gives the field expression close to the source channel junction and therefore is responsible for the forward tunneling phenomenon. The second component of equation (14.14), the field expression at the drain-channel junction, governs the reverse tunneling phenomenon.

$$E_{x1}(x,y) = -\sum_{i=1}^{n} c_{n1} \frac{\cosh((l_1 - x)/\lambda_n)}{\lambda_n \sinh(l_1/\lambda_n)} \sin((y+A)/\lambda_n) = -\sum_{i=1}^{n} K_n \sin((y+A)/\lambda_n)$$

$$(14.15)$$

[Where, $K_n = c_{n1} \frac{\cosh((l_1-x)/\lambda_n)}{\lambda_n \sinh(l_1/\lambda_n)}$, $z = y + A$ (say) and $n = 1, 3, 5, \ldots$]

$$
\begin{aligned}
= -\Big[&K_1 \left\{ (z/\lambda_1) - \frac{(z/\lambda_1)^3}{3!} + \frac{(z/\lambda_1)^5}{5!} - \cdots \right\} \\
&+ K_3 \left\{ (z/\lambda_3) - \frac{(z/\lambda_3)^3}{3!} + \frac{(z/\lambda_3)^5}{5!} - \cdots \right\} \\
&+ K_5 \left\{ (z/\lambda_5) - \frac{(z/\lambda_5)^3}{3!} + \frac{(z/\lambda_5)^5}{5!} - \cdots \right\} + \cdots \Big]
\end{aligned}
$$

$$(14.16)$$

Substituting the condition, $\lambda_1 = n\lambda_n \Rightarrow \lambda_3 = \lambda_1/3, \lambda_5 = \lambda_1/5, \ldots$ in equation (14.15) and further simplification, we got,

$$= -N_1 K_1 \sinh(z/\lambda_1) = -N_1 K_1 \sinh\left(\frac{y+A}{\lambda_1}\right)) - N_1 K_1 \cosh(y/\lambda_1) \qquad (14.17)$$

where N_1 is the multiplication factor depending upon the number of series terms (Gholizadeh and Hosseini 2014). Similarly, from equation (14.14) it is obtained that,

$$E_{x2}(x,y) = -\sum_{i=1}^{n} d_{n2} \frac{\cosh(x/\lambda_n)}{\lambda_n \sinh(l_1/\lambda_n)} \sin((y+A)/\lambda_n) = -\sum_{i=1}^{n} M_n \sin((y+A)/\lambda_n)$$

$$(14.18)$$

[where, $M_n = d_{n2} \frac{\cosh(x/\lambda_n)}{\lambda_n \sinh(l_1/\lambda_n)}$ and $n = 1, 3, 5, \ldots$]. By further simplification, as before, we got,

$$E_{x2}(x,y) = -N_2 M_1 \cosh(y/\lambda_1), \qquad (14.19)$$

where N_2 is the multiplication factor.

Now, the carrier generation rate in TFET device is given by Kane's model (Kane 1960) as follows,

$$G_{BTBT} = A_K E^{D1} \exp(-B_K/E) \qquad (14.20)$$

where A_K & B_K are Kane's parameters. $D_1 = 2$ or 2.5 for direct or indirect band gap materials, respectively. So, the drain current is obtained as,

$$I_{DS} = q \int_v G_{BTBT} dv = q \int_v A_K E \cdot E_{avg}^{D1-1} \exp(-B_k/E_{avg}) \qquad (14.21)$$

where E and E_{Avg} are the local electric field and average electric field respectively.

For heterojunction (source to channel junction), the local electric field, E, is given by equation (14.15) and the average Electric Field (Gholizadeh and Hosseini 2014) is given by, $E_{Avg} = E_{g,eff}/(ql_{s,path}) = E_{g2} - \Delta E_v$, where

$$\Delta E_v = (\chi_2 - \chi_1)/2 + (E_{g2} - E_{g1})/2 \qquad (14.22)$$

$l_{s,path}$ is the tunnelling path, given by $l_{s,path} = l_{s2} - l_{s1}$ $\qquad (14.23)$

l_{s2} and l_{s1} are the lengths along the channel where potential differences are $(E_{g,eff} + \Delta\phi)/q$ and $E_{g,eff}/q$ respectively and $\Delta\phi = E_{VS} - E_{CC}$. Therefore,

$l_{s,\text{path}}$, l_{s2} and l_{s1} are calculated from the following boundary conditions, $\Psi_{s1}(l_{s1},0) - \phi_0 = E_{g,\text{eff}}/q$, $\Psi_{s1}(l_{s2},0) - \phi_0 = (E_{g,\text{eff}} + \Delta\phi)/q$ and $\phi_0 = \Psi_{s1}(0,0)$. The expression for l_{s2} and l_{s1} are shown in appendix.

Therefore, the drain current (source to channel) equation is written from equations (14.21)–(14.23), as follows,

$$
\begin{aligned}
I_{S \to C} &= q \int_{-t_{\text{si}}/2}^{t_{\text{si}}/2} \int_{l_{s1}}^{l_{s2}} A_K N_1 K_1 \cosh\left(\frac{y}{\lambda_1}\right) \left(\frac{E_{g,\text{eff}}}{qx}\right)^{D1-1} \cdot \exp\left(-\frac{B_K q}{E_{g,\text{eff}}} x\right) dx dy \\
&= \frac{q A_K N_1 E_{g,\text{eff}}^{D1-1}}{q^{D1-1}} \int_{-t_{\text{si}}/2}^{t_{\text{si}}/2} \int_{l_{s1}}^{l_{s2}} c_{11} \lambda_1 \frac{\cosh\left((l_1 - x)/\lambda_1\right)}{\sinh\left(l_1/\lambda_1\right)} \cdot \cosh\left(\frac{y}{\lambda_1}\right) \cdot \\
&\quad \exp\left(-\frac{B_K q}{E_{g,\text{eff}}} x\right) dx dy \\
&= \frac{q^{2-D1} A_K N_1 E_{g,\text{eff}}^{D1-1} c_{11} \lambda_1 \exp\left(l_1/\lambda_1\right)}{\sinh\left(l_1/\lambda_1\right) \left(1/\lambda_1 + B_K q/E_{g,\text{eff}}\right)} \cdot S_1\left(l_{s,\text{path}}\right) \cdot \left\{S_2\left(t_{\text{si}}/2\right) - S_2\left(-t_{\text{si}}/2\right)\right\}
\end{aligned}
$$

$$(14.24)$$

Similarly, for drain to channel junction (homo junction) the local electric field, E, is given by equation (14.18), and the average electric field is given by,

$$E_{\text{avg}} = E_g/\left(q l_{d,\text{path}}\right), \tag{14.25}$$

where $l_{d,\text{path}} = l_{d2} - l_{d1}$ $l_{d,\text{path}}$ is the tunneling path at the drain junction and the drain to channel current, $I_{D \to C}$, is calculated from equations (14.21), (14.23), and (14.25), following the same manner as $I_{S \to C}$ is calculated. Hence the total tunneling current i.e. source to drain current, I_{DS} (SILVACO Inc 2010) is given as,

$$I_{\text{DS}} = I_{S \to C} + I_{D \to C} \tag{14.26}$$

Equation (14.26) is the final expression for drain current. The detailed calculation of drain current, I_{DS}, and the expressions of $S_1(x)$ and $S_2(y)$ are shown in Appendix.

14.4   RESULTS AND DISCUSSION

Here, we validated our model data with the simulated ones obtained through the SILVACO ATLAS device simulator (Sarkar et al. 2016). Depending upon the heterostructure materials the band-to-band tunneling mechanism may be a direct or indirect phenomenon and hence Kane's parameters are modified accordingly. All the results in this section are generated by considering GaAsSb-InGaAs heterostructure TFET with the dimensions of 2 and 1.5 nm as the front and back oxide thickness, until and unless stated otherwise. Parameters like gate oxide thickness, gate dielectrics, and heterostructure materials are varied in both modeling and simulation and results are compared to evaluate the merit of the proposed mathematical model.

First, the surface potential profile of the proposed device is computed through both the infinite series method (our proposed model) and Young's parabolic approximation method. Figure 14.6 represents the nature of surface potential computed

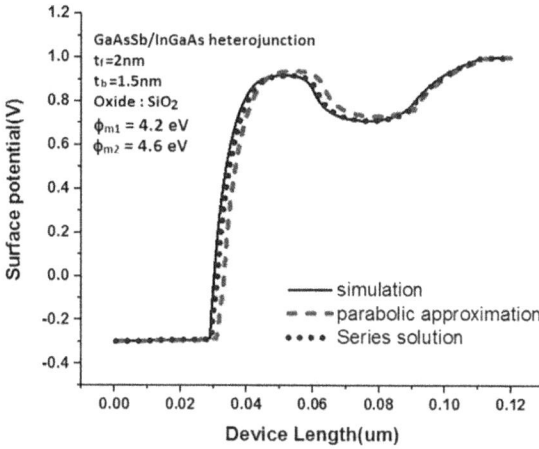

Figure 14.6 Surface potential vs. channel length. Hence, the surface potential derived by variable separation method compared with results of the parabolic approximation method and TCAD simulation.

through the aforementioned methods along with the ATLAS simulation for comparative study. It is clear from the above Figure 14.6 that the proposed model of surface potential fits better with the simulation result rather than that obtained from the parabolic approximated model. The proposed surface potential model developed by the series method can be effectively tuned by incorporating a suitable number of series terms in the calculation in order to get the result matched with the simulation data. Therefore, the infinite series model is considered in this literature for accurate computation of device characteristics. In Figure 14.7 the surface potential profiles of

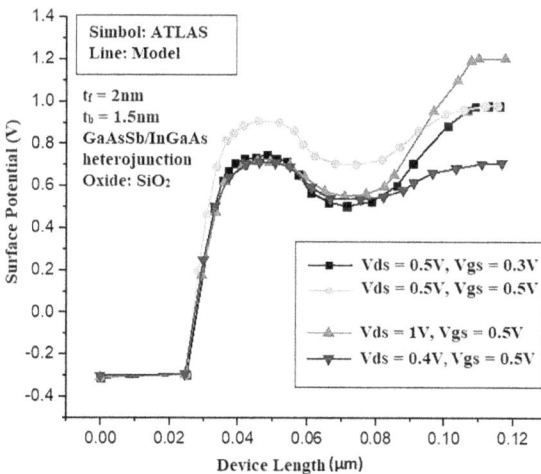

Figure 14.7 Surface potential vs. channel length. Hence, the surface potential for various bias conditions is given.

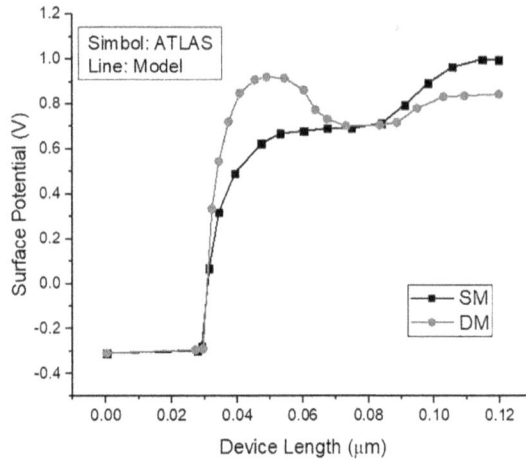

Figure 14.8 Surface potential profile and ($M_1 = 4.2$ eV and $M_2 = 4.6$ eV) with its equivalent SMDG structure ($M = 4.4$ eV) for $V_{GS} = V_{DS} = 0.5$ V.

the proposed device are depicted for the various V_{DS}, while $V_{GS} = 0.5$ V and for various V_{GS}, keeping drain voltage constant ($V_{DS} = 0.5$ V). The said Figure 14.7 shows a good agreement of modeled results with simulation data.

Figure 14.8 presents the surface potentials profile of a Dual Metal Double Gate (DMDG) TFET structure with metal work functions $\Phi_{M_1} = 4.2\,\text{eV}$ and $\Phi_{M_2} = 4.6\,\text{eV}$ respectively and equivalent single metal double gate (SMDG) TFET structure ($\Phi_M = 4.4\,\text{eV}$, average value). The exponential rise of surface potential and its unequal distribution along the device length proves better gate controllability of the DMDG structure over the channel region than that of its counterpart, the SMDG structure. The work function engineering plays the key role here to improve the ON current and reduce the ambipolar conduction.

Figure 14.9 shows the profiles of electric field distributions of the DMDG and SMDG TFET structures. Both, lateral and total electric field profiles are plotted in the same Figure 14.9 for the bias condition of $V_{GS} = V_{DS} = 0.5$ V. The Figure 14.9 clearly depicts the superiority of the field intensity of the DMDG structure in terms of the higher peak at the junction of source and channel and the profile has a decaying nature at the drain and channel junction, whereas the SMDG structure has a slight peak at the said junction, which could degrade its current characteristics.

Figure 14.10 represents the profile of the total electric field along the channel length for varying gate oxide thickness. Here the oxide thickness for the front gate oxide is kept constant at 2 nm and the three different values taken for back oxide thickness are 1.5, 2, and 2.5 nm respectively. The peak of the electric field at the tunneling junction (source channel) is shown in the inset. Lesser the value of the oxide thickness, the more the value of the field intensity and the more it is spread over the channel region. The calculations regarding asymmetric gate oxide structure are incorporated in the proposed analytical model, which best suits the fine-tuning of

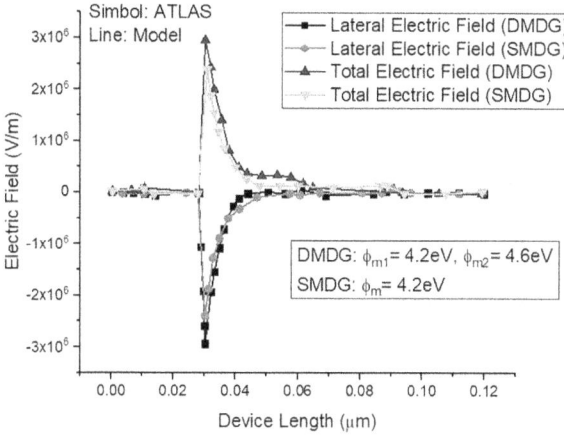

Figure 14.9 Electric field profile (total and lateral) of DMDG structure.

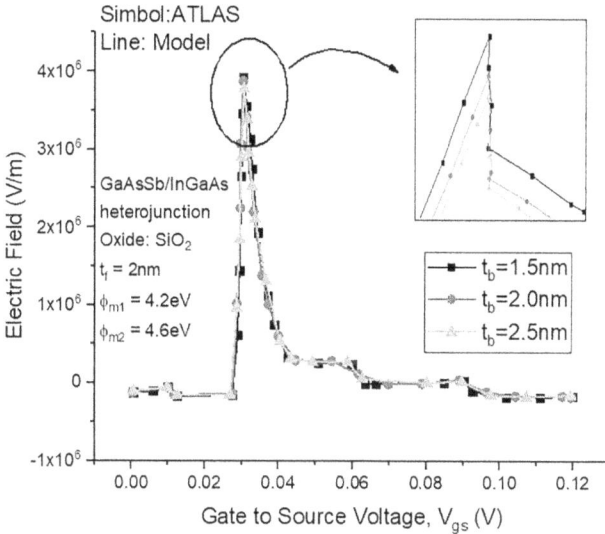

Figure 14.10 Total electric field profile with varying back-gate oxide thickness.

the electric field profile of the device by varying the one-sided oxide thickness, while keeping the other side thickness constant.

Figure 14.11 represents the role of different gate oxide materials over the electric field profiles of the proposed device. Three oxide materials, HfO_2 ($\varepsilon = 25\ \varepsilon_0$), Si_3N_4 ($\varepsilon = 7.25\ \varepsilon_0$), and SiO_2 ($\varepsilon = 3.9\ \varepsilon_0$) are used to study the field profiles of the device. The peak of the field intensity is maximum and widely spread over the channel region for Hafnium dioxide, which is followed by Silicon nitride and Silicon dioxide respectively as shown in Figure 14.11.

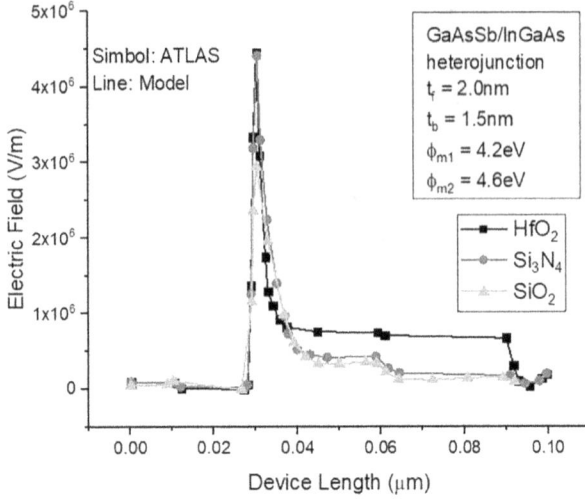

Figure 14.11 Total electric field profile with varying gate oxide materials.

For a range of low values of gate voltage (V_{GS} below threshold voltage), the substrate of the TFET device is occupied with only depletion charges. As the gate voltage increases beyond the threshold voltage, inversion charges start accumulating in the channel region. In our proposed analytical model, the 2-D Poisson's equation is solved for the surface potential, considering only the depletion charges in the substrate of the device. Therefore, we got quite accurate results in computing the I_D-V_{GS} characteristics in the subthreshold domain whereas in the super-threshold region, the calculated drain is over-estimated according to the data obtained from the simulation results.

Figure 14.12 shows a comparative study of the drain current of DMDG-A-HTFET and its counterpart i.e. SMDG-A-HTFET. The only difference between the structures is that the work function engineering is deployed for the DMDG structure, which contributes to achieving higher ON current, lesser subthreshold slope, and reduced ambipolarity for the said structure. In the subthreshold region, the analytical model gets perfectly matched with the simulation data whereas, in the super-threshold region the model couldn't capture the drain current accurately for the device structure. This justifies the fact already explained in the previous paragraph. At the same time, it is worth mentioning that we can neglect this computation error due to its insignificant impact on circuit performance as the device usually operates nearly the threshold voltage while incorporated in VLSI circuits.

Figure 14.12 represents the transfer characteristics of the proposed device structure for three different heterostructure material systems given by Germanium-Silicon (Ge-Si), Galium Antimonide-Indium Arsenide (GaSb-InAs), and Galium Arsenide Antimonide-Indium gallium Arsenide (GaAsSb-InGaAs). The condition for heterojunction at the tunneling region has been incorporated in the analytical model and

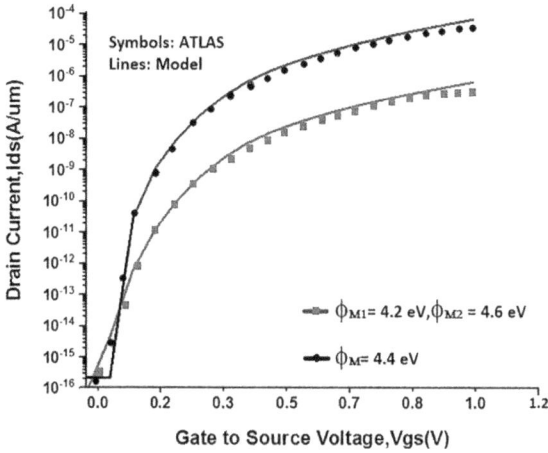

Figure 14.12 I_D Vs. V_{GS} characteristics for DMDG-HTFET and equivalent SMDG-HTFET structures.

therefore the model is verified by using the aforementioned material systems. The proposed model utilizes energy band gap, electron affinity, and hence the band off-set at the junction of source and channel i.e. heterojunction in computing the drain current. The Kane's parameters are also changed for the respective channel material for the device. Figure 14.13 shows the improved drain current performance of the respective structures in terms of high ON current and better ON/OFF current ratio. The drain current model, matched with the simulation data, justifies the accuracy of the proposed model.

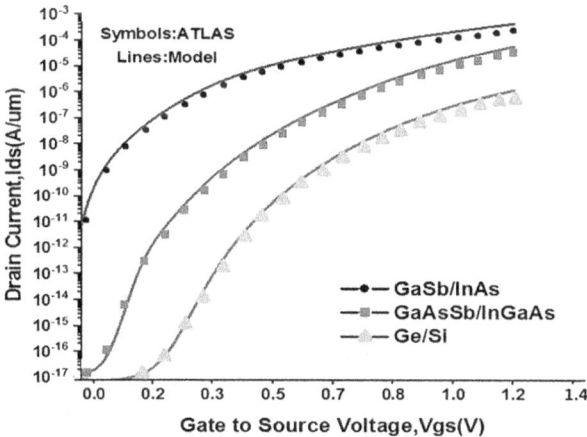

Figure 14.13 I_D Vs. V_{GS} characteristics for DMDG-HTFET and equivalent SMDG-HTFET structures and (b) A-DMDG-HTFET structure with different heterostructure material systems at $V_{DS} = 0.5$ V.

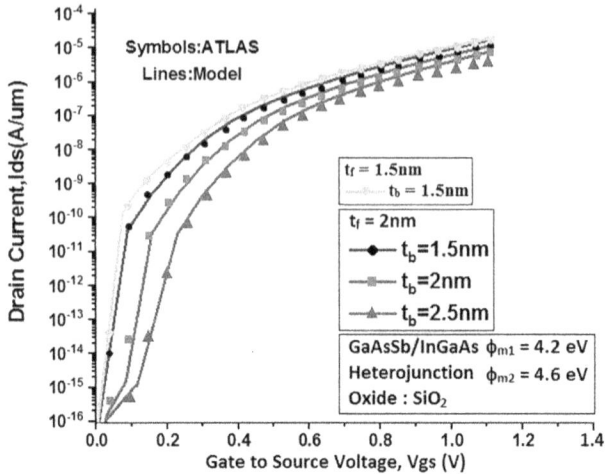

Figure 14.14 I_D vs. V_{GS} characteristics for the different back-gate oxide thickness.

The condition for asymmetric gate oxide thickness is incorporated in the proposed analytical model. The oxide capacitance differs for different thicknesses. This impacts the surface potential model and changes the drain current accordingly. The ON current variations of the structures are not very significant for thickness variations but it highly impacts the subthreshold swing and the leakage current of the devices. Therefore the asymmetric gate oxide thickness can minutely tune the trade-off between the ON current and subthreshold swing and leakage current of the device.

Here Figure 14.14 represents the I_D-V_{GS} characteristics of the proposed device for different back-gate oxide thicknesses. A set of curves are plotted with a variable back oxide thickness of 1.5, 2, and 2.5 nm for a constant value of 2 nm of front oxide thickness. Along with these, the drain current curve is plotted with the same value of front and back oxide thickness of 1.5 nm each. The subthreshold swings and ON currents of different structures change accordingly with the change of thicknesses of oxides, depicted in the said Figure 14.14.

To verify the accuracy of the proposed model, different oxide materials with different permittivity are also used. The higher dielectric material causes a higher capacitive effect over the substrate of the device in order to achieve exponential rise and a higher value of ON current. Figure 14.15 shows the comparison among the drain current curves obtained from the device structures having HfO_2, Si_3N_4, and SiO_2 respectively. The highest drive current and lowest subthreshold swing are achieved from the structure using HfO_2, followed by Si_3N_4 and SiO_2 as oxide materials. Our proposed model shows a good agreement with the simulation data as depicted in both the figures, mentioned above.

Figure 14.16 shows the transfer characteristics of the proposed device for negative gate bias. The electric field at the junction (reverse tunneling) of the drain and channel is computed from the surface potential model from which the generation rate

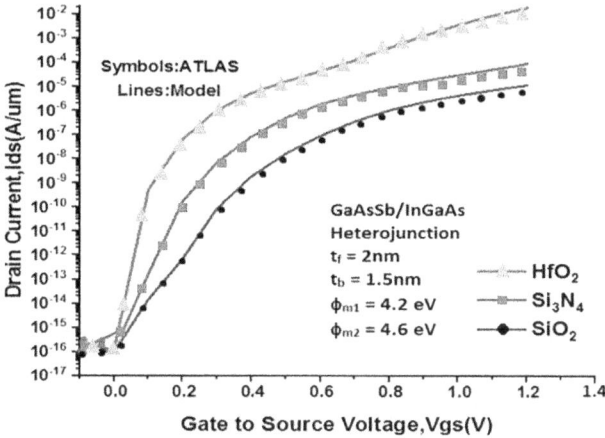

Figure 14.15 I_D vs. V_{GS} characteristics for different gate oxide materials at $V_{DS} = 0.5$ V.

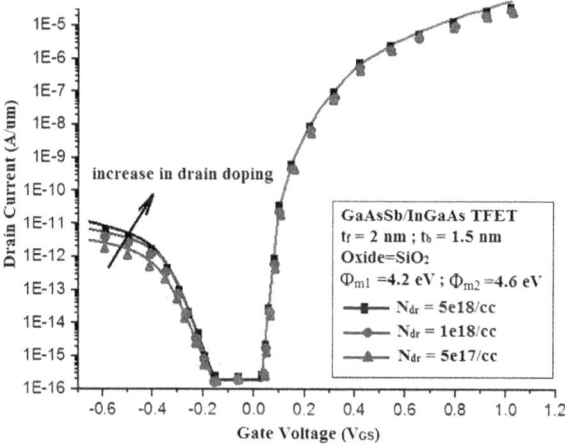

Figure 14.16 Transfer characteristics of the proposed device for negative gate voltages.

is computed at the said junction. Here the carrier generation mechanism is governed by the negative gate bias and ambipolar conduction takes place, which is of a similar nature as the drive current. The said Figure 14.16 shows the set of drain current curves for negative gate voltages for the doping concentration of 5×10^{18}/cc, 1×10^{18}/cc, and 5×10^{17}/cc respectively at the drain region. Figure 14.17 represents the variation of subthreshold swing with respect to the variation of drain current and the results find close agreement with the simulation data and justifies the reliability of the proposed analytical model.

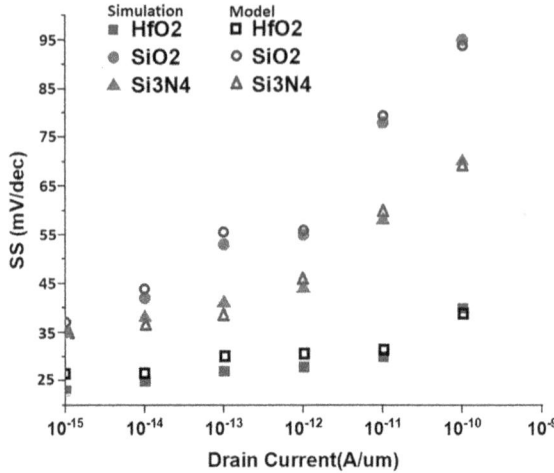

Figure 14.17 Transfer characteristics of the proposed device for subthreshold slope (SS) profile with varying I_{DS} for different gate oxide materials.

14.5 CONCLUSION

In this work, a closed-form solution of an analytical model for asymmetric gate DMDG HTFET has been developed using the separation of variables method. It has been explicitly narrated in the Eigenfunction and Eigenvalue section of this literature about how we deduce the closed-form expression for surface potential, Electric Field, and drain current by converging the corresponding series expressions. Asymmetric gate oxide and Dual Metal gate structure which uses work function engineering have created a significant impact on the device transfer characteristics. We performed best case analysis to obtain enhanced device performance in terms of a higher I_{ON}/I_{OFF} ratio and lesser subthreshold swing (SS). Simulation outputs are also in good agreement with the results of analytical modeling, justifying the accuracy and reliability of the proposed model.

APPENDIX

$$c_{n1} = -2\left\{ (V_s - \beta_1)\lambda_1 \cos\left(\frac{t_{si}}{2\lambda_1} - 1\right) - \left(\frac{qN_{ch}}{\varepsilon_{si}} - V_{gs1}\right)\left(t_{si}^2\lambda_1 + 2\lambda_1^2\right)\cos\left(\frac{t_{si}}{2\lambda_1} - 1\right)\right.$$
$$\left. -\alpha_{11}\left(\lambda_1^2 \sin\left(\frac{t_{si}}{\lambda_1}\right) - \frac{t_{si}}{\alpha_{11}}\right)\right\}\Big/ t_{si}$$

$$c_{n2} = \left\{\frac{c_{n1}}{\sin(l_1/\lambda_n)} - \frac{d_{n2}}{\sin(l_2/\lambda_n)} + \frac{v_{21}}{\sin y_1}\coth\left(\frac{l_1}{\lambda_n}\right)\right\}\Big/\left\{\coth\left(\frac{l_1}{\lambda_n}\right) + \coth\left(\frac{l_2}{\lambda_n}\right)\right\}$$

$$d_{n1} = \left[\frac{v_{21}(y)}{\sin(y_1)}\coth(l_1/\lambda_n) - \left\{\frac{c_{n1}}{\sin(l_1/\lambda)} - \frac{d_{n2}}{\sin(l_2/\lambda_n)}\right\}\right]\Big/\coth\left(\frac{l_1}{\lambda_n}\right) + \coth\left(\frac{l_2}{\lambda_n}\right)$$

$$d_{n2} = \frac{-2\left\{(V_d - \beta_{12})\lambda_1 \cos\left(\frac{t_{si}}{2\lambda_1}\right) + (v_d - \beta_{12})\lambda_1 - \left(\frac{qN_{ch}}{\varepsilon_{si}} - V_{gs1}\right)\right\}}{\left(t_{si}^2\lambda_1/4 + 2\cos\left(\frac{t_{si}}{2\lambda_1}\right)\lambda_1^3\right) - 2\lambda_1 - \alpha_{12}\sin\left(\frac{t_{si}}{2\lambda_1}\lambda_1^2 - \frac{t_{si}}{2}\lambda_1\right)} \Bigg/ t_{si}$$

$$l_{s1} = \lambda_1 \sin\left(\frac{l_1}{\lambda_1}\right)\left\{\frac{E_{g,\text{eff}}}{qc_{11}\sin(A/\lambda_1)} + 1\right\} + l_1 \,\&\, l_{s2}$$

$$= \lambda_1 \sin\left(\frac{l_1}{\lambda_1}\right)\left\{\frac{E_{g,\text{eff}} + \Delta\phi}{qc_{11}\sin(A/\lambda_1)} + 1\right\} + l_1$$

$$I_{S \to C} = q \int_{-t_{si}/2}^{t_{si}/2} \int_{l_{S1}}^{l_{S2}} AK(NK_1 \cosh\left(\frac{y}{\lambda_1}\right) \cdot \left(\frac{E_{g,\text{eff}}}{qx}\right)^{D_1-1}) \cdot \exp\left(\frac{-B_Kq}{E_{g,\text{eff}}}x\right) dxdy$$

$$I_{S \to C} = \frac{qAK(-N)\left(E_{g,\text{eff}}\right)^{D1-1}}{q^{D1-1}} \int_{-t_{si}/2}^{t_{si}/2} \int_{l_{S1}}^{l_{S2}} c_{11}\lambda_1 \cdot \frac{\cosh\left(\frac{(l_1-x)}{\lambda_1}\right)}{\sinh\left(\frac{l_1}{\lambda_1}\right)} \cdot$$

$$\cosh\left(\frac{y}{\lambda_1}\right)\exp\left(\frac{-B_Kq}{E_{g,\text{eff}}}x\right) dxdy$$

$$I_{S \to C} = \frac{qAK(-N)\left(E_{g,\text{eff}}\right)^{D1-1}c_{11}\lambda_1}{q^{D1-1}\sinh\left(\frac{l_1}{\lambda_1}\right)} \int_{-t_{si}/2}^{t_{si}/2}\int_{l_{S1}}^{l_{S2}} c_{11}\lambda_1 \cdot \exp\frac{(l_1-x)}{\lambda_1}\exp\left(\frac{y}{\lambda_1}\right)\frac{\exp\left(\frac{-B_Kq}{E_{g,\text{eff}}}x\right)}{x^{D1-1}}dxdy$$

$$= -\frac{qAK(N)\left(E_{g,\text{eff}}\right)^{D1-1}c_{11}\lambda_1 \exp\left(\frac{l_1}{\lambda_1}\right)}{q^{D1-1}\sinh\left(\frac{l_1}{\lambda_1}\right)}\int_{l_{S1}}^{l_{S2}}\frac{\exp\left(-x\left(\frac{1}{\lambda_1}+\frac{B_Kq}{E_{g,\text{eff}}}\right)\right)}{x^{D1-1}}dx \cdot \int_{-t_{si}/2}^{t_{si}/2}\exp\left(\frac{y}{\lambda_1}\right)dy$$

$$= -\frac{qAKN\left(E_{g,\text{eff}}\right)^{D1-1}c_{11}\lambda_1 \exp\left(\frac{l_1}{\lambda_1}\right)}{q^{D1-1}\sinh\left(\frac{l_1}{\lambda_1}\right)}\left(-\frac{1}{\lambda_1+\frac{B_Kq}{E_{g,\text{eff}}}}\right)[S_1(x)]_{l_{S1}}^{l_{S2}} \cdot \lambda_1 [S_2(y)]_{-t_{si}/2}^{t_{si}/2}$$

$$= -\frac{q^{2-D1}AKN\left(E_{g,\text{eff}}\right)^{D1-1}c_{11}\lambda_1 \exp\left(\frac{l_1}{\lambda_1}\right)}{\sinh\left(\frac{l_1}{\lambda_1}\right)\left(\frac{1}{\lambda_1}+\frac{B_Kq}{E_{g,\text{eff}}}\right)}S_1\left(l_{s,\text{path}}\right) \cdot \left(S_2\left(\frac{t_{si}}{2}\right) - S_2\left(\frac{-t_{si}}{2}\right)\right)$$

REFERENCES

Bagga, N., and Sarkar, S. K. (2015). An analytical model for tunnel barrier modulation in triple metal double gate TFET. *IEEE Transactions on Electron Devices*, 62(7):2136–2142.

Bhattacharyya, A., Chanda, M., and De, D. (2020a). GaAs0.5Sb0.5/In0.53Ga0.47As heterojunction dopingless charge plasma-based tunnel FET for analog/digital performance improvement. *Superlattices and Microstructures*, 142:106522, doi: 10.1016/j.spmi.2020.106522.

Bhattacharyya, A., Chanda, M., and De, D. (2020b). Analysis of noise-immune Dopingless heterojunction bio-TFET considering partial hybridization issue. *IEEE Transactions on Nanotechnology*, 19:769–777, doi: 10.1109/TNANO.2020. 3033966.

Bhattacharyya, A., De, D., and Chanda, M. (2022). Sensitivity measurement for bio-TFET considering repulsive steric effects with better accuracy. *IEEE Transactions on Nanotechnology*, 21:100–109.

Chanda, M., Mal, S., Mondal, A., and Sarkar, C. K. (2018). Design and analysis of a logic model for ultra-low power near threshold adiabatic computing. *IET Circuit, Device and System*, 12(4): 439–446.

Cui, N., Liu, L., Xie, Q., Tan, Z., Liang, R., Wang, J., and Xu, J. (2013). A two – dimensional analytical model for tunnel field effect transistor and its applications. *Japanese Journal of Applied Physics*, 52:044303.

Das, R., Chattopadhyay, A., Chanda, M. et al. (2022). Analytical modeling of sensitivity parameters influenced by practically feasible arrangement of bio-molecules in dielectric modulated FET biosensor. *Silicon*, doi: 10.1007/s12633-021-01617-z.

Dutta, R., and Sarkar, S. K. (2019). Analytical modelling and simulation-based optimization of broken gate TFET structure for low power applications. *IEEE Transactions on Electron Devices*, 66(8):3513–3520.

Frank, D. J., Taur, Y., and Wong, H. S. P. (1998). Generalized scale length for two-dimensional effects in MOSFET's. *IEEE Electron Device Letters*, 19(10): 385–387.

Gholizadeh, M., and Hosseini, S. E. (2014). A 2-D analytical model for double-gate tunnel FETs. *IEEE Transactions on Electron Devices*, 61(5):1494–1500.

Kane, E. O. (1960). Zener tunnelling in semiconductors. *Journal of Physics and Chemistry of Solids*, 12(2):181–188.

Kumar, S., Goel, E., Singh, K., Singh, B., Kumar, M., and Jit, S. (2016). A compact 2-D analytical model for electrical characteristics of double-gate tunnel field-effect transistors with a SiO_2/high-k stacked gate-oxide structure. *IEEE Transactions on Electron Devices*, 63(8):3291–3299.

Kumar, S., Singh, K., Chander, S., Goel, E., Singh, P. K., Baral, K., Singh, B., and Jit, S. (2018). 2-D analytical drain current model of double-gate heterojunction TFETs with a SiO_2/HfO_2 stacked gate-oxide structure. *IEEE Transactions on Electron Devices*, 65(1):331–338.

Lee, M. J., and Choi, W. Y. (2011). Analytical model of single-gate silicon-on-insulator (SOI) tunnelling field-effect transistors (TFETs). *Solid-State Electronics*, 63(1):110–114.

Liang, X., and Taur, Y. (2004). A 2-D analytical solution for SCEs in DG MOSFETs. *IEEE Transactions on Electron Devices*, 51(9):1385–1391.

Liu, L., Mohata, D., and Datta, S. (2012). Scaling length theory of double-gate interband tunnel field- effect transistors. *IEEE Transactions on Electron Devices*, 59(4):902–908.

Mallik, A., and Chattopadhyay, A. (2011). Drain-dependence of tunnel field-effect transistor characteristics: The role of the channel. *IEEE Transactions on Electron Devices*, 58(12):4250–4257.

Mamidala, J. K., Vishnoi, R., and Pandey, P. (2016). *Tunnel Field-Effect Transistors (TFET) Modelling and Simulation.* Wiley: Hoboken, NJ. ISBN: 978-1-119-24629-9.

Mojumder, N. N., and Roy, K. (2009). Band-to-band tunnelling ballistic nanowire FET: Circuit-compatible device modelling and design of ultra-low-power digital circuits and memories. *IEEE Transactions on Electron Devices*, 56(10): 2193–2201.

Safa, S., Noor, S. L., and Khan, M. Z. R. (2017). Physics-based generalized threshold voltage model of multiple material gate tunnelling FET structure. *IEEE Transactions on Electron Devices*, 64(4):1449.

Saha, P., and Sarkar, S. K. (2019). Drain current modeling of proposed dual material elliptical gate-all- around heterojunction TFET for enhanced device performance. *Superlattices and Microstructures,* 130:194–207, doi: 10.1016/j.spmi.2019.04.022, https://www.sciencedirect.com/journal/superlattices-and-microstructures/vol/130/suppl/C.

Sarkar, A., De, S., Chanda, M., and Sarkar, C. K. (2016). Low power VLSI design: Fundamentals. *De Gruyter Oldenbourg*, doi: 10.1515/9783110455298.

Sarkhel, S., Bagga, N., and Sarkar, S. K. (2016). Compact 2D modeling and drain current performance analysis of a work function engineered double gate tunnel field-effect transistor. *Journal of Computational Electronics*, 15:104–114.

SILVACO Inc. (2010) *ATLAS User's Manual*, Santa Clara, CA.

Wu, J., Min, J., and Taur, Y. (2015). Short-channel effects in tunnel FETs. *IEEE Transaction on Electron Devices*, 62(9):3019–3024.

Index

293

For Product Safety Concerns and Information please contact our EU
representative GPSR@taylorandfrancis.com
Taylor & Francis Verlag GmbH, Kaufingerstraße 24, 80331 München, Germany

www.ingramcontent.com/pod-product-compliance
Lightning Source LLC
Chambersburg PA
CBHW060334220326
41598CB00023B/2703